Aquatic Production Systems

Series Editor

Gavin Burnell, School of Biological, Earth and Environmental Sciences, University College Cork, Cork, Ireland

Food production on land is limited by available sites, access to uncontaminated freshwater and a suitable climate range. Catering for an anticipated increase in world population up to 10 billion people by 2050 requires closing three gaps:

A **56 percent food gap** between crop calories produced in 2010 and those needed in 2050.

A **593 million-hectare land gap** between global agricultural land area in 2010 and expected agricultural expansion by 2050.

An **11-gigaton GHG mitigation gap** between expected agricultural emissions in 2050 and the target level needed to hold global warming below 2oC.

The largest increase in production will have to come from marine environments and also from a better use of terrestrial habitats including sustainable freshwater management. This book series will examine some of the key areas where this required growth will take place. Each volume will first examine existing technologies and their limitations. It will then explore the new species and technologies that will be key in moving these sectors forward including details of species life cycles, processes and systems. Finally it will consider societal, governance and economic issues that will have to be solved in order for the new production methods to succeed.

Fearghal O'Donncha • Martin Føre • Jon Grant
Editors

Digital Twin for Marine Fish Farming

Springer

Editors
Fearghal O'Donncha
IBM Research Europe
Dublin, Ireland

Martin Føre
Department of Engineering Cybernetics
NTNU
Trondheim, Norway

Jon Grant
Department of Oceanography
University of Dalhousie
Halifax, NS, Canada

ISSN 2730-7832 ISSN 2730-7840 (electronic)
Aquatic Production Systems
ISBN 978-3-032-10361-1 ISBN 978-3-032-10362-8 (eBook)
https://doi.org/10.1007/978-3-032-10362-8

This Springer imprint is published by the registered company Springer Nature Switzerland AG
The registered company address is: Gewerbestrasse 11, 6330 Cham, Switzerland

This book is dedicated to the late Dr Jon Grant, one of the driving forces behind its conception. Jon was a tireless advocate for aquaculture research and among the first to investigate aquaculture through an oceanographer's lens. His research on aquaculture-related topics is well published and diverse, ranging from carrying capacity, species performance, environmental assessment, benthic ecology, pathogen dispersal, and more. Jon was an avid collaborator, always willing to make new connections to advance research and understanding. He was very student focused and trained dozens of graduate students and future leaders through his laboratory at Dalhousie University in Canada. First and foremost, Jon was a fierce defender of objective science. He was never afraid to admonish a colleague if he felt they were being biased. The aquaculture research community has lost a great champion. It is hoped that in a small way this book can pay homage to Dr Grant's lifetime of accomplishments and commitment to modern aquaculture development.

Introduction

Aquaculture and Intensive Marine Fish Farming

According to the Organisation for Economic Co-operation and Development (OECD), the global Blue Economy, which encompasses aquaculture and fisheries, shipping, offshore wind energy, maritime and coastal tourism, and marine biotechnology, was estimated to be worth USD 2.7 trillion in 2019, representing 3–4% of the world's gross value added (OECD 2025). As the Ocean Economy grows, it's crucial that maritime industries and the utilisation of ocean resources and ecosystems are environmentally sustainable. Economic activities must be aligned with the long-term carrying capacity of the ocean ecosystems to maintain balance (Bhatia et al. 2017).

Aquaculture is a core component of the Blue Economy, valued at $296B in 2022 (FAO 2024). A vibrant fish farming industry is key to combatting food insecurity and poverty into the twenty-first century, and to progressing towards the UN Sustainable Development Goals (SDGs) (FAO 2024). Despite being historically one of the fastest-growing food systems in the world (Klinger and Naylor 2012), aquaculture now faces a set of industrial and environmental challenges that are slowing its expansion. A comprehensive study by Sumaila et al. (2022) found that while global production remains high, the rate of growth has peaked and is declining—for example, the 5-year moving average growth rate for Atlantic salmon has dropped to just 0.9%. These trends suggest that without innovation or intervention, the sector risks stagnation. This development is in direct conflict with the widely stated goal that aquaculture must supply an increasing proportion of the nutritional demands of a growing world population. It is thus important that efforts are made to reverse this situation and thereby enable further industrial growth. This is best achieved by first identifying and then solving the underlying challenges. Solving such challenges will typically require a cross-disciplinary effort combining biology and technology. In this book we focus on how technology, specifically digital twin technology, can play a central role in this landscape.

While these perspectives concern all forms of aquaculture, this book mainly focuses on the application of technology in the intensive marine fish farming segment, for the following reasons:

1. It is the most capital-intensive segment of aquaculture, making it easier to justify investments in new technologies, as these can often be directly linked to increased revenue and profitability.
2. It already makes more extensive use of technology than other aquaculture sectors, facilitating the integration of new tools and systems at both the farm and corporate levels.
3. Intensive marine fish farming takes place in challenging environments—typically farther from shore and more exposed to harsh weather—where manual operations are more difficult. These conditions make the sector especially well-suited to benefit from digital twin technologies.
4. Its production intensity, measured as fish output per farmer, surpasses all other forms of animal husbandry. This makes it difficult—but essential—to maintain a clear overview of animal health and behaviour in order to support ethical and sustainable production.

Before delving into the specifics on digital twin technology and how it can contribute to ensure ethical, sustainable, and continued growth in aquaculture, we first outline some historical aspects and current practices in intensive marine fish farming.

Intensive Marine Fish Farming

The roots of aquaculture go back thousands of years to pond-based production of cyprinids, plants, and other organisms in Asia; however, the history of intensive marine farming of finfish spans only a few decades. Atlantic salmon (*Salmo salar* L.) is one of the most common species in intensive marine farming, both due to the high status of salmon meat on the global market and the economic success of the industry, which surpasses that of other farmed aquatic species. Salmon farming began on a trial basis in the 1960s, but it wasn't until the 1980s and 1990s that it became an established industry in Norway, the UK, and Chile (Laird 1996). In 2022, global salmon production reached approximately 2.86 million tonnes, with a market value of around 19 billion euros (FAO 2024; Salmon 2025). Most salmon production is today concentrated in Norway, Chile, the UK, and Canada.

Current practices in the salmon farming industry are divided into three distinct phases; the first is the freshwater period conducted in land-based facilities. During this phase, the fish are first hatched from fertilised eggs, and then reared up through the larval and parr stages, until they undergo smoltification, a metamorphosis that prepares them for life in seawater. The land-based period is exclusively conducted in tanks of either flow-through configuration or based on various degrees of water recirculation. Here, the fish are inoculated with vaccines and continually inspected to ensure that they are ready for life at sea. When the fish are found to be smoltified and deemed mature enough for transfer (usually after 12–18 months) to marine farming sites and released into net pens, thereby entering the second phase of their lives.

In the sea-based phase, the fish are usually reared in fish farms incorporating 15–20 pens, a feeding barge and a submerged mooring system keeping the farm stationary. Each pen consists of a net enclosure suspended from a flexible floating collar attached to the mooring system. The lower end of the net enclosure is attached to point weights or sinker tubes that extend the enclosure so that it maintains an internal volume that can house the fish. A centralised feeding system placed at the barge is connected to feeding tubes extending to all cages and attached to feed spreaders in each pen. This allows the system to deliver formulated feed to all the cages. It also enables customising or adjusting the feeding schedule for each cage. In addition to feeding, the most common farming operations during this phase include inspection of farm components and fish, lice counts, parasite/pathogen treatments and collection of dead fish. During the sea-based phase, salmon undergo their most significant growth, increasing in weight from smolts of 100–1000 g to fully grown adults of 5–6 kg, typically over a period of 18 months. When the fish are deemed to have reached their final weight, they undergo a period of feed retention (7–14 days) before being transferred to well-boats—specialised vessels equipped with tanks for transporting live fish—which carry them to onshore processing facilities. Their arrival at the plant finalises the production cycle in that the fish are then sedated, euthanised, processed, and packed, thereafter being shipped to the markets.

Over the past three decades, there has been a significant increase in the number of fish used in salmon aquaculture, with Scottish aquaculture alone transferring 47 million fish to sea in 2018, up from 25 million in 1990 (Munro and Wallace 2018). However, as the salmon industry has intensified, there is an increasing demand for insight regarding feed usage, environmental control and response, and operational decision-making. This has stimulated developmental trends towards new technological solutions designed to assist farmers in making farming operations more efficient, safe, and sustainable. Examples of such innovations include technologies for welfare assessments (Barreto et al. 2022), feeding systems that automatically consider feeding activity (Folkedal et al. 2022), and remote-controlled/autonomous vehicles used in maintenance and inspection tasks (Kelasidi and Svendsen 2023). Although usually developed for salmon farming, these tools are often transferrable to other forms of intensive pen-based fish farming.

Despite the emergence of new innovations and tools, many of the operations in intensive salmon farming still entail manual tasks and subjective assessments. Moreover, the indications of stagnated growth seen in production numbers for the salmon industry (Sumaila et al. 2022) show that even though the sector is considered profitable, new measures need to be taken to enable continued growth. Increased digitisation and introduction of new technologies are probably necessary to achieve this.

Digital Twin Technology

A digital twin is a virtual replica of a physical object, process, or system that is designed to provide a virtual representation of the real-world object or system that can be used to monitor, analyse, or optimise its performance (Tao and Qi 2019). This variety in applications can be linked with the capability levels of a digital twin implementation, which range from monitoring, estimating, and presenting present system states, through the ability to predict and forecast future system states (e.g., answering "what if" questions), to the active feedback control of the system or asset in question (What is digital-twin technology? | McKinsey 2024; Rasheed et al. 2020). Digital twins achieve these capability levels by combining real-world data, simulation, and machine learning models with data analysis, a concept that is rooted in the field of cyber-physical systems (i.e., the integration of physical systems with computational and communication technologies).

Digital twins are currently adopted within a wide range of industrial, societal, and public sectors, including manufacturing (Soori et al. 2023), transportation and logistics (Barykin et al. 2021), shipping (Katsoulakos et al. 2024; Timms et al. 2024; Timms et al. 2025), healthcare (Sun et al. 2023), urban planning (Deng et al. 2021), energy management (Yu et al. 2022), retail (Shoji et al. 2022), and defence (Mendi et al. 2021). Many of these sectors already leverage digital twin capabilities to detect problems and increase efficiency. This broad adoption across various domains highlights one of the key strengths of digital twin technology: its inherent interoperability. Specific examples include using digital twins to simulate and test different scenarios to enhance efficiency and reduce downtime in manufacturing plants (Soori et al. 2023). In the automotive sector, Volvo utilises digital twin solutions to predict component stress and potential failure (Scurria et al. 2022), enabling more effective predictive monitoring during the operational life of vehicles. In a less commercial use-case, NASA employs digital twins to monitor the status of their spacecrafts (Glaessgen and Stargel 2012) to maintain a more detailed overview during missions.

Since purely model-based simulations and digital twins both utilise digital models to replicate the internal processes and mechanisms of the system of interest, it might not always be apparent how a digital twin can be distinguished from a simulator. However, where models often seek to simulate a single system, a digital twin will usually represent a virtual environment that also covers how the system interacts with other assets or systems. This enables richer studies of the system or systems in question. Further differences also exist. For example, while simulations usually don't benefit from availability of real-time data, most digital twins entail some integration with real-time (and possibly historical) data. Hence, one can categorise different levels of digital simulations into three distinct tiers (Kritzinger et al. 2018):

1. *Digital Simulations*: These represent the most basic level, where a model or simulation is used to replicate specific processes or systems in a controlled environment. Digital simulations are often used to analyse and optimise processes,

predict outcomes, or test designs without needing to involve real-world systems. However, they are generally static, meaning they are not continuously updated with real-time data. Instead, they rely on predefined inputs to generate results based on specific scenarios. Their accuracy is thus solely based on model fidelity. Some digital twin definitions include the concept of standalone digital twins that in reality are digital simulations (Rasheed et al. 2020).
2. *Digital Shadows*: This intermediate level involves a one-way connection between the physical and digital environments. In a process often termed data assimilation, data from physical systems are captured and transmitted to a digital model for analysis. However, the flow of information is unidirectional—while the digital model reflects changes in the physical system, it does not influence or control it. In terms of capability levels (Rasheed et al. 2020), this would enable descriptive, diagnostic, predictive, and prescriptive capabilities.
3. *Digital Twins*: At the highest level, digital twins go beyond simulations and digital shadows by establishing a bi-directional flow of information between the physical system and its digital counterpart. This means not only does the digital twin receive real-time data from sensors and IoT devices embedded in the physical system but send feedback, control signals, or adjustments back to the physical system to influence its operations. This bi-directional interaction allows for dynamic optimisation, predictive maintenance, and autonomous decision-making (corresponding to an autonomous capability level (Rasheed et al. 2020)).

These levels represent a progression in digital modelling capabilities, moving from static, isolated simulations to highly integrated, real-time digital twins that support predictive analytics and decision-making. By having better and constantly updated data related to a wide range of areas, combined with the added computing power that accompanies a virtual environment, digital twins can study more issues from far more vantage points than standard simulations can—with greater ultimate potential to improve products and processes.

Digital Twins and Fish Farming

Considering the nature of the challenges facing the fish farming industry, it is apparent that there are significant benefits in adapting digital twin technology to aquaculture. However, while fish farming is not the most technology intensive sector, the industry has over the last decades had a focus on increased digitisation. This has led to the adaptation of different data-driven paradigms in aquaculture management with the intent of improving production efficiency, reducing costs, and enhancing environmental sustainability. Some of the data-driven paradigms that have been adopted include:

- Precision aquaculture (Føre et al. 2018; O'Donncha and Grant 2019), which is an approach that uses real-time data from sensors to monitor and optimise various

aspects of fish production, including feed management, water quality, and disease control.
- Big data analytics (Ba et al. 2023), which is the use of advanced data analytics tools and techniques to analyse large and complex datasets generated from different sources. By applying this approach to aquaculture and analysing the data, patterns, and trends, insights can be gained informing decision-making and improving farm performance.
- Internet of Things (IoT) (O'Donncha et al. 2021b) which involves the use of interconnected devices, sensors, and systems to collect and share data across organisational and conceptual boundaries. By leveraging IoT, aquaculture producers can monitor and manage different aspects of their operations in real-time, including water quality, feed management, and environmental conditions, thereby adding value across the aquaculture value chain.

A common thread across these paradigms are scalable frameworks for collecting and managing data generated on farms, as well as AI-backed processing to extract actionable insights from these sources. This serves as a solid foundation on which to develop future digital twin solutions. A digital twin approach would extend these with interconnected processing and simulation capabilities and a data-informed, interactive, AI-backed, virtual representation of the various operational processes and environmental dynamics that occur on fish farms. This virtual representation serves as the backbone for stakeholders to monitor farm conditions, explore different management and control scenarios, detect anomalous or exceptional circumstances, and extract optimal decision cognisant of possible scenarios.

The interoperability of digital twins opens new possibilities for value creation across the aquaculture value chain by enabling common dataspace solutions and integrated virtual representations and simulation. This means that digital twins can consider all aspects of farm product cultivation, harvesting, sustainable management, and sales in treating these as being interconnected, which is the case in the real world. They can thus enhance decision-making in aquaculture, particularly when a decision may have impacts on multiple stages in the value chain. Examples of this include how to steer essential water quality variables to prevent future disease outbreaks, how to automate feeding to optimise fish growth while minimising waste, how to manage fish welfare and reduce mortalities while maintaining good growth rates (O'Donncha et al. 2021a), and how to streamline veterinary interventions such that downstream production is minimally impaired. Additionally, digital twin technologies can help mitigate exposure to external stressors such as pathogens/diseases, parasites, chemical pollutants, and microplastics, and assist with harvesting decision-making and financial management based on economic yield forecasts and input controls (O'Donncha et al. 2024). Realising digital twins that can achieve these benefits depends on technological components that monitor, simulate, and create virtual replicas of complex farm systems in the marine environment.

In summary, we envision that realising digital twin technology for aquaculture can lead to several industrial trends that can shape the future of the sector:

- Although intensive aquaculture operations, such as salmon production, will be the first segment to benefit from digital twin solutions, it may be possible to adapt the solutions to other production forms and species. Digital twins can thus lead to increased technological exchange between different segments in the industry.
- A robust integration of these technological components will offer the industry a more objective, precise, and accurate understanding of current conditions in marine fish farms. This empowers producers to make better-informed decisions that enhance production efficiency.
- In the short term, digital twin technology can lead to safer, more sustainable, and efficient farm-level management.
- In long term, this will stimulate the industry towards more holistic and sustainable practice, positioning it to meet the growing demand for food. This shift will be crucial in bridging the gap between current food production and future demands.

Rationale and Outline of This Book

This book aims to pave the way for adopting digital twin technology as a core tool in transforming aquaculture practices. The book will describe the current state-of-the-art within the enabling technologies required to realise digital twins for aquaculture, how these elements can be synthesised into digital twin solutions, and how the resulting digital twins may be used as tools by the industry. This book is structured as follows:

- We begin by outlining the state of the art in marine aquaculture, starting with an overview of the latest advances in fish monitoring (Chap. 2) followed by a review of robotics and autonomous vehicles used in commercial pen-based production (Chap. 3).
- Subsequent chapters will shift to the virtual realm, exploring the current applications of machine learning (Chap. 4) and mechanistic/physics-based modelling (Chap. 5). We will also discuss how these methodologies can be integrated using hybrid analysis techniques (Chap. 6).
- These foundational chapters collectively define the core components necessary to develop effective digital twins for aquaculture. Chap. 7 presents our vision for synthesising these elements into a cohesive digital twin framework.
- The final three chapters will focus on practical applications, addressing how digital twin technology can tackle critical industry challenges, such as
  - Adapting to climate change (Chap. 8)
  - Streamlining supply chain integration in aquaculture (Chap. 9)
  - Designing future solutions for intensive pen-based fish production (Chap. 10)

Although many chapters contain examples and case studies from various production systems and species (e.g., closed production units/RAS, seaweed production and pond-based production), the book is largely focused on practices related to intensive marine finfish production. The editorial team decided on this limitation early on based on the extreme diversity that is seen in aquaculture with respect to species, production systems, and degree of intensity. This diversity renders the act of making an exhaustive industry-wide study of the development and application of digital twins difficult and would also result in a more complex storyline that would feature either a larger number of chapter or considerably longer chapters.

The choice of intensive marine finfish farming as a main focus for this book was primarily motivated by two factors:

1. Marine finfish farming is the most technologically advanced, capital intensive, and profitable segment in the sector today, potentially rendering the time horizon for industrial application of digital twins shorter than for less capital-intensive production forms.
2. Compared with many farmed species (e.g., seaweed, mussels), finfish tend to exhibit complex behaviours and lead complex lifestyles, meaning that the benefits of targeted use of technology to improve observability and controllability of production are likely to be higher.

In conclusion, we hope that this book can provide roadmap for the future realisation of digital twins in this sector, and that the industry eventually will reap the benefits of doing so. However, we also hope that it can serve as motivation for similarly exploring the potential of adopting digital twin technology for other segments within aquaculture.

References

Ba A, O'Donncha F, Ploennigs J, Azmat M (2023) Efficient extraction of insights at the edges of distributed systems. In: 2023 IEEE international conference on big data (big data). IEEE, pp 1610–1619. Accessed 19 May 2025. [Online]. Available: https://ieeexplore.ieee.org/abstract/document/10386230/?casa_token=pijQXS_5AWYAAAAA:bnvm_fH4fYwMO-wBlEiRsKENpvramLcZvuTe2AteDMDG6sXtnMRSry4YyTDBpaDDwosXmBHs

Barreto MO, Rey Planellas S, Yang Y, Phillips C, Descovich K (2022) Emerging indicators of fish welfare in aquaculture. Rev Aquac 14(1):343–361. https://doi.org/10.1111/raq.12601

Barykin SY, Bochkarev AA, Dobronravin E, Sergeev SM (2021) The place and role of digital twin in supply chain management. Acad Strateg Manag J 20:1–19

Bhatia R et al (2017) Sustainable ocean economy, innovation and growth: a G20 initiative for the 7th largest economy in the world. Policy Area 2030 Agenda Sustain Dev

Deng T, Zhang K, Shen Z-JM (2021) A systematic review of a digital twin city: a new pattern of urban governance toward smart cities. J Manag Sci Eng 6(2):125–134

FAO (2024) The state of world fisheries and aquaculture 2024 – blue transformation in action. Food and Agriculture Organization of the United Nations. [Online]. Available: https://doi.org/10.4060/cd0683en

Folkedal O et al (2022) Deployment of hydroacoustic feeding control in salmon sea-cages; biological and technical considerations. Aquaculture 561:738700

Føre M et al (2018) Precision fish farming: a new framework to improve production in aquaculture. Biosyst Eng 173:176–193

Glaessgen E, Stargel D (2012) The digital twin paradigm for future NASA and US Air Force vehicles. In: 53rd AIAA/ASME/ASCE/AHS/ASC structures, structural dynamics and materials conference 20th AIAA/ASME/AHS adaptive structures conference 14th AIAA, p 1818

Katsoulakos T, Tsiochantari G, O'Donncha F, Kaklamanis E, Maccari A, Mucharski M (2024) Shipping digital twin landscape. In: State-of-the-art digital twin applications for shipping sector decarbonization. IGI Global, pp 1–25. Accessed 19 May 2025. [Online]. Available: https://library.oapen.org/bitstream/handle/20.500.12657/90121/9781668498491.pdf?sequence=1#page=21

Kelasidi E, Svendsen E (2023) Robotics for sea-based fish farming. In: Zhang Q (ed) Encyclopedia of smart agriculture technologies. Springer International Publishing, Cham, pp 1–20. https://doi.org/10.1007/978-3-030-89123-7_202-1

Klinger D, Naylor R (2012) Searching for solutions in aquaculture: charting a sustainable course. Annu Rev Environ Resour 37:247–276

Kritzinger W, Karner M, Traar G, Henjes J, Sihn W (2018) Digital twin in manufacturing: a categorical literature review and classification. Ifac-Pap 51(11):1016–1022

Laird LM (1996) History and applications of salmonid culture. In: Pennell W, Barton BA (eds) Principles of salmonid culture. Elsevier

Mendi AF, Erol T, Doğan D (2021) Digital twin in the military field. IEEE Internet Comput 26(5):33–40

Munro L, Wallace I (2018) Marine Scotland science: Scottish fish farm production survey. [Online]. Available: https://www.gov.scot/publications/scottish-fish-farm-production-survey-2018/

O'Donncha F, Grant J (2019) Precision aquaculture. IEEE Internet Things Mag 2(4):26–30

O'Donncha F et al (2021a) Data driven insight into fish behaviour and their use for precision aquaculture. Front Anim Sci 2:695054

O'Donncha F, Akhriev A, Eck B, Burke M, Filgueira R, Grant J (2021b) Deployment and management of time series forecasts in ocean industry. In: 2021 IEEE international conference on big data (big data). IEEE, pp 4091–4096. Accessed 19 May 2025. [Online]. Available: https://ieeexplore.ieee.org/abstract/document/9671877/?casa_token=sl7nDnfNgGUAAAAA:SOxd4GqWaixjKCaFaGdJ6riID2Yfoh2DORax3LwYYhReKbP4vYDu561hapWpAVa14_cAKqtP

O'Donncha F, Akhriev A, Fragoso B, Icely J (2024) Forecasting closures on shellfish farms using machine learning. Aquac Int 32(5):5603–5623. https://doi.org/10.1007/s10499-024-01438-y

OECD (2025) The ocean economy to 2050. OECD Publishing, Paris. Accessed 19 May 2025. [Online]. Available: https://doi.org/10.1787/a9096fb1-en

Rasheed A, San O, Kvamsdal T (2020) Digital twin: values, challenges and enablers from a modeling perspective. IEEE Access 8:21980–22012

Salmon – main producers see record-breaking exports. Globefish. Accessed 19 May 2025. [Online]. Available: https://www.fao.org/in-action/globefish/news-events/news/news-detail/Salmon%2D%2D-Main-producers-see-record-breaking-exports/en

Scurria L, Risaliti E, Buss D, Kubo P, Tamarozzi T, Cornelis B (2022) Executable digital twin-prevent the early failure of a truck anchorage using smart virtual sensors. SAE Int J Adv Curr Pract Mobil 4(2022-01–0767):1309–1317

Shoji K, Schudel S, Onwude D, Shrivastava C, Defraeye T (2022) Mapping the postharvest life of imported fruits from packhouse to retail stores using physics-based digital twins. Resour Conserv Recycl 176:105914. https://doi.org/10.1016/j.resconrec.2021.105914

Soori M, Arezoo B, Dastres R (2023) Digital twin for smart manufacturing, a review. Sustain Manuf Serv Econ:100017

Sumaila UR et al (2022) Aquaculture over-optimism? Front Mar Sci 9. Accessed 25 Feb 2023. [Online]. Available: https://www.frontiersin.org/articles/10.3389/fmars.2022.984354

Sun T, He X, Li Z (2023) Digital twin in healthcare: recent updates and challenges. Digit Health 9:20552076221149651. https://doi.org/10.1177/20552076221149651

Tao F, Qi Q (2019) Make more digital twins. Nature 573(7775):490–491. https://doi.org/10.1038/d41586-019-02849-1

Timms A, Langbridge A, O'Donncha F (2024) Agentic anomaly detection for shipping. Presented at the NeurIPS 2024 workshop on open-world agents. Accessed 4 Feb 2025. [Online]. Available: https://openreview.net/forum?id=mmBrJRoppK

Timms A, Langbridge A, Antonopoulos A, Mygiakis A, O'Donncha F (2025) Agentic AI for digital twin. In: AAAI conference on artificial intelligence, Philadelphia. Accessed 5 Feb 2025. [Online]. Available: https://research.ibm.com/publications/agentic-ai-for-digital-twin

What is digital-twin technology? | McKinsey. Accessed 23 Sept 2024. [Online]. Available: https://www.mckinsey.com/featured-insights/mckinsey-explainers/what-is-digital-twin-technology

Yu W, Patros P, Young B, Klinac E, Walmsley TG (2022) Energy digital twin technology for industrial energy management: classification, challenges and future. Renew Sustain Energy Rev 161:112407. https://doi.org/10.1016/j.rser.2022.112407

Contents

About the Editors

Martin Føre is an Associate Professor at the Norwegian University of Science and Technology (NTNU), Department of Engineering Cybernetics. He holds an MSc (2006) and a PhD (2011) within Engineering Cybernetics from NTNU, and before rejoining NTNU in 2019, he worked as a Research Scientist/Senior Research Scientist at SINTEF Ocean. Throughout his research career, he has specialised towards the application of cybernetic methods and control systems components in aquaculture situations, and he had a central role in the development of the Precision Fish Farming (PFF) concept for finfish aquaculture.

Jon Grant was a Professor of Oceanography at Dalhousie University (Canada). He earned his BSc in Zoology from Duke University in 1976 and his PhD in Biology from the University of South Carolina in 1981. He joined Dalhousie as a NATO Postdoctoral Fellow in 1981, became a Full Professor of Oceanography in 1997, and held multiple positions that reflected his international leadership in aquaculture research.

He was an outstanding scientist, publishing over 200 scientific papers and supervising more than 50 master's and PhD students.

Professor Grant passed away during the final stages of this book's preparation. This book is dedicated to his memory.

Fearghal O'Donncha is a Research Scientist and Technical Lead at IBM Research Europe—Ireland. His work focuses on applying AI, hybrid cloud, accelerated discovery, and quantum computing to real-world challenges. A core research interest is the development of human-centred digital twin technologies for environmental systems such as ships, ports, and fish farms. These virtual models allow rapid exploration of different management and environmental scenarios, helping to identify optimal strategies.

He is a Chartered Engineer and holds a PhD from the National University of Ireland, Galway. He has published over 80 scientific articles and holds more than 50 patents. He leads global patenting in AI Engineering at IBM.

Abbreviations

ABM	Animal-Based Measures
AI	Artificial Intelligence
CNN	Convolutional Neural Network
DDM	Data-Driven Model
DT	Digital Twin
ECG	Electrocardiogram
EDITO	European Digital Twin of the Ocean
EU GFL	European Union General Food Law
FCR	Feed Conversion Ratio
FMEA	Failure Modes and Effects Analysis
GAN	Generative Adversarial Network
GDST	Global Dialogue on Seafood Traceability
H_2S	Hydrogen Sulphide
HAM	Hybrid Analysis and Modelling
HMS	Health, Safety and Work Environment
IoT	Internet of Things
KBM	Knowledge-Based Model
LLM	Large Language Model
LSTM	Long Short-Term Memory
MSC	Marine Stewardship Council
NLP	Natural Language Processing
OT	Optimal Transport
OWI	Operational Welfare Indicator
PE-rings	Polyethylene Rings
PFF	Precision Fish Farming
QoL	Quality of Life
QR Code	Quick Response Code
RAS	Recirculating Aquaculture System
RFID	Radio Frequency Identification
RNN	Recurrent Neural Network
SDG	Sustainable Development Goal

SEA-TRACES	EU Project on Seafood Traceability
SSP	Sustainable Shrimp Partnership
TRL	Technology Readiness Level
WI	Welfare Indicator
WOAH	World Organisation for Animal Health
YOLO	You Only Look Once

Part I
On the Farm

Chapter 1
Fish Health and Welfare Monitoring for Potential Digital Twins

Chris Noble, René Alvestad, Pablo Arechavala-López, Finn Olav Bjørnson, Nina Bloecher, Meredith Burke, Petr Císař, Lynne Falconer, Martin Føre, Gaute A. N. Helberg, David Izquierdo-Gomez, Kristbjörg Edda Jónsdóttir, Sunil Kadri, Jelena Kolarevic, Santhosh K. Kumaran, Thomas Larsson, Ingrid Måge, Jonatan Nilsson, Samuel Ortega, Sonia Rey Planellas, Lars Erik Solberg, Lars H. Stien, Bjørn-Steinar Sæther, Linda Tschirren, Elisabeth Ytteborg, and Lucas Zena

Abstract Measuring, monitoring, assessing, auditing and safeguarding fish welfare in aquaculture is of the utmost importance to ensure the overall sustainability of the sector. Welfare is key to many short- and long-term operational farming decisions,

---

C. Noble (✉) · R. Alvestad · D. Izquierdo-Gomez · S. K. Kumaran · T. Larsson · I. Måge · S. Ortega · L. E. Solberg · L. Tschirren · E. Ytteborg · L. Zena
Nofima, Tromsø, Norway
e-mail: chris.noble@nofima.no

P. Arechavala-López
Mediterranean Institute of Advanced Studies (IMEDEA-CSIC/UIB), Esporles, Spain

F. O. Bjørnson · N. Bloecher · K. E. Jónsdóttir
SINTEF Ocean, Trondheim, Norway

P. Císař
Faculty of Fisheries and Protection of Waters, CENAKVA, University of South Bohemia in České Budějovice, Nové Hrady, Czech Republic

M. Burke · L. Falconer
Institute of Aquaculture, University of Stirling, Stirling, Scotland, UK

S. Rey Planellas
Welfare Unit, IRTA, Girona, Spain

Institute of Aquaculture, University of Stirling, Stirling, Scotland, UK

M. Føre
Department of Engineering Cybernetics, NTNU, Trondheim, Norway

G. A. N. Helberg
The Norwegian College of Fishery Science, UiT The Arctic University of Norway, Tromsø, Norway

Lerøy Seafood Group ASA, Bergen, Norway

F. O'Donncha et al. (eds.), *Digital Twin for Marine Fish Farming*,
Aquatic Production Systems, https://doi.org/10.1007/978-3-032-10362-8_1

but the strength of a given decision depends upon the quality and context of the information the farmer has. However, it is often difficult to measure and monitor fish welfare, especially when farming large numbers of fish and when available tools are not as robust or as operational as needed. Animal welfare is measured and monitored using welfare indicators (WIs), and while there are a wide array of welfare indicators available, these may not be suitable for all species, life stages, husbandry routines, rearing systems or budgets. Digitalising this process is a way to address and resolve some of the challenges a stakeholder faces regarding documenting and auditing input- and outcome-based WIs. The following chapter outlines what welfare is and a range of technologies that one can use to measure and monitor fish welfare, with emphasis on real-time online monitoring. It also provides examples of how AI can add value to these technologies and considers how to integrate welfare indicators into a digital twin.

Keywords Aquaculture · Health · Welfare · Digital twin

1.1 Chapter Aims and Scope

In this chapter, we give the reader an introduction to the concept of fish welfare and the different types of indicators that are used to measure it. We will then outline some of the digital technologies available for measuring and monitoring different types of welfare indicators (WIs) to give the reader an overview of what is available, particularly in terms of real-time online monitoring capabilities that are central to holistic digital twins. Potential challenges, pitfalls and benefits will be addressed in relation to different sensing and data interpretation methods. Additionally, case studies will be used to outline potential scenarios where digitalisation is applicable or already in use and will outline what technologies are real time online. Lastly, the prospects and challenges that artificial intelligence (AI) and machine learning (ML) present for automating and streamlining the measuring and monitoring process will be discussed, with the aim of helping the reader consider the integration of welfare indicators into a digital twin. The chapter will not propose a digital twin or twins for fish health and welfare; it will specifically outline the type of data that should be

S. Kadri
RINA, Research Institute for Environment and Livelihoods, Charles Darwin University, Brinkin, NT, Australia

J. Kolarevic · B.-S. Sæther
Nofima, Tromsø, Norway

The Norwegian College of Fishery Science, UiT The Arctic University of Norway, Tromsø, Norway

J. Nilsson · L. H. Stien
Animal Welfare Research Group, Institute of Marine Research, Matre, Norway

collected to support potential welfare measurement and monitoring technologies for incorporating into future digital twins. For information on the steps that are being taken toward the development of fish health and welfare digital twin or twins, and what should be considered in their construction, we direct the reader to works by, e.g. Budaev et al. (2024, 2025) and Giske et al. (2025a).

1.2 Fish Welfare and How to Measure It

Animal welfare has been a fundamental theme in many aquaculture sectors for a number of years, and a wide array of stakeholders, including farmers, draw attention to this and promote its optimisation as part of the wider sustainable food production system framework (e.g. Broom 2018; European Commission 2020). Welfare is key to the Sustainable Development Goals (SDGs) of the UN 2030 Agenda for Sustainable Development (Keeling et al. 2019; Gonzalez 2023). Despite its unquestioned relevance, the definition of animal and, in particular, fish welfare is less clear.

1.2.1 Fish Welfare and Welfare Indicators

Most definitions of animal welfare acknowledge that animals have an intrinsic value and do not exist solely for our benefit. While it is not in the scope of this chapter to go into the background of each definition, we would like to introduce the readership to a few commonly used definitions that can then help frame fish welfare in aquaculture settings and facilitate a structured approach to measuring, monitoring and assessing it. One long-standing welfare definition that is gaining increasing traction in terrestrial and aquatic farming considers the animal's quality of life (QoL), defining welfare as "the quality of life as perceived by the animal itself". This definition was proposed by Bracke et al. (1999) and was adopted for fish in the early 2010s by Stien et al. (2013). Meanwhile, the World Organisation for Animal Health (WOAH) adopts a similar approach, defining welfare as "the physical and mental state of an animal in relation to the conditions in which it lives and dies" (World Organisation for Animal Health 2024). Both definitions can incorporate the positive and negative experiences of the fish and how these can affect their overall welfare status at any given moment. These definitions can also be considered in relation to the One Health-One Welfare concept, where One Health can be defined as "an integrated, unifying approach that aims to sustainably balance and optimize the health of people, animals and ecosystems" (World Health Organisation 2024) and One Welfare can be defined as "the interrelationships between animal welfare, human wellbeing and the physical and social environment" (e.g. Garcia Pinillos 2018). One Health has been considered in many earlier articles in relation to aquaculture (e.g. Stentiford et al. 2020). While it does consider both the mental and physical health of the animal in the farming system, the One Welfare concept builds upon and strengthens considerations in relation to the animal's mental state.

Another conceptualisation of good welfare is to see it as the balance between quality and quantity of stress in the context of allostasis. Allostasis refers to the physiological and behavioural process by which the body actively maintains stability through change in response to various stressors (Sterling 2012). These processes, which aim to restore balance after exposure to stressors and demands, are costly, and their sum is referred to as the allostatic load, a consolidating approach that may be of particular use in digital twins (Giske et al. 2025a). Conditions that result in allostatic loads that are persistently too high or far too low impair animal health and welfare (Korte et al. 2007 and references therein). Alternatively, the welfare status of an animal can also be defined as a function of the fulfilment of its welfare needs. The Five Domains model, originally proposed in 1994 and updated most recently in 2020, collates numerous welfare needs into four general domains related to (i) nutrition, (ii) the environment the animal is subjected to, (iii) the health status of the animal, and (iv) its behavioural interactions that then influence a fifth domain, (v) the animal's mental state, and therefore its welfare status (Mellor et al. 2020). In this chapter, we will consider fish health and welfare both independently and from an integrated perspective. This separation, coupled with integration, is important, because health is a central tenet of the five domains model for welfare (Mellor et al. 2020) and is also central to other welfare needs frameworks such as the earlier Five Freedoms model (Farm Animal Welfare Council 1993). However, good health alone does not predicate good welfare (Segner et al. 2012), and similarly, poor health doesn't always lead to poor welfare if an animal has an asymptomatic health problem that it is unaware of (e.g. Veasey 2017). This outlining of the definition of animal welfare status as a cumulation of (un)fulfilled welfare needs is the basis for understanding when an animal has a good welfare status, which is mostly done by characterising signs of good or poor welfare, so-called welfare indicators.

Welfare indicators are used to monitor and document the fulfilment of welfare needs which a stakeholder can then use to draw conclusions regarding welfare. An indicator is "... a measurement or value which gives you an idea of what something is like" (Collins Dictionary © Collins 2024) and WIs must address the fulfilment of, or change in, one or more welfare needs of the animal. They can then be used to provide supporting evidence for documenting whether welfare needs are being met and can help diagnose welfare problems or challenges. Indicators have varying utilities within a welfare assessment framework. They can be used to (i) measure welfare, via "... the process of quantifying the characteristics of an animal's welfare ..."; (ii) monitor welfare, which "... involves observing and checking the progress or quality of welfare over a period, maintaining regular surveillance, and systematically reviewing and reporting on the findings", and also changes in welfare; (iii) audit welfare, which involves "an on-site verification activity, such as an inspection or examination, to ensure compliance with established welfare standards and requirements"; and (iv) assess welfare "... the process of judging or deciding on the nature and quality of an animal's state of welfare based on available evidence" (citations reproduced from Wilkins et al. 2024).

WIs can be input-based, monitoring factors such as the environmental conditions the fish are subject to, or outcome-based, which specifically address the animals themselves (also termed animal-based measures, ABM's, e.g. EFSA 2012). In a farm setting, outcome-based indicators can be applied at a group level (e.g. mortality, appetite, schooling behaviour or the disease status of the fish population) or an individual level (e.g. physiological parameters such as plasma cortisol levels, injuries such as skin wounds and fin damage, malformations, etc., see Stien et al. 2020). Farm-friendly WIs are termed operational welfare indicators (OWIs); these must be relatively simple and easy to use in farming situations and be both reliable and comparable (e.g. Noble et al. 2012). Indicators that are sampled on the farm or from farming situations but are more complex and need further analysis (e.g. in a laboratory before a result is obtained) are termed Laboratory-based Welfare Indicators (LABWIs) see Noble et al. (2018). Alternatively, indicators can also be categorised as to whether they involve invasive or non-invasive methods in their monitoring and collection. Invasive methods "include those that physically disturb the fish by handling or capture and/or affect the physical integrity of the fish and/or had an impact on fish fitness or behaviour" while non-invasive methods "may refer to methods which avoid any direct physical contact including handling, capture and restraint" (see Barreto et al. 2022 and references therein). The monitoring of input-based indicators is generally non-invasive, especially with regard to monitoring the rearing environment the fish are subjected to. The monitoring of numerous outcome-based indicators, such as behaviour, appetite, or mortality, can also be termed non-invasive as they do not involve handling the fish to collect data. However, many OWI and LABWI quantification and monitoring methods at the individual level, such as those used for manually documenting the physiological or injury status of the animal, involve capturing, physically handling and sampling the fish and are therefore invasive. A key principle of measuring and monitoring welfare is that data collection should have a minimal or even no impact on the welfare of the animal, and where possible, non- or minimally invasive methods should therefore be utilised (Barreto et al. 2022). Based on the metrics resulting from measuring the welfare indicators, the fulfilment of the welfare needs, and therefore the welfare status, is then assessed.

1.2.2 Fish Welfare Assessments

The assessment of fish welfare should be comprehensive and include indicators that address all, or the majority of all known welfare needs of the fish (e.g. Stien et al. 2013; Tschirren et al. 2021). However, knowledge of relevant welfare needs for various species and life stages may not be available (Saraiva et al. 2019; Franks et al. 2021), and it may be challenging to use all the indicators, or even the most relevant ones, that are appropriate for a given welfare need. If this is the case, a stakeholder should be aware of the potential constraints and limitations this will place upon their welfare assessment and, consequently, upon their welfare-related

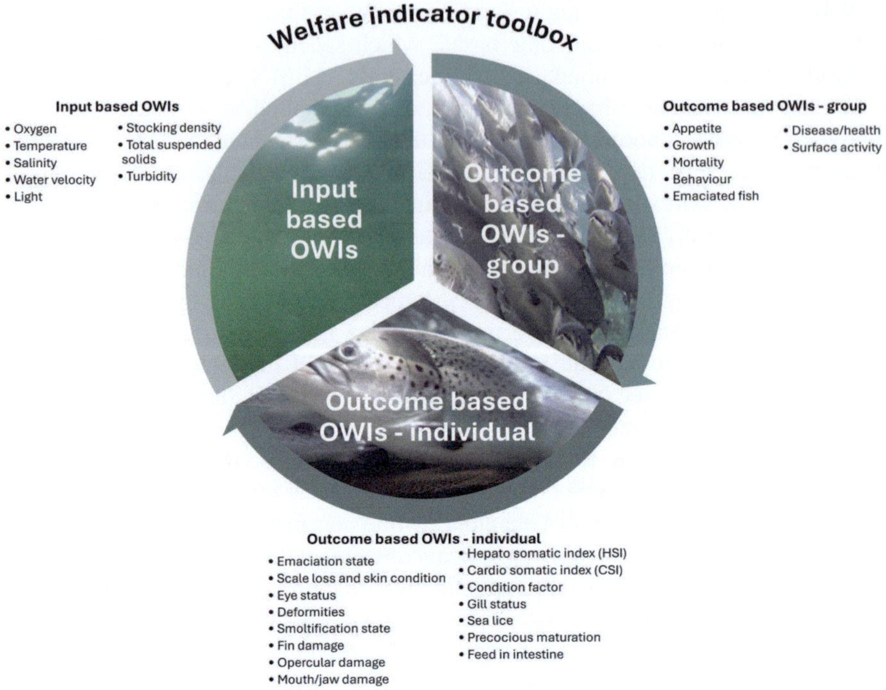

Fig. 1.1 Summarising OWIs that are appropriate for Atlantic salmon reared in marine net pens, including input- and outcome-based OWIs. (Adapted with permission from Noble et al. 2018. Photos: Lars H. Stien. Illustration: Gunhild Seljehaug Johansson and Chris Noble)

farming decisions. Further, it should be noted that *no single welfare indicator can address all welfare needs for fish*. Therefore, a selection of indicators, often termed a welfare indicator toolbox (e.g. Noble et al. 2018), must be used to satisfy the quantification, monitoring and assessment requirements for all, or the majority of all, known welfare needs for each species and life stage, to build an evidence-base for conducting a comprehensive welfare assessment. In addition to the species and life stage, such toolboxes must also be suitable for the different rearing systems and operations. Recent toolboxes for salmonids in open marine net pens and land-based systems have been described by Noble et al. (2018, 2020) and by others for species such as European seabass (*Dicentrarchus labrax*) and gilthead seabream (*Sparus aurata*) (Saraiva et al. 2022; Pavlidis and Samaras 2020; Pavlidis et al. 2024). An example of an indicator-based toolbox for measuring, assessing and monitoring the welfare of Atlantic salmon reared in marine, open commercial pens is outlined above (Fig. 1.1).

There are many ways that each indicator can be monitored. As far as the authors are aware, there are no standardised methods for the application of the toolboxes, although steps in this direction have been taken in some sectors and production regions (see, e.g. the Laksvel Welfare Scoring Protocol for Atlantic salmon produced in marine net pens, Nilsson et al. 2022, the Norwegian Standard 9417 "*Salmon and Rainbow Trout—Terminology and Methods for Documentation of Production*" NS 9417 2022, and also the RSPCA welfare standards, RSPCA 2020, 2024). This variability on how to implement and apply a framework for the monitoring, measuring, assessing and auditing of operational fish welfare indicators can directly affect the quality and utility of the data that is used to make a welfare assessment. While the missing standards may create uncertainty of application, variation in results or hurdles for comparability, they also allow for adaptability and specification of the toolbox. One promising advancement is the digitalisation of the aquaculture sector, allowing for more standardised measurements through automation, while the use of artificial intelligence (AI) allows for adaptation and specificity and has use for future digital twins in fish farming.

1.3 Precision Fish Farming and Fish Welfare

This section introduces the reader to some of the existing or emerging optical, acoustic and biosensor-based technologies for monitoring fish welfare in a farming environment, which may pave the way for the future use of digital twins. While the technologies can be grouped and classified in different ways, we organise them in relation to whether they are suitable for monitoring input- and/or outcome-based indicators. Further, the potential benefits and challenges of using each technology will be highlighted in relation to each sensing and data interpretation method, and the potential for using artificial intelligence (AI) to improve their utility will be discussed.

1.3.1 Precision Fish Farming

Precision Fish Farming (PFF) is a term adapted from Precision Livestock Farming (PLF, Berckmans 2004) by Føre et al. (2018a). PFF aims to "(1) improve accuracy, precision and repeatability in farming operations; (2) facilitate more autonomous and continuous biomass/animal monitoring; (3) provide more reliable decision support and; (4) reduce dependencies on manual labour and subjective assessments, and thus improve staff safety" and will thus "improve animal health and welfare while increasing the productivity, yield and environmental sustainability in commercial intensive aquaculture" (Føre et al. 2018a). PFF can improve the type of data that is collected and the way that data is collected and utilised, and this is especially important when considering the accuracy and utility of digital twin solutions. To highlight the utility of PFF in measuring and monitoring fish health and welfare, we will primarily focus on Atlantic salmon and the monitoring technologies applicable for different rearing systems (recirculating aquaculture systems on land, RAS and net pen production systems used during the marine ongrowing phase). This focus is mostly due to the rapid digitalisation and technologically intensive nature of the salmon industry relative to other aquaculture sectors (Moe Føre et al. 2022), but we acknowledge that these other sectors are also accelerating the development and application of PFF frameworks.

When considering PFF and technologies that can be used to improve our monitoring of the fish and their environment, we will consider AI as an integral part of PFF in driving a transformative shift in aquaculture, and improving how we monitor, manage, and optimise fish farming operations (Føre et al. 2018a; Mandal and Ghosh 2024). This will in turn help create a more comprehensive digital twin that encapsulates a wider range of factors and dynamics relevant for fish health and welfare. At the core of this transformation lies the ability to collect and analyse vast amounts of data from aquaculture systems. Automated systems, equipped with sensors and cameras, enable continuous monitoring of crucial input- and outcome-based welfare indicators, including water temperature, dissolved oxygen saturation, fish behaviour and injury status (e.g. Bui et al. 2024; Liu et al. 2023a; Lindholm-Lehto 2023). This approach can outdo traditional manual methods by ensuring data consistency, accuracy and real-time insights (Arabelli and Bernatin 2022; Zhang et al. 2022; Tolentino et al. 2020), thereby reducing the risk of missing important variability that typically arises from low-frequency sampling. Such systems can help facilitate the early detection of anomalies or environmental changes, allowing for faster responses and mitigating risks to fish health and welfare (Petkovski and Shehu 2023: Tolentino et al. 2020; Mandal and Ghosh 2024). Additionally, by eliminating human error and standardising data collection processes, automation enhances data reliability and enables sophisticated analysis, and by harnessing the power of AI and machine learning, this data can be transformed into actionable insights (Mandal and Ghosh 2024). It is further important to consider what data can be made available real time, online from the physical measurement and monitoring technologies via the Internet of Things (IoT), paradigm (i.e. the use of intelligent sensors with innate capabilities to analyse and/or transmit their measurements) to create a responsive and synchronised digital twin that is truly reflective of the physical twin and the multitude of factors and variables it is subjected to (Arulmozhi et al. 2024).

However, building effective automation and digital twins still faces many challenges. We need better data, matured technology and skills before these tools can truly reach their full potential. While automation offers numerous advantages, it's crucial to consider potential drawbacks, especially concerning human-animal interactions. Reducing human contact could diminish our understanding of an animal's needs, potentially impacting welfare. Therefore, responsible implementation of automation necessitates careful consideration of its ethical implications.

1.3.2 Digital Twins and Fish Welfare Assessments

A digital twin is, a virtual replica of a physical object, a process or system that can be used for monitoring, analysis and optimisation across a range of capability levels (Tao and Qi 2019; Rasheed et al. 2020; McKinsey 2024) and may be used as twin of a fish or a farm to advance the model capabilities from descriptive to diagnostic and even predictive, prescriptive and autonomous (Føre et al. 2024). It is linked to the physical farm (a physical twin) through the exchange of real-time data (Arulmozhi et al. 2024). As fish and fish farms are highly complex systems, these digital twins will rely on data of high enough quality and quantity to reliably and accurately describe current states, diagnose past problems and predict future outcomes. This prerequisite poses a challenge for today's fish welfare assessments as many indicator measurements are still manual processes. Therefore, the automation and digitalisation of measuring welfare indicators are key to creating the data needed for digital twins.

1.4 Monitoring of Input-Based Welfare Indicators

The monitoring and control of input-based WIs in a production system are integral to ensuring that different water quality parameters meet the welfare needs of the fish. Such a monitoring programme must consider how, where and how often relevant parameters are examined to provide enough data of sufficient quality for the creation of an accurate digital twin, and where possible, the generated data should be real time and online (Føre et al. 2024). Further, the scope, diversity and level of monitoring and control are largely dependent upon the type of production system and the production intensity. As different production systems offer different challenges and opportunities for monitoring and control, two cases are detailed below, first, (semi-) closed recirculation systems on land, and second, open marine net pens.

1.4.1 Case Study: Recirculating Aquaculture Systems

In recirculating aquaculture systems (RAS), a high level of control of the water environment can be achieved due to their (semi-)closed water system. This requires certain water quality parameters to be automatically and continuously maintained at desired and predetermined levels set by operators. Water flows, temperature, dissolved oxygen saturations, pH and ORP (oxidation reduction potential) are some of the water environment parameters that can be controlled and adjusted. This is commonly done via the use of supervisory control and data acquisition (SCADA) systems, where networked data are collected by remote terminal units, such as real-time sensors, programmable logic controllers (PLC) and other controllers (temperature, flow and pH controllers), and the data is presented to operators via a human machine interface (HMI). The HMI allows the operator to monitor system operation in real time, adjust set points, observe trends, respond and acknowledge alarms and access recorded system data (Timmons et al. 2018). In addition to the level of control that is achieved via industrial automation systems, a number of important characteristics of RAS water such as carbon dioxide, nitrogen compounds and organics can only be monitored, and their levels are largely dependent on operational conditions, feed, fish number and metabolic state and biofilter function in the system. These parameters can be monitored (i) continuously via real-time sensors and instruments (inline/online), (ii) by taking point measurements at certain locations in the system using portable instruments and probes, or (iii) by taking water samples that are subsequently analysed in a lab on site or at offsite labs.

Continuous monitoring via real-time sensors can ensure rapid detection of changes in the water environment and adequate and timely responses. An IoT approach can also facilitate the effective flow of this data between physical and digital twins. For example, a fast response time (minutes) is required if there are electrical issues in the system, problems with the water levels in the tanks or the oxygenation/aeration systems. In contrast, a moderate response time (hours) is required if there are changes in temperature, carbon dioxide and pH that are not yet at levels that can pose a welfare risk. Changes in alkalinity and concentration of nitrogenous compounds (ammonia and nitrite) normally occur slowly and require a response time within days (Timmons et al. 2018). Multiple real-time sensors also allow for simultaneous monitoring of water quality at different places within the system. However, the real-time sensors used in RAS are expensive, and these sensors may sometimes originate from the wastewater industry where water quality ranges and detection requirements differ markedly from farming conditions. Ammonia real-time sensors are a good example, as accurate and cost-efficient continuous measurement of this nitrogenous compound is still lacking for aquaculture. However, a new data assimilation method has recently been suggested for measuring ammonia that combines information from simulation models to enhance real-time estimates in RAS (Jamal et al. 2024).

Given the importance of water quality parameters in RAS, clear and science-based standards must be available for all stakeholders. Some steps toward this are the Norwegian Standard 9417 "Salmon and Rainbow Trout—Terminology and Methods for Documentation of Production" where some limited sampling points for tank-based measurements are addressed, such as monitoring outflows 5cm after the tank drain, or at 1/3 of the water depth within the tank (NS 9417 2022). Further, a range of water quality parameters that should be complied with in RAS are outlined in species-specific welfare standards and certification schemes, such as the RSPCA welfare standards for rainbow trout and Atlantic salmon (RSPCA 2020, 2024). However, more standardised and comprehensive monitoring plans should be applied at each RAS facility in the future to ensure fish health and welfare (Lindholm-Lehto 2023). For instance, Åtland et al. (2020) suggest three main system locations for monitoring different water qualities: (i) the inlet or untreated water; (ii) production water or after water treatment, before it enters the fish tank; and (iii) the tank outlet water or the water that is discharged from the fish tank. The first location monitors potential variations in inlet water quality, due to seasonal changes in temperature, salinity or pathogen levels, which may influence the water treatment before it even enters the system, e.g. through nanofiltration to reduce sulphate and pathogen levels

(Fernandes et al. 2024). The second location reflects water treatment efficiency, which provides information on the water quality that the fish will receive in the tanks and is important for monitoring biofilter function. And the third location is the most representative when it comes to documenting the water quality fish have been exposed to. Furthermore, other measurements of water quality, such as velocity and light conditions, must be measured in the tanks. Overall, current challenges like the high costs of real-time sensors and labour-intensive point measurements must be overcome, and a more automated, cost-effective and time-efficient monitoring of input-based welfare indicators should be established to facilitate the creation of digital twins in RAS.

1.4.2 Case Study: Open Marine Net Pens

In open marine net pens, where the possibility to control the water environment is limited, systematic monitoring of central factors such as oxygen, temperature and salinity (and often turbidity and water velocity) can help evaluate potential welfare risks on specific sites and help steer daily husbandry operations. Problems with dissolved oxygen saturation within the system require a rapid response if levels approach those deemed undesirable or critical. Changes in ambient temperatures can also pose a risk if the change is rapid, prolonged, the fish are not acclimated to the change or if the temperatures approach critical thresholds. Water velocity can affect both water quality and the behaviour of the fish, while salinity and turbidity are generally low risk if levels are not critical for welfare. However, it should be noted that a farmer can influence some water quality parameters within a pen such as dissolved oxygen saturation levels, which can be increased or maintained with supplementary oxygenation technologies, or by improving water exchange. Fish in open systems can be exposed to seasonal changes in the water environment in addition to fluctuations caused by tidal currents and differing current speeds, plankton blooms, the upwelling of colder water with potentially lower DO levels, storms, freshwater runoffs or even exposure to hydrogen sulphide from deep waters (e.g. Noble et al. 2018).

As Atlantic salmon aquaculture is a leader in production technologies (Iversen et al. 2020), the sector has adopted varying scales of technological solutions for monitoring input-based OWIs in marine pens. These can range from the use of a basic Secchi disc for measuring turbidity, hand-held sensors for point sampling, e.g. once a day, to real-time probes that measure temperature, oxygen, salinity, chlorophyll-a, pH and turbidity continuously over the course of a day. While both are important, continuous monitoring is crucial for a successful digital twin, to capture the oceanographic-driven variations in water quality that may impact upon the health, welfare and growth of the fish (Burke et al. 2021). It is also important to monitor the physical parameters of a sea farm's surroundings, particularly ocean currents, waves and tidal action (e.g. Su et al. 2023). Many farms will incorporate weather stations as well as current meters such as Acoustic Doppler Current

Profilers (ADCPs), though these can be quite expensive and the data can be difficult to manage. Novel single-point current sensors (ADVs) can be placed at certain depths and use a compass and tilt instead. These are more conducive to being incorporated into an IoT monitoring approach, where a network of sensors allows communication between them to measure and analyse the data in situ using cloud or edge-based servers (Rastegari et al. 2023).

Novel technologies are also being developed to facilitate mitigation through monitoring a range of emerging exogenous environmental issues that are increasingly driven by climate change. For example, plankton blooms and gelatinous plankton (jellyfish) are becoming a growing threat for fish produced in open marine net pens in some regions (Bosch-Belmar et al. 2020; Clinton et al. 2021; León-Muñoz et al. 2018). Many mitigation methods depend on accurate predictions of algae or jellyfish occurrence to allow timely deployment. Plankton forecasting using predictive models is supported by monitoring, conducted using different remote sensing approaches relying on satellites or unmanned aerial vehicles preforming, e.g. hyperspectral imaging (reviewed in, e.g. Pitt et al. 2024 and Zahir et al. 2024), while traditional water sampling conducted daily at farm sites (e.g. Weeks et al. 2022) relies on assessment by trained personnel and is thus labour intensive and costly. Assessment of water samples based on environmental DNA (eDNA), increasingly used for pathogen surveillance (Bass et al. 2023). While the method has limitations associated with, e.g. spatial or temporal precision, it may work as an early warning tool to prompt/initiate physical sampling of the culture species. Monitoring can also be carried out using cameras; this can include cameras such as feed cameras or cameras mounted onto moving underwater vehicles and drones, allowing the surveillance of larger areas. In addition to in-water cameras, photos taken by aerial drones and hyperspectral and multispectral imaging are utilised from satellites, aircraft and UAVs to quantify water quality (Ma et al. 2023). Moreover, aerial drones can be used for physical water sampling, expediting sampling acquisition over large areas and allowing validation of image-based analyses and models (Kimura et al. 2019; Horricks et al. 2022). Other optical sensors (e.g. based on cell imaging and fluorescence) that detect and identify planktonic organisms in the water column can be deployed from stationary buoys, providing real-time data from crucial sampling points. The retrieval of water quality metrics based on optical properties offers advantages that traditional methods, which rely on collecting and analysing water samples, cannot provide. Key water quality attributes measurable using hyperspectral imaging include chlorophyll-a, transparency and total suspended solids (e.g. Cao et al. 2023). While hyperspectral technology presents a promising solution for measuring water quality, several challenges remain. These include limitations related to the instrumentation and the algorithmic methods used to estimate various water quality parameters from optical signals. Finally, acoustic technology such as echosounders can observe gelatinous plankton (e.g. Magalhães et al. 2020), although species identification is still challenging, yet likely to improve with advancement of sonar technology.

Given the importance of water quality parameters in net pens, clear and science-based standards must be available for all stakeholders. Some steps toward this are the Norwegian Standard 9417 "Salmon and Rainbow Trout—Terminology and Methods for Documentation of Production", which states that water quality should be measured in the middle of the net pen at 3, 5 and 15 metres deep and at the maximum pen depth (NS 9417 2022). Further, the UK has similar requirements, and some of these are currently being revised; see Scottish Environment Protection Agency (SEPA) https://www.sepa.org.uk/regulations/water/aquaculture/regulatory-framework/. Additionally, species-specific information regarding the type and frequency of water quality monitoring in net pens is also addressed in the RSPCA welfare standards (RSPCA 2020, 2024).

1.4.3 Real-Time Measuring of Input-Based Welfare Indicators

Independent of differing production systems, using continuous real-time (inline/online) vs. single-point water quality sensors has various aspects that need to be considered. This includes not only current advantages or disadvantages of available technologies but also the anticipation of future solutions, such as emerging measuring methods and the integration of AI. Table 1.1 outlines differing input-based WIs, how they are monitored and the systems they are relevant and applicable for.

Single-point sensors and measurements, achieved through handheld instruments, probes or lab kits, remain a cost-effective and often reliable method for data collection. However, they can be time-consuming and require standardisation based on specific factors such as location, time of day and production conditions (including light and feeding). To ensure good data quality, proper calibration and maintenance of the probes are essential. Additionally, following appropriate water sampling and storage protocols is critical when using lab measurements to monitor the water environment. Further, it is important to understand the limitations of lab kits and to consider potential interference from other water quality parameters.

The main advantage of using real-time sensors is for continuous monitoring of the water environment, giving the stakeholder the possibility to observe trends over time (e.g. Falconer et al. 2025) during production and use recorded data for modelling and prediction using machine learning and digital twin technology. For example, while vertical and horizontal profiling of input-based water quality indicators using point measurements (Gorle et al. 2018) or mobile systems (Stien et al. 2008) has utility, only the continuous data collection of multiple sensors at varying depths will provide the detailed information that has most utility for a digital twin (Su et al. 2023). However, as real-time sensors are continuously immersed, biofouling and consequent drift in measured values are a common issue (Delgado et al. 2021). Therefore, regular cleaning of the probes, either manual or automatic with pressurised air or water, is necessary to ensure good data quality; automated quality control or quality assurance/quality control (e.g. Timms et al. 2011; Dempsey et al. 2024), based on set acceptable variations or deviations, e.g. using control charts, can serve as a failsafe to detect and mitigate or allay possible consequences of erroneous

Table 1.1 Outline of key input-based welfare indicators to monitor in different types of production systems, including (i) the monitoring methods available (stationary/continuous inline and online systems or point source instruments either lab-based or portable), (ii) relevance to different production systems and (iii) source material, primarily related to Atlantic salmon production

	Monitoring method			Production systems				
		Point source instruments						
Input-based welfare indicators	Continuous systems	Lab-based	Portable	RAS	FTS	OMNP	SCCS	References
Alkalinity	×	×		×			×	Lazado et al. (2022) and Jafari et al. (2024)
Carbon dioxide (CO_2)	×	×	×	×	×		×	Nilsen et al. (2017), Höglund et al. (2023), Lazado et al. (2022) and Jafari et al. (2024)
Dissolved oxygen saturation (DO)	×		×	×	×	×	×	Fivelstad et al. (1999), Nilsen et al. (2017), Gorle et al. (2018), Solstorm et al. (2018) and Burke et al. (2021)
Electrical conductivity (EC)	×		×	×	×		×	Davidson et al. (2017) and Lazado et al. (2022)
Hydrogen sulphide (H_2S)	×			×				Lien et al. (2022)
Light intensity and quality			×	×	×	×	×	Bui et al. (2013), Handeland et al. (2013), Noble et al. (2018) and Qiu et al. (2015)
Nitrite-N, nitrate-N	×	×	×	×			×	Davidson et al. (2017) and Lazado et al. (2022)
Noise	×		×	×	×	×	×	Radford and Slater (2019) and Oppedal et al. (2024)
Oxidative reduction potential (ORP/redox)	×		×	×				Davidson et al. (2017)
Plankton (algae, jellyfish)	×		×		×	×	×	Pitt et al. (2024) and Zahir et al. (2024)
pH	×		×	×			×	Kolarevic et al. (2014) and Lazado et al. (2022)

Salinity	×	×	×	×	×	×	×	Johansson et al. (2009), Ytrestøyl et al. (2020) and Lazado et al. (2022)
Temperature	×	×	×	×	×	×	×	Johansson et al. (2009), Kolarevic et al. (2014), Nilsen et al. (2017) and Chatziantoniou et al. (2023)
Total ammonia nitrogen (TAN)		×		×			×	Davidson et al. (2017) and Lazado et al. (2022)
Total organic carbon (TOC)	×	×		×				Kolarevic et al. (2014)
Turbidity	×	×	×	×		×	×	Treasurer et al. (2003), Kolarevic et al. (2014) and Lazado et al. (2022)
Water velocity, current direction, waves	×		×	×	×	×	×	Plew et al. (2015), Gorle et al. (2018), Nilsen et al. (2019) and Johannesen et al. (2022)

Indicators listed alphabetically

RAS recirculating aquaculture systems, *FTS* flow-through systems on land, *OMNP* open marine net pens, *SCCS* semi-closed containment systems in the sea measurements. Further, the ability to calibrate real-time sensors is desirable to check the accuracy of measurements. Choosing factory-calibrated instruments could lead to uncertainties with data quality and limit monitoring capabilities during the production cycle. Additionally, the price of real-time sensors can limit the implementation in many smaller farms or less capital-intensive sectors.

As a side note, outcome-based welfare indicators such as changes in fish behaviour can be used as a warning system for poor input-based WIs like water quality. For example, erratic swimming behaviour can indicate increasing H_2S levels (Ciani et al. 2024) or gasping at the surface can be caused by low dissolved oxygen saturation levels (Noble et al. 2018). It is further worth reiterating that no single WI can address all welfare needs and that consequently several different WIs must be measured and put in context and in relation to each other (see next section on outcome-based indicators for further information).

When casting an eye to future monitoring possibilities for input-based WIs, an emerging approach for monitoring the rearing environment is passive acoustic monitoring, PAM (e.g. Lindseth and Lobel 2018). PAM is a non-invasive method for monitoring the acoustic signals that are present in the environment, which until recently has primarily focused on natural environments. Acoustic signals, or sounds, are made by a wide range of components in an aquaculture setting, from the equipment and personnel to the fish itself. The combination of sounds that are present at a specific place can be referred to as the soundscape, which can provide information about the noise levels the fish are subjected to (Popper and Hawkins 2019), providing an input-based OWI that is currently rarely monitored. Sound has two elements: particle motion (the kinetic energy, which can be measured in particle acceleration levels (PAL), dB re 1 $\mu m/s^2$), which all fish species can detect, and sound pressure (the potential energy, measured in sound pressure levels (SPL), dB re 1 μPa), which can only be detected by fish species with a swim bladder close to, or connected to, the ear (Nedelec et al. 2016; Popper and Hawkins 2018). Particle motion would likely be the best to monitor for most species from a welfare perspective; however, technology for monitoring sound pressure is currently much more accessible. Sound pressure can be measured using hydrophones, with most solutions being developed for monitoring oceanic sounds. To our knowledge, hydrophones that are specifically designed for aquaculture purposes are not yet available. However, hydrophones have been used to investigate and describe sound pressure levels for several production environments, including RAS tanks (Bart et al. 2001; Craven et al. 2009; Oppedal et al. 2024; Helberg et al. 2024), net pens (Radford and Slater 2019; Oppedal et al. 2024), and ponds (Bart et al. 2001; Radford and Slater 2019). A recent report and article on the soundscapes in salmon aquaculture in Norway describes a lot of variation in net pen soundscapes, particularly related to time of day and different operations, while land-based tanks often have higher but more consistent and predictable background noise levels (Oppedal et al. 2024; Barrett and Oppedal 2025). While exact thresholds for undesirable sound levels might be difficult to establish, it is much easier to look at trends and deviations. Sudden sounds in particular will often cause a behavioural response in the fish indicating stress (Oppedal et al. 2024). PAM could be used for continuous monitoring, gathering data on normal sound levels and identifying any rapid changes or deviations. In addition to monitoring noise, PAM could be used to monitor system operation and the performance of equipment that is used either in, or close to, the production systems. Equipment, such as pumps, oxygenation technologies and feeding systems, produces acoustic signals which could indicate whether they are performing as intended or not (Craven et al. 2009; Helberg et al. 2024). A deviation in the soundscape could indicate equipment malfunction; however, the specific sounds made by different equipment must be characterised and identified to link a change in the soundscape to a specific piece of equipment. Similarly, live video streams may find increased use in the future for system and function surveillance, be it through manual observation or computer vision algorithms.

Whether for existing or emerging technologies, AI may be used both to expand the measurement methods of input-based WIs through AI-driven data processing and to augment the analysis of data from various measurement sources. In the first case, AI can be involved where more complex information needs processing to deliver WI measurement, such as computer vision-based processing of images and videos or multidimensional acoustical data. In the second case, AI can be used for improved pattern recognition in single-source data, such as water quality sensors or for advanced analysis of data from multiple sources to perform analytics on the descriptive, diagnostic and predictive level. This may allow the future inclusion of input-based WIs in the daily operation that are currently LABWIs, such as the detection and prediction of harmful algal blooms (McIlwaine and Casado 2021) or jellyfish occurrences (Zahir et al. 2024). A particular implementation of this advanced AI-based analytics would be digital twins of fish or farms. However, currently, there are still considerable disadvantages to the use of AI-based WI measurement and analytics. For a start, an IoT approach to monitoring farms necessitates electricity and internet availability, which can be difficult in more remote areas (Rastegari et al. 2023). Further, the inherently demanding environment on fish farms poses challenges for the endurance and maintenance of hardware. Additionally, existing hardware and software solutions are expensive, and investment risks are considerable for farms. Lastly, the technology is still in its infancy, and solutions and features are not yet robust and widely applicable.

1.5 Monitoring Outcome-Based Welfare Indicators

The monitoring of outcome-based WIs is critical to making sure that the system and the operations the fish are subjected to consider and meet the welfare needs of the fish. The range of monitoring and surveillance possibilities is diverse and is influenced by the type of WIs that require documenting, the rearing system, capital availability and monitoring needs of the producer and sector. When considering the monitoring of outcome-based OWIs at the individual and group level, visual, acoustic and biosensor-based monitoring technologies are becoming more commonplace in aquaculture, especially in net pen production (Li et al. 2020a, b). While these are not the only technologies that are available to the farmer for enhancing the monitoring of WIs, this subchapter will focus on outlining their application, and its outputs are briefly summarised in Table 1.2 at the end of the section.

1.5.1 Optical-Based Monitoring

Over the last two decades, camera-based monitoring of outcome-based WIs has become more commonplace due to the increased accessibility and affordability of cameras, enhanced quality of recordings and advancements in processing techniques (Antonucci and Costa 2020; Mandal and Ghosh 2024). Camera-based monitoring in aquaculture can utilize both over- or underwater cameras, and these can be single or multiple cameras per rearing system (Biazi and Marques 2023). Overwater systems provide long-term benefits, as they are free from biofouling and not affected by rapid light attenuation, resulting in clearer images (Hao et al. 2024). However, they are limited to surface-level observations, including behaviours like jumping or feeding at the surface (Li et al. 2024), but they can also identify fish facing health challenges, such as salmon smolts that isolate themselves in response to stressors (Stradmeyer 1994; Handeland et al. 1996). Underwater cameras, which may be static or mobile, offer different benefits and challenges. Mobile cameras allow for a wider viewing range but increase the risk of fish collisions, especially if the equipment moves too quickly or if environmental conditions or other stressors impact the behavioural control and mobility of the fish. Specific technologies, including near-infrared cameras, have been used for monitoring fish behaviour including feeding (Zhou et al. 2017), and estimating biomass (Saberioon and Cisar 2018; Lopez-Tejeida et al. 2022). Further, visible spectrum cameras are most commonly used to monitor fish behaviour (Pinkiewicz et al. 2011; Zion 2012; Liu et al. 2022, feeding behaviour (Li et al. 2020b), biomass (Li et al. 2020a), injuries (Li et al. 2022a, b, c) and certain input-based water quality parameters (Antonini et al. 2024; Chiu et al. 2022; Wu et al. 2022). Additionally, 3D cameras and stereo vision are used for fish position monitoring in tanks (Saberioon and Cisar 2016) and biomass estimation in net pens (León-Muñoz et al. 2018), see also Zhao et al. (2025). Furthermore, although technology readiness levels for in situ hyperspectral imaging remain somewhat limited, primarily due to challenges in developing underwater imaging equipment, a breakthrough study by Pettersen et al. (2019) showcased the potential of an underwater hyperspectral imaging system to efficiently detect and distinguish between differing life stages of parasitic salmon lice attached to Atlantic salmon under laboratory conditions based on their unique spectral signatures.

Cameras offer stakeholders a non-invasive method for monitoring various outcome-based WIs applicable at both individual and group levels (Barreto et al. 2022). Particularly in net pen aquaculture, these cameras are currently primarily employed for feeding management, allowing personnel to directly observe feed distribution and fish responses in real time (Zhou et al. 2018). Advancements

Table 1.2 Outlining how differing technologies can be used to monitor key outcome-based welfare indicators (and also some input-based welfare indicators) and potential reference materials

Technology		Outcome-based welfare indicator						
		Individual level		Group level		Additional input-based OWIs		
		Welfare indicator	References	Welfare indicator	References	Welfare indicator	References	IoT
Optical	Overwater	Presence/absence of some injuries (e.g. to the dorsal side of the fish)	C. Noble, pers. obs.	Behaviour including those associated with negative and positive welfare	Li et al. (2022a, b, c) and Cui et al. (2024)	Presence of algal blooms, jellyfish, other planktonic threats	Kimura et al. (2019) and Martin-Abadal et al. (2020)	Yes
		Presence of loser fish	C. Noble, pers. obs.	Appetite	Zhao et al. (2017)			Yes
		Tail beat frequency	Sasaki et al. (2024)					Yes
	Underwater	Emaciation state	Banno et al. (2024a)	Mortality (approximate or precise numbers)	Ranjan et al. (2023)	Visibility (Turbidity)	Antonini et al. (2024) and Chiu et al. (2022)	Yes
		Injuries to snout, jaw, eye, opercula, skin (wound), fins Deformities of: spine, jaw, fin	Ellis et al. (2019), Chen et al. (2022) and Noble, pers. obs.	Behaviour including those associated with negative and positive welfare states	Pinkiewicz et al. (2011)	Presence of algal blooms, jellyfish, other planktonic threats	McIlwaine and Casado (2021)	Yes
		Ectoparasite level	Bui et al. (2024)	Appetite	Li et al. (2020b)			Yes
		Maturation state	Vambeseth (2022)					Yes
		Ventilatory frequency/mouth opening frequency	Zheng et al. (2014) and Hokin et al. (2024)					Yes
		Presence of loser fish	Banno et al. (2024a)					Yes

(continued)

Table 1.2 (continued)

Technology		Outcome-based welfare indicator						
		Individual level		Group level		Additional input-based OWIs		
		Welfare indicator	References	Welfare indicator	References	Welfare indicator	References	IoT
Hyperspectral	Overwater					Presence of algal blooms, visibility (turbidity)	Cao et al. (2023)	Yes
	Underwater	Ectoparasite level	Pettersen et al. (2019)					Yes
	During sampling	Fin haemorrhaging	Lindberg et al. (2023)					No
		Smoltification state	Svendsen et al. (2021a)					No
		Liver fat status	Ortega et al. (2024)					No
Acoustic	Active			Appetite	Folkedal et al. (2022)			Yes
				Behaviour including those associated with negative and positive welfare states	Banno et al. (2024b)			
	Passive			Feeding activity	Rosten et al. (2023), Helberg et al. (2024), Mallekh et al. (2003) and Lagardère et al. (2004)	Noise	Craven et al. (2009), Oppedal et al. (2024) and Radford and Slater (2019)	Yes
				Stress	Jónsdóttir et al. (2024)	System performance	Helberg et al. (2024)	Yes
				Surface activity	Rountree et al. (2018), Helberg et al. (2024)			Yes

Biosensors	Acoustic transmitters	Behaviour including those associated with negative and positive welfare states	Føre et al. (2011, 2018b), Kolarevic et al. (2016), Stehfest et al. (2017), Stockwell et al. (2021) and Ulvund et al. (2021)	–	–	Water temperature	Stehfest et al. (2017)	Yes
						Depth	Stehfest et al. (2017)	
						Dissolved oxygen levels	Stehfest et al. (2017)	
						Salinity and/or conductivity	Mitamura et al. (2017)	
	Bio-loggers	Behaviour including those associated with negative and positive welfare states	Kolarevic et al. (2021), Korus et al. (2024) and Morgenroth et al. (2024)	–	–	Water temperature	Korus et al. (2024)	No
		Respiratory frequency	Kolarevic et al. (2021)			Depth	Korus et al. (2024)	
		Heart rate	Gamperl et al. (2021), Hvas et al. (2020), Korus et al. (2024), Morgenroth et al. (2024), Warren-Myers et al. (2021) and Yousaf et al. (2022)					
LABWIs—e.g. histology, ECGs, system hydrodynamics		Skin health status; heart health status	Sveen et al. (2021) and Zena et al. (2024)	–	–	Rearing system hydrodynamics	Gorle et al. (2018)	No

in computer vision techniques have enhanced the capability of cameras to automatically make observations of both the fish and their behaviour.

For example, object detection facilitates continuous and accurate measurements of fish, including biomass estimations and mortality counts, or the monitoring of falling feed pellets, which can serve as a proxy for appetite (Li et al. 2020a; Liu et al. 2023b; Tanaka et al. 2019; Lines et al. 2001; Saberioon and Cisar 2018; Tian et al. 2024; Hu et al. 2021; Barreto et al. 2022; Georgopoulou et al. 2024). Further, automated feeding systems that assess pellet wastage and fish feeding behaviours can adjust ration sizes and feeding rates dynamically. When combined with environmental monitoring, these technologies provide valuable insights into input- and outcome-based WIs, as changes in feeding behaviours can signal emerging health problems (Martins et al. 2012). Similarly, the tracking of objects enables the assessment of swimming speed, direction or social interactions at both individual and group levels (Zion 2012). Additionally, these monitoring techniques can detect abnormal swimming behaviours, physical deformities, injuries, stunted growth, growth anomalies and parasites (Banno et al. 2024a; Bui et al. 2024; Li et al. 2022a, b, c; Mandal and Ghosh 2024) providing early detection of diseases that may require timely intervention (Li et al. 2022a, b, c; Yu et al. 2023). Overall, algorithms can be trained to recognize specific patterns indicative of health issues and signs of disease, stress, stunted growth or abnormal behaviour enabling proactive measures (Banno et al. 2024a; Palaiokostas 2021; Wang et al. 2023; Mandal and Ghosh 2024; Li et al. 2022a, b, c). Despite their advantages, limitations such as poor lighting, occlusion, background noise and restricted fields of view remain, particularly in larger production systems (Yang et al. 2021; Mandal and Ghosh 2024).

In addition to AI-driven solutions, remote monitoring technologies such as drones, ROVs (remotely operated vehicles) and other remote sensing tools enable stakeholders to monitor risks to fish health and welfare (Ubina et al. 2021; Akram et al. 2022; English et al. 2024; Wu et al. 2022). These technologies enhance overall monitoring opportunities and allow for the collection and analysis of data in real time. Furthermore, AI can support the identification of individual fish within a group, paving the way for individualised treatment and management practices (Cisar et al. 2021). By integrating these innovative monitoring systems, aquaculture practices can improve fish welfare, optimize feeding strategies and ultimately enhance operational efficiency.

1.5.2 Acoustic-Based Monitoring

Acoustic systems, including both passive and active technologies such as single and multibeam sonar systems, play a crucial role in underwater mapping in various industries and are increasingly used in aquaculture. By integrating AI, these systems can significantly enhance their utility, improving data accuracy, interpretation and operational efficiency for fish farming, particularly by being applied to measure outcome-based WIs at both the group and individual levels.

Active acoustics in the form of echosounders, in its simplest form, consists of emitting a sound pulse and measuring the backscatter of this pulse. Backscattering is created when the emitted sound hits an object that has a change in density compared to the medium the sound was emitted in. This echo is what is used as an information source. Echosounders are typically used to monitor the seabed or detect fish schools in the open ocean and estimate the biomass of the school. A direct application in an aquaculture context for estimating fish biomass has proven difficult due to high fish densities. Echosounders are therefore mainly used to monitor the fish school's distribution within the net pen, either vertically (e.g. Johannesen et al. 2020; Folkedal et al. 2022) or horizontally if multiple echosounders are deployed (e.g. Klebert et al. 2023).

The simplest and most common form of echosounders is the single beam echosounder. These consist of one unit which both transmits the sound pulse and receives the echo. They are made to observe the water column and therefore deployed either from the surface looking down or submerged in or beneath the net pen looking upward. A submerged echosounder has the advantage of less motion compared to a surface deployed one, given that its motion has been decoupled from the net pen. It also has a clear reference point with the surface as it creates a strong echo. Deploying from the surface is however an easier operation and can therefore be more cost efficient.

In recent years, multibeam echosounders have also been tested in an aquaculture setting. These are in simple terms a sensor consisting of several beams, which can be electronically steered in different directions and thereby provide high-resolution images (Kristmundsson et al. 2023; Banno et al. 2024b). This combined with their backscattering strength allows monitoring of spatial distribution and swimming behaviour, with the possibility of tracking individual fish within its visual field. The high resolution of the multibeam sonar allows target detection within the pen but also beyond the net of the pen (Banno et al. 2024b).

Single beam echosounders were not originally designed for net pens; hence, they are rarely used for biomass estimations as the high density has the potential to cause a shadowing effect, i.e. high fish density obscuring the fish beyond (Eilertsen et al. 2021). To minimize this issue, the entire water column can be integrated, or data beyond the highest density point can be ignored. Overall, single beam echosounders have been used to monitor feeding behaviour (Folkedal et al. 2022) and preferred position when exposed to high waves and strong currents (Johannesen et al. 2020). Single beam echosounders are well suited to provide information on the group's preferences, while multibeam echosounders provide the additional possibility of detecting singular fish within the school.

However, multibeam sonar data is also expected to be influenced by shadowing effects (Banno et al. 2024b), and issues have been encountered with deterioration of the signal after 10 m due to the high biomass (Kristmundsson et al. 2023). But close to the sonar, they are capable of detecting targets within the school. This feature of multibeam sonars enables many new possibilities, but a significant challenge is the large amount of data it generates, and labour intensity of manual analysis of backscatter data (Kristmundsson et al. 2023). Even data from single beam echosounders can be challenging to work with for similar reasons. Although the data volume is smaller than that of multibeam sonars, the manual analysis of backscatter data is labour intensive, and the data size can become challenging when using several echosounders over long periods of time.

Automatic data analysis using AI or ML can be a tool to address these challenges. There are already examples of AI and ML applied on sonar data in marine net pens, with promising results. Måløy (2020) could detect a change in behaviour 1 month prior to disease being detected within the fish population, using neural networks and single beam echosounder data. Kristmundsson et al. (2023) used machine vision techniques to detect targets in multibeam images in a net pen. Furthermore, combining data from single beam echosounders with current individual-based fish models (Føre et al. 2009) and ML (hybrid analysis and modelling) has proven to be a potential tool for a more comprehensive future behavioural monitoring system (Su et al. 2023; Zhang and Su 2025).

As stated earlier, passive acoustics refers to the use of listening devices to detect and analyse sound waves in the environment without generating any sound or signals of its own. This technology captures environmental sounds produced by organisms and infrastructure. In addition to monitoring environmental noise within the rearing system, passive acoustic monitoring (PAM) could also be used to monitor outcome-based WIs by directly monitoring sounds produced by the fish.

Feeding activity appears to have received the most attention so far; for instance, Atlantic salmon feeding has been linked to changes in the soundscape in both net pens (Rosten et al. 2023) and in RAS tanks (Helberg et al. 2024). Feeding sounds have been found for different species such as brown trout, rainbow trout, turbot (Lagardère et al. 2004) and Atlantic salmon (Helberg et al. 2024). Jónsdóttir et al. (2024) have also used PAM to monitor the stress levels of salmon in marine net pens during a crowding operation.

There is currently no established gold standard for which methods to use when analysing the soundscape with the intent of monitoring fish. Mallekh et al. (2003) focused on specific acoustic signals produced by turbot while feeding; Jónsdóttir et al. (2024) used an automated detector to count instances of a specific acoustic signal; Rosten et al. (2023) used 10-min averages of sound intensity levels, while Helberg et al. (2024) demonstrated the potential of using acoustic indices. Unfortunately, it might be difficult to differentiate between the different sounds that are present. Sounds of wind, rain, currents, birds, other fish, farming operations and particularly vessel activity affect the soundscapes of net pens creating a variable soundscape which can mask the sounds of interest (Barrett and Oppedal 2025). In tanks, sounds from the system operation (pumps, oxygenation, feeding system), and reflections from the tank wall, could make it difficult to distinguish fish sounds (and vice versa). This is further challenged by the limited knowledge that is available about which sounds are produced by the fish, as well as why and in which situations they make them. While there is limited knowledge about how to minimise disturbing noise in net pens, there has been some research on reducing noise in tanks. Davidson et al. (2007) have described measures to minimise noise in fibreglass tank, such as ensuring the inlet pipe does not come into contact with the rearing tank; making sure effluent piping infrastructure is insulated and is not connected to a collective drain line (Petersen and Jurevicius 2009); elevating tanks on cement blocks; and seating tanks on insulated padding. With regard to enhancing the utility of the PAM data using AI/ML, machine learning based on spectrograms has been tested to classify feeding response from recorded audio (Du et al. 2023; Zeng et al. 2023).

1.5.3 Biosensor-Based Monitoring

Monitoring fish behaviour and physiology helps stakeholders understand how the animal is affected by and copes with their rearing environment (e.g. Macaulay et al. 2021; Martins et al. 2012; Saraiva et al. 2022). Advances in animal tracking technologies based on biosensors, in particular acoustic telemetry and bio-logging, can provide valuable insights for responsible aquaculture management, allowing for monitoring of not only outcome-based WIs at individual and group level but by extension also some input-based WIs. Acoustic transmitters and bio-loggers are inserted in, or fixed on, animals to record aspects of environment the animal is subject to, or aspects of its, e.g. behaviour or physiology (Whitford and Klimley 2019; Barreto et al. 2022). There are a wide variety of biotelemetry and bio-logging tools and techniques available that are applicable to and offer utility in aquaculture settings (Whitford and Klimley 2019; Macaulay et al. 2021). However, the expected welfare costs and benefits of these tools and methods to the animals should be evaluated, given the unavoidable invasive nature of in situ biosensors (Macaulay et al. 2021; Barreto et al. 2022).

Acoustic telemetry technology consists of battery-powered tags that transmit registered data via hydro-acoustic signals to external receivers (Føre et al. 2017; Edwards et al. 2019), which are generally submerged near or within the fish tank or pen. Acoustic transmitters can be equipped with different sensors for measuring, e.g. pressure, temperature (internal or external), salinity and/or conductivity and accelerometers. Receivers, which can also be equipped with environmental sensors, capture data and process into exploitable data. This data is then saved locally or uploaded to the cloud (Føre et al. 2017). Since acoustic tags are attached externally or surgically implanted in individual animals, only a subset of individuals within a larger group are tagged and act as sentinels to obtain relevant information. Fish swimming activity, trajectories, spatial use and feeding behaviour are among the most common outcome-based WIs used to monitor fish welfare with acoustic telemetry. These behavioural parameters have been monitored (on e.g. salmonid and cleaner fish species) with acoustic transmitters in different aquaculture settings in order to assess acclimation and fish responses to husbandry procedures (operational routines, density, feeding strategy) and environmental conditions (lighting, atmospheric and/or water conditions) (Bégout Anras and Lagardère 2004; Brooker et al. 2020; Leclercq et al. 2018; Stockwell et al. 2021; Stehfest et al. 2017). The use of accelerometers (acoustic acceleration transmitters) has gained special attention in fish tracking during the last decade in aquaculture, allowing a more detailed assessment of salmonid behaviour in both tanks (Gesto et al. 2020; Kolarevic et al. 2016; Zupa et al. 2021) and net pens (Føre et al. 2011, 2018b; Ulvund et al. 2021).

Bio-loggers are battery-powered devices equipped with sensors that store registered information in an internal memory. Similar to transmitters, loggers are attached externally or surgically implanted in the fish to record components of their environment (e.g. Korus et al. 2024), physiology (e.g. Hvas et al. 2020) or behaviour (e.g. Kolarevic et al. 2021). However, loggers do not transmit the recorded information, and data must be retrieved directly from the device. This implies that the tagged fish (or the logger) must be recovered, and data can be downloaded only after the tag has been removed from the animal. This can be challenging in commercial settings as (i) it can be hard to re-capture the tagged fish, (ii) the sensors may also detach and be lost out of the rearing system (Barreto et al. 2022), and (iii) the data does not fit with real-time digital twins. Nevertheless, data storage tags (DST) or bio-loggers have been used to assess the effects of environmental and husbandry conditions in several fish species and aquaculture settings (including Atlantic salmon in net pens, Johansson et al. 2009; Korus et al. 2024). Bio-loggers can utilise different types of sensors (e.g. depth and temperature sensors, accelerometers, magnetometers, gyroscopes) allowing for the documentation and examination of various outcome-based WIs, such as swimming activity, locomotion, trajectories or the spatial distribution of farmed fish. A miniaturised biosensor equipped with a tri-axial accelerometer, attached to the fish operculum, allows the assessment of respiratory frequency and body tail movement of fish in captivity (Martos-Sitcha et al. 2019). This device has been tested in several species, including Atlantic salmon and rainbow trout (Calduch-Giner et al. 2022; Kolarevic et al. 2021), but only at experimental scale. Bio-loggers can be also equipped with electrodes to obtain physiological metrics, such as heart rate and electrocardiograms (ECG), in free-swimming fish. In the last decade, ECG loggers have been widely used to assess the effects of environmental conditions and husbandry operations upon the potential stress response of the procedure (included stunning and slaughtering) on Atlantic salmon (Gamperl et al. 2021; Hvas et al. 2020; Korus et al. 2024; Morgenroth et al. 2024; Warren-Myers et al. 2021; Yousaf et al. 2022) and rainbow trout (Brijs et al. 2018, 2019a, b). ECG loggers equipped with other sensors can give relevant information of both physiological and behavioural responses of captive fish, and therefore, a wider assessment of their response to stress and welfare, but the size of the fish is a limiting factor in these cases. In addition, the suitable attachment of ECG loggers is a challenge, since they must be inserted into the abdominal cavity, close to the pericardium, and a high level of expertise is required to avoid potential undesirable side effects on health and welfare in relation to their implantation (e.g. Brown et al. 2011).

Some authors have combined acoustic transmitters and bio-loggers on Atlantic salmon individuals within the same net pen (Føre et al. 2017; Svendsen et al. 2021b), highlighting the complementary nature of these technologies. Acoustic tags deliver data in real time, whereas bio-loggers do not allow real-time applications but provide relevant information on a number of behavioural and physiological WIs.

In this sense, some prototypes have been already tested in farmed fish where the same device is equipped with data storage and transmitters, which will increase the potential for use and applicability of these technologies to monitor output-based WIs in commercial farms. Despite current limitations, biosensors and tracking technologies are useful observational tools, allowing the quantification of relevant output-based and input-based WIs, which combined with mathematical modelling conform a powerful tool for better interpretation, presentation and refinement of the measurement data, improving the decision support systems and management plans (Føre et al. 2024).

1.5.4 Ex Situ Monitoring

Automation and digitalisation may make the measurement of complex welfare indicators, which are currently considered laboratory-based WIs (LABWIs), more accessible for on-site farm use. These indicators currently involve either more complicated measuring methods or include collecting samples for lab/off site analysis. They can be more time-consuming and costly and hence are used less frequently and often in smaller sample sizes. Future AI-based technological advances may be applied both at the moment of sampling and in the processing or analysis of samples to make these WIs more operational. An example of the first could be the use of hyperspectral imaging during routine handling, and the augmentation of tissue analysis could serve as an illustration for the latter.

Hyperspectral imaging has been successfully validated as a method to determine the smoltification status of juvenile Atlantic salmon, distinguishing whether they are in the freshwater stage (parr) or have transitioned to the seawater stage (smolt) through spectral analysis (Svendsen et al. 2021a). This technique could potentially reduce the risks of, e.g. transferring poorly smolted fish to sea. It also offers much promise as a high-throughput technology for measuring and monitoring injury-based OWIs during fish handling and sampling. Lindberg et al. (2023) demonstrated the effectiveness of a hyperspectral imaging system for identifying and measuring the severity of haemorrhaging in the dorsal fins of Atlantic salmon (see Fig. 1.2). The method utilises the measurement of specific absorption peaks of haemoglobin within the visible light spectrum to detect blood in and on the fins and has the potential to surpass human capabilities in identifying active injuries. A recent study has explored stress responses in lumpfish by non-invasively analysing their biofluorescence using hyperspectral imaging (Juhasz-Dora et al. 2024). The findings indicate that the fluorescence of lumpfish noticeably changes when they are stressed, and fluorescence may be a valuable emerging welfare indicator. Hyperspectral imaging has also been demonstrated as an effective technique for estimating liver fat content (Ortega et al. 2024). This method has been tested on three aquaculture species: Atlantic salmon, European seabass, and Atlantic cod. The results indicate that hyperspectral imaging is a rapid and cost-effective alternative for quantifying fat content, making it potentially valuable for applications in aquaculture settings.

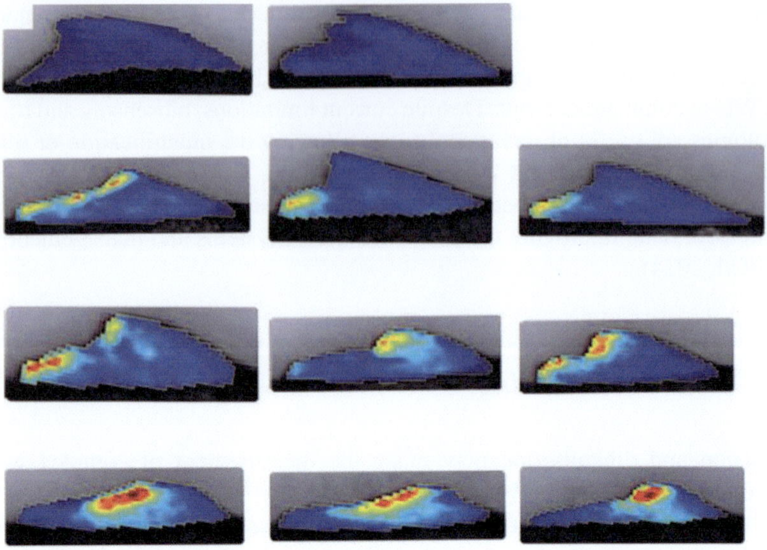

Fig. 1.2 Summarising how hyperspectral imaging can be used to detect fin haemorrhaging within the dorsal fin of Atlantic salmon, where blood intensity has been visualised via a blue-red colour scale. (Figure Copyright © 2023 Lindberg, Durland, Heia, Noble, Alvestad and Difford, and the figure has been adapted from Lindberg et al. 2023, an open-access article distributed under the terms of the Creative Commons Attribution License (CC BY))

Overall, this research showcases hyperspectral imaging as a promising method for enhancing welfare monitoring within aquaculture settings, offering a significant tool for the aquaculture sector.

For lab-based technologies such as immunohistology and gene expression analysis, AI-based advances may greatly enhance their usability for aquaculture in the future. First, AI algorithms can automate image analysis (e.g. Sveen et al. 2021), leading to quicker and more accurate identification of cells and biomarkers while reducing human error. Additionally, AI may enable the integration of large datasets from various sources, enhancing insights into correlations between gene expression, phenotypic outcomes and environmental conditions. Moreover, AI can streamline workflows by automating repetitive tasks like sample preparation or data entry and improve quality control in the laboratories. While the incorporation of AI into these lab-based technologies can potentially boost efficiency, accuracy and effectiveness in complex outcome-based WIs and make them more affordable, applicable and usable, AI may also, e.g. generate more data to interpret and make the process more time-consuming. They may also make mistakes, and the algorithm may need to be updated to increase and maintain accuracy. Further, these approaches and indicators do not offer the stakeholder an opportunity for continuous online monitoring, which is essential for digital twin (s), but future PFF developments may allow for this transition.

Another noteworthy lab-based technology is electrocardiogram (ECG) analysis for real-time cardiac monitoring, which provides valuable insights into underlying heart disease by tracking changes in heart rate and rhythm in fish. Diagnosing ECG abnormalities and arrhythmias requires distinguishing normal from abnormal heartbeats and classifying them accordingly. However, since ECG screening for diagnostic purposes is a relatively new method in farmed fish (Zena et al. 2024), identifying disease-related abnormalities can be challenging and time-consuming. AI-powered ECG interpretation holds promise for enhancing the accuracy and efficiency of detecting cardiac abnormalities in farmed fish.

1.6 Summary: Monitoring Fish and Their Environment and Integrating This Data into Digital Twins

This chapter first introduced the reader to the concept of animal welfare and ways to measure and monitor fish welfare and the environment they are subjected to, via the use of welfare indicators. Thereafter, examples of differing types of technologies and instruments that can add utility to the measurement, assessment, monitoring and auditing of animal welfare are presented, including how they might be incorporated into an IoT approach. The following section outlines how the data from the physical sensors can be considered in relation to aquacultural digital twin solutions with further added value being realised via the integral application of artificial intelligence and machine learning. As we stated in the introduction, we do not propose digital twins for fish health and welfare in this chapter (these are not available at time of writing), but we do strongly suggest the reader refers to articles by, e.g. Budaev et al. (2024, 2025) and Giske et al. (2025a) for an in-depth overview of how fish health and welfare should be considered in digital twins, as they outline the components and mechanisms of potential twin models.

1.6.1 Fish Health and Welfare Data into Digital Twins

A digital twin is a virtual depiction of a physical object, process or system that is used to simulate, predict and optimize performance. Aquacultural digital twins may be applied at different levels, such as at the level of an individual fish, rearing unit or site. In either case, a comprehensive digital twin should consider and address key attributes of an animal's biology that relates to their health and welfare to strengthen their knowledge base and enhance their credibility and utility. While an animal's health and welfare state are a complex interplay of a multitude of factors, digital twins are inherently designed to consider these complexities in a holistic fashion. Budaev et al. (2024) and Giske et al. (2025a) have reviewed the need and knowledge required for developing and designing a digital twin for individual Atlantic salmon.

They outline the utility that the potential a salmon digital twin has for early warning, diagnostic and predictive capabilities in relation to health and welfare and outlined modelling frameworks for potential digital twins (Budaev et al. 2024; Giske et al. 2025a). Specific models and frameworks related to the mathematical modelling of input-based WIs such as dissolved oxygen conditions within a rearing system (Føre et al. 2024), outcome-based WIs such as appetite (Budaev et al. 2025) and behaviour (Su et al. 2023; Zhang and Su 2025) have also been developed.

While the commercial application of digital twins in aquaculture is still in its infancy (e.g. Su et al. 2023; Ubina et al. 2023), the field is rapidly evolving with new possibilities likely entering the market at regular intervals. The technology level of the sector is high enough for the consideration of digital twins in aquaculture, and the possibilities of real-time data collection can be integrated into their construct (Føre et al. 2024).

The validity of models such as digital twins depends on the quality and quantity of the data they receive. With regard to animal welfare, the selected indicators that provide data to the twin must provide information on the fulfilment of the majority (preferably all) of the known welfare needs of the species and life stage to maximise the twin's utility. The indicators that are outlined in this chapter, and the automation of their assessment, have a mostly function-based focus (Segner et al. 2012) with the exception of, e.g. behaviour, heart rate and some physiological indicators. The link between, e.g. behaviour and welfare, through decision-making, facilitates a more holistic assessment of welfare through behavioural observation (Giske et al. 2025b). This emphasises the importance of including outcome-based welfare indicators based on behaviour into a welfare assessment and into digital twins (Giske et al. 2025b). While fundamental research is progressing on emotions and positive welfare in farm animals (Rault et al. 2025), the basis of current fish welfare assessments relies on data from more scientifically substantiated welfare indicators that have an, e.g. morphological, behavioural and physiological focus (e.g. Segner et al. 2012). Bridging this gap between the physical and mental state of an individual can enhance the utility of both existing welfare assessments and also future digital twins on fish health and welfare.

Digital twins are reliant on real-time health and welfare data for maintaining their utility and relevance to the real-life situations the fish can encounter on a farm. This chapter outlines examples of a range of input- and outcome-based welfare indicators that can be currently monitored in real time and online. If properly set up, a digital twin system can provide real-time insight into the biomass which can be used for detecting and potentially preventing or limiting disease outbreaks, as well as optimising feed plans and feed distribution, thus enhancing farm management practices and improving efficiency and sustainability.

However, stakeholders must be vigilant about introducing bias to the development and application of digital twins (e.g. Mihai et al. 2022), especially where certain systems, sectors or regions have more comprehensive monitoring than others. This can mean that digital twin solutions designed for salmon in a specific production system would not be suitable for other species or systems where monitoring technologies are not equally utilised. While some of the required monitoring technologies could potentially be adapted from other species/production systems, potential knowledge gaps would need to be identified first, which can present opportunities for cross-industry applications of monitoring solutions.

1.6.2 Collecting Data for Digital Twins

Technological and instrumentation possibilities for measuring and monitoring fish health and welfare in aquaculture are diversifying in their breadth and scope and many offer online/inline capabilities that will satisfy the real-time requirements of incorporating input- and outcome-based WIs into digital twins.

Digital twins of specific sites are achievable today, but a major obstacle to collecting data into a digital twin is the lack of standards. Most sensors come with their own protocol, their own cloud and their own API. Another obstacle when it comes to aquaculture and modelling beyond a specific site is collecting data between sites and companies. This is particularly relevant for representing disease risks in a digital twin, since the twin shouldn't only consider its local environment but rather the whole ecosystem to achieve the best models and predictions. This leads to a balancing act between preserving company competition advantages and providing data for the creation of common value (NCE Seafood Innovation 2023).

With the introduction of the European Commission's Data Act in 2024, "connected products will have to be designed and manufactured in a way that empowers users (businesses or consumers) to easily and securely access, use and share the generated data" this will work in favour of digital twins which can utilize different connected products to gather and integrate different data sets within sites and companies. The next big challenge will be to ensure sharing of data across companies so digital twins can consider factors outside their own sites. Here there is a significant gap between technology suppliers who suggest this can be done within 1–2 years and fish farmers who suggest this needs 5–10 years (NCE Seafood Innovation 2023).

For digital twins of specific pens or sites, work has already begun in this direction for a limited number of farmed fish species (e.g. Su et al. 2023; Lima et al. 2023). The integration of the monitoring of the fish and their environment will also allow for the development of more holistic digital twins and improve their utility for the stakeholder.

The potential absence of accurate, repeated and standardised measurements of fish welfare can make monitoring a production system via a digital twin difficult as continuous real-time monitoring is a key facet of their utility (Føre et al. 2024). Opportunities for PFF and AI to non-invasively monitor fish health, disease patterns, growth, behaviour, etc. are available (Rather et al. 2024). However, diagnosing fish diseases is complex, since diseases are often multifactorial, and it may be difficult to provide precise descriptions, and/or the symptoms may be difficult to detect. Thus, more accurate diagnostic tools are still required, and these tools are often invasive. The analysis and processing of the data these tools generate often requires a high level of expertise to interpret the results, but this can be streamlined with possibilities such as digital histopathology (Sveen et al. 2021). To advance in the field, a better insight and understanding of the links between environment, fish behaviour, signs and symptoms of diseases and characterisation of early-stage disease markers are needed.

1.6.3 Integrating Data in Digital Twins

There are several ways data can travel from the edge sensors to a digital twin that needs to be considered when designing a digital twin system. The first major architectural decision is whether to base the system on a pull- or push-based approach. Many sensor providers provide automatic transfer to their own proprietary cloud and then provide API access to this service. A digital twin service can then access this system at regular intervals to ask for updated data. Since the data has to be pulled from the service without knowing when the data was last updated, this approach cannot achieve true real-time access to data, although it can approach it through rapid requests to the cloud service.

An alternative to the pull-based architecture is push based. Central to this architecture is the concept of a publish-subscribe, an architectural pattern that provides a framework for exchanging messages between publishers and subscribers in an asynchronous manner. Within this pattern, there exists multiple different IoT protocols that can handle machine-to-machine communication for a digital twin. These have different trade-offs that need to be considered for the given system, but the two major approaches to choose between are the hub and spokes architecture, connecting edge sensors to a central application, or a peer-to-peer architecture which is optimised for distributed processing. The push-based approach delivers new sensor data to the digital twin system as soon as it is registered on the sensor.

Thus, when designing the data integration part of a digital twin, a major thing to consider is what the finished digital twin will be used for and whether it requires access to real-time data or how close to real-time data is necessary for making good decisions. This will influence the choice of underlying integration architecture.

1.6.4 Hurdles for Operational Feasibility

There are many ways to implement a framework for monitoring fish welfare and applying a digital twin that incorporates fish health and welfare data, and different systems and sites may pose specific challenges that require tailored solutions. For instance, a reliable communication backbone may allow for most or all of storage and computation to occur in the "cloud" (even though edge computing solutions may still have advantages). However, common to all solutions involving a digital twin are a set of sensors monitoring the fish and the environment on the farm, storage capacity for data and computation capacity for providing decision support or other services. With the sensor technologies discussed earlier, one key challenge concerns the amount of data generated, which is generally substantial for both optic- and acoustics-based technologies. This in turn quickly places high demands on storage, computation and bandwidth for communications. However, for a digital twin, it is likely that such raw streams are not used directly but rather in a pre-processed form where "useful" information has been extracted. Therefore, in most cases, when the processing can be done at the farm at the expense of local computing capacity, the amount of data involved is more amenable to storage and significantly reduces the communication load. Such an edge-computing approach presumes the raw sources may be discarded which is most likely true for a digital twin, but less so for manual overview, where necessary; however, even this is usually only used on the fly, as reviews of long histories of, for instance, video streams are likely not commonplace. One consequence of discarding raw data is that it precludes retrospective studies for applying more targeted models, for instance, to learn new patterns explaining or predicting specific events. While this may not be a common need for fish farms, and often may be done in prepared studies, a compromise may be to build in triggers to store particularly relevant subsets of data. Overall, there will emerge a range of solutions ranging from mature, simple technologies that can operate locally with moderate storage and computing capacity along with low requirements for technical and biological competency, to the more involved cases where larger firms with many farms could participate in developing improved models for AI and digital twins. While the former may lag in terms of development, it is still likely that the "low-hanging fruit" will provide commercial gains for many.

Some solutions will evolve to be robust and mature, while others may remain more difficult to apply before they mature. From a stakeholder's perspective, taking advantage of these new solutions may require a level of technical competency for understanding how and if new developments fit into the farm, and biological competency for interpreting signals and incorporating them into appropriate decision support. This added complexity and need of competency has implications for the IT infrastructure of the farm (see, e.g. Føre et al. 2024) as it may require some form of maintenance depending on if the solution is provided for by an external company or if the infrastructure is built in-house: Servers may need to be reset; fault situations (such as power failures and backup issues) may need to be resolved, analysing the output of the tools in terms of the events happening on the farm.

Acknowledgments This work was supported by funding from the Nofima projects 'DigitalAqua' grant number 12749, 'ML4AKVA' grant number 14224, 'AI-WELL' grant number 13937, the Research Council of Norway SIS project 'Next-Level-Welfare', grant number 14473, the Institute of Marine Research project 16191 'SatsForVel', and the UK Research and Innovation Future Leaders Fellowship (MR/V021613/1). Part of the work was also supported by VVI CENAKVA Research Infrastructure (LM2023038, MEYS CR, 2023–2026) and the Research Council of Norway Innovation Project for the Industrial Sector 'Shoaling Collective Health Observation and Outbreak Likelihood System (SCHOOLS)' grant number 355949.

We also thank the reviewers of an earlier draft of this chapter for their constructive and insightful comments that have helped improve this final version.

References

Akram W, Casavola A, Kapetanović N, Miškovic N (2022) A visual servoing scheme for autonomous aquaculture net pens inspection using ROV. Sensors 22(9):3525. https://doi.org/10.3390/s22093525

Antonini G, Pearce JM, Berruti F, Santoro D (2024) A novel camera-based sensor for real-time wastewater quality monitoring. Water Pract Technol 19(9):3778–3793

Antonucci F, Costa C (2020) Precision aquaculture: a short review on engineering innovations. Aquacult Int 28:41–57. https://doi.org/10.1007/s10499-019-00443-w

Arabelli R, Bernatin T (2022) Water quality monitoring and controlling systems for aquaculture. In: International conference on data science, machine learning and applications. Springer, Singapore, pp 57–67. https://doi.org/10.1007/978-981-99-2058-7_7

Arulmozhi E, Deb NC, Tamrakar N, Kang DY, Kang MY, Kook J, Basak JK, Kim HT (2024) From reality to virtuality: revolutionizing livestock farming through digital twins. Agriculture 14(12):2231. https://doi.org/10.3390/agriculture14122231

Åtland Å, Dale T, Bloecher N (2020) Environmental considerations in aquaculture health management. In: Aquaculture health management. Academic. ISBN:978-0-12-813359-0, pp 235–280. https://doi.org/10.1016/B978-0-12-813359-0.00008-7

Banno K, Gonçalves FMF, Sauphar C, Anichini M, Hazelaar A, Sperre LH, Stolz C, Aas GH, Gansel LC, da Silva Torres R (2024a) Identifying losers: Automatic identification of growth-stunted salmon in aquaculture using computer vision. Mach Learn Appl 16:100562. https://doi.org/10.1016/j.mlwa.2024.100562

Banno K, Gao S, Anichini M, Stolz C, Tuene SA, Gansel LC (2024b) Expanded vision for the spatial distribution of Atlantic salmon in sea cages. Aquaculture 588:740879. https://doi.org/10.1016/j.aquaculture.2024.740879

Barreto MO, Rey Planellas S, Yang Y, Phillips C, Descovich K (2022) Emerging indicators of fish welfare in aquaculture. Rev Aquac 14(1):343–361. https://doi.org/10.1111/raq.12601

Barrett LT, Oppedal F (2025) Characterisation of the underwater soundscape at Norwegian salmon farms. Aquaculture 602:742334. https://doi.org/10.1016/j.aquaculture.2025.742334

Bart AN, Clark J, Young J, Zohar Y (2001) Underwater ambient noise measurements in aquaculture systems: a survey. Aquac Eng 25(2):99–110. https://doi.org/10.1016/S0144-8609(01)00074-7

Bass D, Christison KW, Stentiford GD, Cook LS, Hartikainen H (2023) Environmental DNA/RNA for pathogen and parasite detection, surveillance, and ecology. Trends Parasitol 39(4):285–304. https://doi.org/10.1016/j.pt.2022.12.010

Bégout Anras M-L, Lagardère JP (2004) Measuring cultured fish swimming behaviour: first results on rainbow trout using acoustic telemetry in tanks. Aquaculture 240:175–186. https://doi.org/10.1016/j.aquaculture.2004.02.019

Berckmans D (2004) Automatic on-line monitoring of animals by precision livestock farming. In: Livestock production and society. Academic, Wageningen, pp 287–294. https://doi.org/10.3920/9789086865673_023

Biazi V, Marques C (2023) Industry 4.0-based smart systems in aquaculture: a comprehensive review. Aquac Eng:102360. https://doi.org/10.1016/j.aquaeng.2023.102360

Bosch-Belmar M, Milisenda G, Basso L, Doyle TK, Leone A, Piraino S (2020) Jellyfish impacts on marine aquaculture and fisheries. Rev Fish Sci Aquac 29(2):242–259. https://doi.org/10.1080/23308249.2020.1806201

Bracke MBM, Spruijt BM, Metz JHM (1999) Overall animal welfare assessment reviewed. Part 1: Is it possible? Neth J Agric Sci:279–291

Brijs J, Sandblom E, Axelsson M, Sundell K, Sundh H, Huyben D, Broström R, Kiessling A, Berg C, Gräns A (2018) The final countdown: continuous physiological welfare evaluation of farmed fish during common aquaculture practices before and during harvest. Aquaculture 495:903–911. https://doi.org/10.1016/j.aquaculture.2018.06.081

Brijs J, Sandblom E, Axelsson M, Sundell K, Sundh H, Kiessling A, Berg C, Gräns A (2019a) Remote physiological monitoring provides unique insights on the cardiovascular performance and stress responses of freely swimming rainbow trout in aquaculture. Sci Rep 9:9090. https://doi.org/10.1038/s41598-019-45657-3

Brijs J, Sandblom E, Rosengren M, Sundell K, Berg C, Axelsson M, Gräns A (2019b) Prospects and pitfalls of using heart rate bio-loggers to assess the welfare of rainbow trout (*Oncorhynchus mykiss*) in aquaculture. Aquaculture 509:188–197. https://doi.org/10.1016/j.aquaculture.2019.05.007

Brooker AJ, Davie A, Leclercq E, Zerafa B, Migaud H (2020) Pre-deployment acclimatisation of farmed ballan wrasse (*Labrus bergylta*) to sea-cage conditions promotes behaviour analogous to wild conspecifics when used as cleaner fish in Atlantic salmon (*Salmo salar*) farms. Aquaculture 520:734771. https://doi.org/10.1016/j.aquaculture.2019.734771

Broom DM (2018) The scientific basis for action on animal welfare and other aspects of sustainability. In: Farming, food and nature. Routledge, pp 93–100

Brown C, Laland K, Krause J (2011) Fish cognition and behavior. Wiley

Budaev S, Dumitru ML, Enberg K, Handeland SO, Higginson AD, Kristiansen TS, Opdal AF, Railsback SF, Rønnestad I, Vollset KW, Mangel M (2024) Premises for a digital twin of the Atlantic salmon in its world: agency, robustness, subjectivity and prediction. Aquaculture, fish and fisheries 4(1):e153. https://doi.org/10.1002/aff2.153

Budaev S, Cusimano G, Rønnestad I (2025) Fishmet: a digital twin framework for appetite, feeding decisions and growth in salmonid fish. In: Feeding decisions and growth in salmonid fish. https://doi.org/10.1002/aff2.70064

Bui S, Oppedal F, Korsøen ØJ, Sonny D, Dempster T (2013) Group behavioural responses of Atlantic salmon (*Salmo salar* L.) to light, infrasound and sound stimuli. PLoS One 8(5):e63696. https://doi.org/10.1371/journal.pone.0063696

Bui S, Folkedal O, Nerbø IS, Barrett LT (2024) Standard operasjonsprosedyre for manuell lusetelling til validering av automatisk telling. Rapport fra havforskningen.. https://imr.brage.unit.no/imr-xmlui/handle/11250/3172564

Burke M, Grant J, Filgueira R, Stone T (2021) Oceanographic processes control dissolved oxygen variability at a commercial Atlantic salmon farm: application of a real-time sensor network. Aquaculture 533:736143. https://doi.org/10.1016/j.aquaculture.2020.736143

Calduch-Giner J, Holhorea PG, Ferrer MÁ, Naya-Català F, Rosell-Moll E, Vega García C, Prunet P, Espmark ÅM, Leguen I, Kolarevic J, Vega A, Kerneis T, Goardon L, Afonso JM, Pérez-Sánchez J (2022) Revising the impact and prospects of activity and ventilation rate bio-loggers for tracking welfare and fish-environment interactions in salmonids and Mediterranean farmed fish. Front Mar Sci 9:854888. https://doi.org/10.3389/fmars.2022.854888

Cao Q, Yu G, Qiao Z (2023) Application and recent progress of inland water monitoring using remote sensing techniques. Environ Monit Assess 195(1):125. https://doi.org/10.1007/s10661-022-10690-9

Chatziantoniou A, Papandroulakis N, Stavrakidis-Zachou O, Spondylidis S, Taskaris S, Topouzelis K (2023) Aquasafe: a remote sensing, web-based platform for the support of precision fish farming. Appl Sci 13(10):6122. https://doi.org/10.3390/app13106122

Chen JC, Chen TL, Wang HL, Chang PC (2022) Underwater abnormal classification system based on deep learning: a case study on aquaculture fish farm in Taiwan. Aquac Eng 99:102290. https://doi.org/10.1016/j.aquaeng.2022.102290

Chiu MC, Yan WM, Bhat SA, Huang NF (2022) Development of smart aquaculture farm management system using IoT and AI-based surrogate models. J Agric Food Res 9:100357

Ciani E, Kvæstad B, Stormoen M, Mayer I, Gupta S, Ribičić D, Netzer R (2024) Early warning through video monitoring: dissolved hydrogen sulphide (H2S) affects Atlantic salmon swimming behavior in recirculating aquaculture systems. Aquaculture 581:740201. https://doi.org/10.1016/j.aquaculture.2023.740201

Cisar P, Bekkozhayeva D, Movchan O, Saberioon M, Schraml R (2021) Computer vision based individual fish identification using skin dot pattern. Sci Rep 11(1):16904. https://doi.org/10.1038/s41598-021-96476-4

Clinton M, Ferrier DE, Martin SA, Brierley AS (2021) Impacts of jellyfish on marine cage aquaculture: an overview of existing knowledge and the challenges to finfish health. ICES J Mar Sci 78(5):1557–1573. https://doi.org/10.1093/icesjms/fsaa254

Collins Dictionary (2024) © Collins 2024. Available at: https://www.collinsdictionary.com/dictionary/english/indicator. Accessed 1 Dec 2024

Craven A, Carton AG, McPherson CR, McPherson G (2009) Determining and quantifying components of an aquaculture soundscape. Aquac Eng 41(3):158–165. https://doi.org/10.1016/j.aquaeng.2009.07.003

Cui M, Liu X, Liu H, Zhao J, Li D, Wang W (2024) Fish tracking, counting, and behaviour analysis in digital aquaculture: a comprehensive review. arXiv preprint arXiv:2406.17800

Davidson J, Frankel AS, Ellison WT, Summerfelt S, Popper AN, Mazik P, Bebak J (2007) Minimizing noise in fiberglass aquaculture tanks: noise reduction potential of various retrofits. Aquac Eng 37:125–131. https://doi.org/10.1016/J.AQUAENG.2007.03.003

Davidson J, Good C, Williams C, Summerfelt ST (2017) Evaluating the chronic effects of nitrate on the health and performance of post-smolt Atlantic salmon *Salmo salar* in freshwater recirculation aquaculture systems. Aquac Eng 79:1–8. https://doi.org/10.1016/j.aquaeng.2017.08.003

Delgado A, Briciu-Burghina C, Regan F (2021) Antifouling strategies for sensors used in water monitoring: review and future perspectives. Sensors 21(2):389. https://doi.org/10.3390/s21020389

Dempsey DP, Torrie NL, Reid GK, Lewis-McCrea LM, McKee A, Woodside R (2024) The makings of a high resolution coastal monitoring program. In: OCEANS 2024-Halifax. IEEE, pp 1–10. https://doi.org/10.1109/OCEANS55160.2024.10753863

Du Z, Cui M, Wang Q, Liu X, Xu X, Bai Z, Sun C, Wang B, Wang S, Li D (2023) Feeding intensity assessment of aquaculture fish using Mel Spectrogram and deep learning algorithms. Aquac Eng 102:102345. https://doi.org/10.1016/j.aquaeng.2023.102345

Edwards JE, Pratt J, Tress N, Hussey NE (2019) Thinking deeper: uncovering the mysteries of animal movement in the deep sea. Deep Sea Res Part Oceanogr Res Pap 146:24–43. https://doi.org/10.1016/j.dsr.2019.02.006

EFSA Panel on Animal Health and Welfare (AHAW) (2012) Statement on the use of animal-based measures to assess the welfare of animals. EFSA J 10(6):2767

Eilertsen E, Asderis M, Papandroulakis N, Caharija W, Frank K, Kapelonis Z, Schellewald C (2021) Quantification of acoustic target shadowing for biomass estimation in aquaculture net-pens. In: OCEANS 2021: San Diego–Porto. IEEE, pp 1–7. https://doi.org/10.23919/OCEANS44145.2021.9706121

Ellis T, Rimmer GSE, Parker SJ, Joiner C, Sebire M, Verner-Jeffreys DW, Lines J (2019) In-tank underwater cameras can refine monitoring of laboratory fish. Anim Welf 28(2):191–203. https://doi.org/10.7120/09627286.28.2.191

English G, Lawrence MJ, McKindsey CW, Lacoursière-Roussel A, Bergeron H, Gauthier S, Wringe BF, Trudel M (2024) A review of data collection methods used to monitor the associations of wild species with marine aquaculture sites. Rev Aquac 16(3):1160–1185. https://doi.org/10.1111/raq.12890

European Commission (2020) Farm to fork strategy: for a fair, healthy and environmentally-friendly food system. Communication from the Commission to the European Parliament, the Council, the European Economic and Social Committee and the Committee of the Regions, COM/2020/381 final. https://eur-lex.europa.eu/legal-content/EN/TXT/?uri=CELEX%3A52020DC0381

Falconer L, Halstensen S, Rinø SF, Noble C, Dale T, Alvestad R, Ytteborg E (2025) Marine aquaculture sites have huge potential as data providers for climate change assessments. Aquaculture 595:741519. https://doi.org/10.1016/j.aquaculture.2024.741519

Farm Animal Welfare Council (1993) Second report on priorities for research and development in farm animal welfare. London, DEFRA, pp 187–205

Fernandes PM, Aalto SL, Åsnes HØ, Rojas-Tirado P, Åtland Å, Letelier-Gordo CO (2024) Sulphate removal by membrane filtration minimizes the risk of hydrogen sulphide formation in fixed bed biofilters. Aquac Eng 106:102426. https://doi.org/10.1016/j.aquaeng.2024.102426

Fivelstad S, Bergheim A, Kløften H, Haugen R, Lohne T, Olsen AB (1999) Water flow requirements in the intensive production of Atlantic salmon (*Salmo salar* L.) fry: growth and oxygen consumption. Aquac Eng 20(1):1–15. https://doi.org/10.1016/S0144-8609(99)00002-3

Folkedal O, Macaulay G, Fosseidengen JE, Mikkelsen G, Myrland J, Søvegjarto B, Klepaker TO, Fernö A, Dempster T, Oppedal F, Stien LH (2022) Deployment of hydroacoustic feeding control in salmon sea-cages; biological and technical considerations. Aquaculture 561:738700. https://doi.org/10.1016/j.aquaculture.2022.738700

Føre M, Dempster T, Alfredsen JA, Johansen V, Johansson D (2009) Modelling of Atlantic salmon (*Salmo salar* L.) behaviour in sea-cages: a Lagrangian approach. Aquaculture 288:196–204. https://doi.org/10.1016/j.aquaculture.2008.11.031

Føre M, Alfredsen JA, Gronningsater A (2011) Development of two telemetry-based systems for monitoring the feeding behaviour of Atlantic salmon (*Salmo salar* L.) in aquaculture sea-cages. Comput Electron Agric 76:240–251. https://doi.org/10.1016/j.compag.2011.02.003

Føre M, Frank K, Dempster T, Alfredsen JA, Høy E (2017) Biomonitoring using tagged sentinel fish and acoustic telemetry in commercial salmon aquaculture: a feasibility study. Aquac Eng 78:163–172. https://doi.org/10.1016/j.aquaeng.2017.07.004

Føre M, Frank K, Norton T, Svendsen E, Alfredsen JA, Dempster T, Eguiraun H, Watson W, Stahl A, Sunde LM, Schellewald C (2018a) Precision fish farming: a new framework to improve production in aquaculture. Biosyst Eng 173:176–193. https://doi.org/10.1016/j.biosystemseng.2017.10.014

Føre M, Svendsen E, Alfredsen JA, Uglem I, Bloecher N, Sveier H, Sunde LM, Frank K (2018b) Using acoustic telemetry to monitor the effects of crowding and delousing procedures on farmed Atlantic salmon (*Salmo salar*). Aquaculture 495:757–765. https://doi.org/10.1016/j.aquaculture.2018.06.060

Føre M, Alver MO, Alfredsen JA, Rasheed A, Hukkelås T, Bjelland HV, Su B, Ohrem SJ, Kelasidi E, Norton T, Papandroulakis N (2024) Digital Twins in intensive aquaculture – challenges, opportunities and future prospects. Comput Electron Agric 218:108676. https://doi.org/10.1016/j.compag.2024.108676

Franks B, Ewell C, Jacquet J (2021) Animal welfare risks of global aquaculture. Sci Adv 7(14):eabg0677. https://doi.org/10.1126/sciadv.abg0677

Gamperl AK, Zrini ZA, Sandrelli RM (2021) Atlantic Salmon (*Salmo salar*) cage-site distribution, behavior, and physiology during a newfoundland heat wave. Front Physiol 12:719594. https://doi.org/10.3389/fphys.2021.719594

Garcia Pinillos R (ed) (2018) One welfare: a framework to improve animal welfare and human well-being. Wallingford/Boston, CAB international. ISBN: 9781786393845 (hardback), 9781786393852 (paperback)

Georgopoulou DG, Vouidaskis C, Papandroulakis N (2024) Swimming behavior as a potential metric to detect satiation levels of European seabass in marine cages. Front Mar Sci 11:1350385. https://doi.org/10.3389/fmars.2024.1350385

Gesto M, Zupa W, Alfonso S, Spedicato MT, Lembo G, Carbonara P (2020) Using acoustic telemetry to assess behavioral responses to acute hypoxia and ammonia exposure in farmed rainbow trout of different competitive ability. Appl Anim Behav Sci 230:105084. https://doi.org/10.1016/j.applanim.2020.105084

Giske J, Dumitru ML, Enberg K, Folkedal O, Handeland SO, Higginson AD, Opdal AF, Rønnestad I, Salvanes AGV, Vollset KW, Zennaro FM (2025a) Premises for digital twins reporting on Atlantic salmon wellbeing. Behav Proces:105163. https://doi.org/10.1016/j.beproc.2025.105163

Giske J, Budaev S, Eliassen S, Higginson AD, Jørgensen C, Mangel M (2025b) Vertebrate decision making leads to the interdependence of behaviour and wellbeing. Anim Behav 221:123101. https://doi.org/10.1016/j.anbehav.2025.123101

Gonzalez TJ (2023) 'Positive' animal welfare in aquaculture as a cardinal principle for sustainable development. Front Anim Sci 4:1206035

Gorle JMR, Terjesen BF, Mota VC, Summerfelt S (2018) Water velocity in commercial RAS culture tanks for Atlantic salmon smolt production. Aquac Eng 81:89–100. https://doi.org/10.1016/j.aquaeng.2018.03.001

Handeland SO, Jarvi T, Ferno A, Stefansson SO (1996) Osmotic stress, antipredator behaviour, and mortality of Atlantic salmon (*Salmo salar*) smolts (1996). Can J Fish Aquat Sci 53(12):2673–2680. https://doi.org/10.1139/f96-227

Handeland SO, Imsland AK, Ebbesson LO, Nilsen TO, Hosfeld CD, Baeverfjord G, Espmark Å, Rosten T, Skilbrei OT, Hansen T, Gunnarsson GS (2013) Low light intensity can reduce Atlantic salmon smolt quality. Aquaculture 384:19–24. https://doi.org/10.1016/j.aquaculture.2012.12.016

Hao Y, Yuan Y, Zhang H, Zhang Z (2024) Underwater optical imaging: methods, applications and perspectives. Remote Sens 16(20):3773. https://doi.org/10.3390/rs16203773

Helberg GAN, Anichini M, Kolarevic J, Sæther BS, Noble C (2024) Soundscape characteristics of RAS tanks holding Atlantic salmon (*Salmo salar*) during feeding and feed withdrawal. Aquaculture 593:741325. https://doi.org/10.1016/j.aquaculture.2024.741325

Höglund E, Folkedal O, Aerts J, Hvas M, Øverli Ø, Jensen AM, Vindas MA, Nilsson J, Kristiansen T (2023) Welfare effects of environmental hypercapnia quantified by indicators based on morphology and allostatic load in Atlantic salmon (*Salmo salar*). Aquaculture 572:739512. https://doi.org/10.1016/j.aquaculture.2023.739512

Hokin M, Miyata T, Kumon K, Oshima SI (2024) Noninvasive health monitoring of fish using optical ventilation signals. Remote Sens Appl Soc Environ 33:101105. https://doi.org/10.1016/j.rsase.2023.101105

Horricks RA, Bannister C, Lewis-McCrea LM, Hicks J, Watson K, Reid GK (2022) Comparison of drone and vessel-based collection of microbiological water samples in marine environments. Environ Monit Assess 194(6):439. https://doi.org/10.1007/s10661-022-10095-8

Hu X, Liu Y, Zhao Z, Liu J, Yang X, Sun C, Chen S, Li B, Zhou C (2021) Real-time detection of uneaten feed pellets in underwater images for aquaculture using an improved YOLO-V4 network. Comput Electron Agric 185:106135. https://doi.org/10.1016/j.compag.2021.106135

Hvas M, Folkedal O, Oppedal F (2020) Heart rate bio-loggers as welfare indicators in Atlantic salmon (Salmo salar) aquaculture. Aquaculture 529:735630. https://doi.org/10.1016/j.aquaculture.2020.735630

Iversen A, Asche F, Hermansen Ø, Nystøyl R (2020) Production cost and competitiveness in major salmon farming countries 2003–2018. Aquaculture 522:735089. https://doi.org/10.1016/j.aquaculture.2020.735089

Jafari L, Montjouridès MA, Hosfeld CD, Attramadal K, Fivelstad S, Dahle H (2024) Biofilter and degasser performance at different alkalinity levels in a brackish water pilot scale recirculating aquaculture system (RAS) for post-smolt Atlantic salmon. Aquac Eng 106:102407. https://doi.org/10.1016/j.aquaeng.2024.102407

Jamal A, Nasser A, van Rijn J (2024) Real-time ammonia estimation in recirculating aquaculture systems: a data assimilation approach. Aquac Eng:102432. https://doi.org/10.1016/j.aquaeng.2024.102432

Johannesen Á, Patursson Ø, Kristmundsson J, Dam SP, Klebert P (2020) How caged salmon respond to waves depends on time of day and currents. PeerJ 8:e9313. https://doi.org/10.7717/peerj.9313

Johannesen Á, Patursson Ø, Kristmundsson J, Dam SP, Mulelid M, Klebert P (2022) Waves and currents decrease the available space in a salmon cage. PLoS One 17(2):e0263850. https://doi.org/10.1371/journal.pone.0263850

Johansson D, Ruohonen K, Juell JE, Oppedal F (2009) Swimming depth and thermal history of individual Atlantic salmon (*Salmo salar* L.) in production cages under different ambient temperature conditions. Aquaculture 290(3-4):296–303. https://doi.org/10.1016/j.aquaculture.2009.02.022

Jónsdóttir KE, Mathiassen JR, Svendsen E, Rosten CM, Bloecher N, Føre M, Olsen RE, Finstad B, Volent Z (2024) Detection of a stress related acoustic signature by passive acoustic monitoring in Atlantic salmon farming. Aquac Eng:102472. https://doi.org/10.1016/j.aquaeng.2024.102472

Juhasz-Dora T, Lindberg SK, Karlsen A, Ortega S (2024) Biofluorescent response in lumpfish *Cyclopterus lumpus* to a therapeutic stressor as assessed by hyperspectral imaging. Sci Rep 14(1):2982. https://doi.org/10.1038/s41598-024-53562-7

Keeling L, Tunón H, Olmos Antillón G, Berg C, Jones M, Stuardo L, Swanson J, Wallenbeck A, Winckler C, Blokhuis H (2019) Animal welfare and the United Nations sustainable development goals. Front Vet Sci 6:336. https://doi.org/10.3389/fvets.2019.00336

Kimura F, Morinaga A, Fukushima M, Ishiguro T, Sato Y, Sakaguchi A, Kawashita I, Yamamoto KT (2019) Early detection system of harmful algal bloom using drones and water sample image recognition. Sens Mater 31(12):4155–4171. https://doi.org/10.18494/SAM.2019.2417

Klebert P, Su B, Nissen O, Kvæstad B (2023) Response of Atlantic salmon in flexible sea cages to waves and currents: measurements at farm-level and modeling. Appl Ocean Res 138:103599. https://doi.org/10.1016/j.apor.2023.103599

Kolarevic J, Baeverfjord G, Takle H, Ytteborg E, Reiten BKM, Nergård S, Terjesen BF (2014) Performance and welfare of Atlantic salmon smolt reared in recirculating or flow through aquaculture systems. Aquaculture 432:15–25. https://doi.org/10.1016/j.aquaculture.2014.03.033

Kolarevic J, Aas-Hansen Ø, Espmark Å, Baeverfjord G, Terjesen BF, Damsgård B (2016) The use of acoustic acceleration transmitter tags for monitoring of Atlantic salmon swimming activity in recirculating aquaculture systems (RAS). Aquac Eng 72–73:30–39. https://doi.org/10.1016/j.aquaeng.2016.03.002

Kolarevic J, Calduch-Giner J, Espmark ÅM, Evensen T, Sosa J, Pérez-Sánchez J (2021) A novel miniaturized biosensor for monitoring Atlantic salmon swimming activity and respiratory frequency. Animals 11:2403. https://doi.org/10.3390/ani11082403

Korte SM, Olivier B, Koolhaas JM (2007) A new animal welfare concept based on allostasis. Physiol Behav 92(3):422–428. https://doi.org/10.1016/j.physbeh.2006.10.018

Korus J, Filgueira R, Grant J (2024) Influence of temperature on the behaviour and physiology of Atlantic salmon (Salmo Salar) on a commercial farm. Aquaculture 589:740978. https://doi.org/10.1016/j.aquaculture.2024.740978

Kristmundsson J, Patursson Ø, Potter J, Xin Q (2023) Fish monitoring in aquaculture using multibeam echosounders and machine learning. IEEE Access 11:108306–108316. https://doi.org/10.1109/ACCESS.2023.3320949

Lagardère JP, Mallekh R, Mariani A (2004) Acoustic characteristics of two feeding modes used by brown trout (*Salmo trutta*), rainbow trout (*Oncorhynchus mykiss*) and turbot (*Scophthalmus maximus*). Aquaculture 240:607–616. https://doi.org/10.1016/j.aquaculture.2004.01.033

Lazado CC, Stiller KT, Shahzad K, Reiten BKM, Marchenko Y, Gerwins J, Radonjic FS, Eckel B, Berge A, Espmark ÅM (2022) Health and welfare of Atlantic Salmon in FishGLOBE V5 – a novel closed containment system at sea. Front Anim Sci *3*:871433. https://doi.org/10.3389/fanim.2022.871433

Leclercq E, Zerafa B, Brooker AJ, Davie A, Migaud H (2018) Application of passive-acoustic telemetry to explore the behaviour of ballan wrasse (*Labrus bergylta*) and lumpfish (*Cyclopterus lumpus*) in commercial Scottish salmon sea-pens. Aquaculture 495:1–12. https://doi.org/10.1016/j.aquaculture.2018.05.024

León-Muñoz J, Urbina MA, Garreaud R, Iriarte JL (2018) Hydroclimatic conditions trigger record harmful algal bloom in western Patagonia (summer 2016). Sci Rep 8(1):1330. https://doi.org/10.1038/s41598-018-19461-4

Li D, Hao Y, Duan Y (2020a) Nonintrusive methods for biomass estimation in aquaculture with emphasis on fish: a review. Rev Aquac 12:1390–1411. https://doi.org/10.1111/raq.12388

Li D, Wang Z, Wu S, Miao Z, Du L et al (2020b) Automatic recognition methods of fish feeding behavior in aquaculture: a review. Aquaculture 528:735508. https://doi.org/10.1016/j.aquaculture.2020.735508

Li D, Li X, Wang Q, Hao Y (2022a) Advanced techniques for the intelligent diagnosis of fish diseases: a review. Animals 12(21):2938. https://doi.org/10.3390/ani12212938

Li D, Wang G, Du L, Zheng Y, Wang Z (2022b) Recent advances in intelligent recognition methods for fish stress behavior. Aquac Eng 96:102222. https://doi.org/10.1016/j.aquaeng.2021.102222

Li X, Hao Y, Akhter M, Li D (2022c) A novel automatic detection method for abnormal behavior of single fish using image fusion. Comput Electron Agric 203:107435. https://doi.org/10.1016/j.compag.2022.107435

Li G, Yao Z, Hu Y, Lian A, Yuan T, Pang G, Huang X (2024) Deep learning-based fish detection using above-water infrared camera for deep-sea aquaculture: a comparison study. Sensors 24:2430. https://doi.org/10.3390/s24082430

Lien E, Valsvik G, Nordstrand JV, Martinez V, Rogne V, Hafsås O, Queralt S, Fathi BS, Aga M (2022) The SeaRAS AquaSense™ system: real-time monitoring of H2S at sub μg/L levels in recirculating aquaculture systems (RAS). Front Mar Sci 9:894414. https://doi.org/10.3389/fmars.2022.894414

Lima AC, Royer E, Bolzonella M, Pastres R (2023) Digital twins for land-based aquaculture: a case study for rainbow trout (*Oncorhynchus mykiss*). Open Res Eur 2:16

Lindberg SK, Durland E, Heia K, Noble C, Alvestad R, Difford GF (2023) Digital scoring of welfare traits in Atlantic salmon (*Salmo salar* L.) – a proof of concept study quantifying dorsal fin haemorrhaging via hyperspectral imaging. Front Anim Sci 4:1162384. https://doi.org/10.3389/fanim.2023.1162384

Lindholm-Lehto P (2023) Water quality monitoring in recirculating aquaculture systems. Aquac Fish Fish 3(2):113–131. https://doi.org/10.1002/aff2.102

Lindseth AV, Lobel PS (2018) Underwater soundscape monitoring and fish bioacoustics: a review. Fishes 3(3):36. https://doi.org/10.3390/fishes3030036

Lines JA, Tillett RD, Ross LG, Chan D, Hockaday S, McFarlane NJB (2001) An automatic image-based system for estimating the mass of free-swimming fish. Comput Electron Agric 31(2):151–168. https://doi.org/10.1016/S0168-1699(00)00181-2

Liu J, Bienvenido F, Yang X, Zhao Z, Feng S, Zhou C (2022) Nonintrusive and automatic quantitative analysis methods for fish behaviour in aquaculture. Aquac Res 53(8):2985–3000. https://doi.org/10.1111/are.15828

Liu C, Wang Z, Li Y, Zhang Z, Li J, Xu C, Du R, Li D, Duan Q (2023a) Research progress of computer vision technology in abnormal fish detection. Aquac Eng:102350. https://doi.org/10.1016/j.aquaeng.2023.102350

Liu H, Ma X, Yu Y, Wang L, Hao L (2023b) Application of deep learning-based object detection techniques in fish aquaculture: a review. J Mar Sci Eng 11(4):867. https://doi.org/10.3390/jmse11040867

Lopez-Tejeida S, Soto-Zarazua GM, Toledano-Ayala M, Contreras-Medina LM, Rivas-Araiza EA, Flores-Aguilar PS (2022) An improved method to obtain fish weight using machine learning and NIR camera with Haar cascade classifier. Appl Sci 13(1):69. https://doi.org/10.3390/app13010069

Ma T, Zhang D, Li X, Huang Y, Zhang L, Zhu Z, Sun X, Lan Z, Guo W (2023) Hyperspectral remote sensing technology for water quality monitoring: knowledge graph analysis and Frontier trend. Front Environ Sci 11:1133325. https://doi.org/10.3389/fenvs.2023.1133325

Macaulay G, Warren-Myers F, Barrett LT, Oppedal F, Føre M, Dempster T (2021) Tag use to monitor fish behaviour in aquaculture: a review of benefits, problems and solutions. Rev Aquac 13:1565–1582. https://doi.org/10.1111/raq.12534

Magalhães C, Martins A, Dos Santos A (2020) New approaches to study jellyfish: from autonomous apparatus to citizen science. In: Zooplankton ecology. CRC Press, pp 227–251

Mallekh R, Lagardère JP, Eneau JP, Cloutour C (2003) An acoustic detector of turbot feeding activity. Aquaculture 221:481–489. https://doi.org/10.1016/s0044-8486(03)00074-7

Måløy H (2020) EchoBERT: A transformer-based approach for behavior detection in echograms. *IEEE Access 8*:218372–218385. https://doi.org/10.1109/ACCESS.2020.3042337

Mandal A, Ghosh AR (2024) Role of artificial intelligence (AI) in fish growth and health status monitoring: a review on sustainable aquaculture. Aquac Int 32(3):2791–2820. https://doi.org/10.1007/s10499-023-01297-z

Martin-Abadal M, Ruiz-Frau A, Hinz H, Gonzalez-Cid Y (2020) Jellytoring: real-time jellyfish monitoring based on deep learning object detection. Sensors 20(6):1708. https://doi.org/10.3390/s20061708

Martins CI, Galhardo L, Noble C, Damsgård B, Spedicato MT, Zupa W, Beauchaud M, Kulczykowska E, Massabuau JC, Carter T, Planellas SR (2012) Behavioural indicators of welfare in farmed fish. Fish Physiol Biochem 38:17–41. https://doi.org/10.1007/s10695-011-9518-8

Martos-Sitcha JA, Sosa J, Ramos-Valido D, Bravo FJ, Carmona-Duarte C, Gomes HL, Calduch-Giner JÀ, Cabruja E, Vega A, Ferrer MÁ, Lozano M, Montiel-Nelson JA, Afonso JM, Pérez-Sánchez J (2019) Ultra-low power sensor devices for monitoring physical activity and respiratory frequency in farmed fish. Front Physiol 10:667. https://doi.org/10.3389/fphys.2019.00667

McIlwaine B, Casado MR (2021) JellyNet: the convolutional neural network jellyfish bloom detector. Int J Appl Earth Observ Geoinf 97:102279. https://doi.org/10.1016/j.jag.2020.102279

McKinsey (2024) What is digital-twin technology? | McKinsey. [Online]. Available: https://www.mckinsey.com/featured-insights/mckinsey-explainers/what-is-digital-twin-technology. Accessed 23 Sept 2024

Mellor DJ, Beausoleil NJ, Littlewood KE, McLean AN, McGreevy PD, Jones B, Wilkins C (2020) The 2020 five domains model: Including human–animal interactions in assessments of animal welfare. Animals 10(10):1870. https://doi.org/10.3390/ani10101870

Mihai S, Yaqoob M, Hung DV, Davis W, Towakel P, Raza M, Karamanoglu M, Barn B, Shetve D, Prasad RV, Venkataraman H (2022) Digital twins: a survey on enabling technologies, challenges, trends and future prospects. IEEE Commun Surv Tutor 24(4):2255–2291. https://doi.org/10.1109/COMST.2022.3208773

Mitamura H, Thorstad EB, Uglem I, Økland F (2017) In situ measurement of salinity during seaward migration of Atlantic salmon post-smolts using acoustic transmitters with data-storage capabilities and conventional acoustic transmitters. Anim Biotelem 5:1–10. https://doi.org/10.1186/s40317-017-0120-4

Moe Føre H, Thorvaldsen T, Osmundsen TC, Asche F, Tveterås R, Fagertun JT, Bjelland HV (2022) Technological innovations promoting sustainable salmon (*Salmo salar*) aquaculture in Norway. Aquaculture Reports 24:101115. https://doi.org/10.1016/j.aqrep.2022.101115

Morgenroth D, Kvaestad B, Økland F, Finstad B, Olsen R-E, Svendsen E, Rosten C, Axelsson M, Bloecher N, Føre M, Gräns A (2024) Under the sea: How can we use heart rate and accelerometers to remotely assess fish welfare in salmon aquaculture? Aquaculture 579:740144. https://doi.org/10.1016/j.aquaculture.2023.740144

Nedelec SL, Campbell J, Radford AN, Simpson SD, Merchant ND (2016) Particle motion: the missing link in underwater acoustic ecology. Methods in Ecology and Evolution 7(7):836–842. https://doi.org/10.1111/2041-210X.12544

Nilsen A, Nielsen KV, Næss A, Bergheim A (2017) The impact of production intensity on water quality in oxygen enriched, floating enclosures for post-smolt salmon culture. Aquac Eng 78:221–227. https://doi.org/10.1016/j.aquaeng.2017.06.001

Nilsen A, Hagen Ø, Johnsen CA, Prytz H, Zhou B, Nielsen KV, Bjørnevik M (2019) The importance of exercise: Increased water velocity improves growth of Atlantic salmon in closed cages. Aquaculture 501:537–546. https://doi.org/10.1016/j.aquaculture.2018.09.057

Nilsson J, Gismervik K, Nielsen KV, Iversen MH, Noble C, Kolarevic J, Frotjold H, Nilsen K, Wilkinson E, Klakegg B, Hauge HS (2022) Laksvel-Standardisert operasjonell velferdsovervåking for laks i matfiskanlegg. Rapport fra havforskningen, Rapport fra havforskningen. https://www.hi.no/hi/nettrapporter/rapport-fra-havforskningen-2022-14#sec-4-3

Noble C, Cañon Jones HA, Damsgård B, Flood MJ, Midling KØ, Roque A, Sæther BS, Cottee SY (2012) Injuries and deformities in fish: their potential impacts upon aquacultural production and welfare. Fish Physiol Biochem 38:61–83. https://doi.org/10.1007/s10695-011-9557-1

Noble C, Gismervik K, Iversen MH, Kolarevic J, Nilsson J, Stien LH, Turnbull JF (eds) (2018) Welfare indicators for farmed Atlantic salmon: tools for assessing fish welfare. Nofima, Tromsø. ISBN 978-82-8296-556-9. 351 pp. www.nofima.no/fishwell/english

Noble C, Gismervik K, Iversen MH, Kolarevic J, Nilsson J, Stien LH, Turnbull JF (eds) (2020) Welfare Indicators for farmed rainbow trout: tools for assessing fish welfare. Nofima, Tromsø. ISBN 978-82-8296-620-7. 310 pp. https://nofima.no/fishwell/trout/

NS 9417:2022 Salmon and rainbow trout – terminology and methods for documentation of production. 2022-10-28. © Standard Norge 2022

Oppedal F, Vindas M, Sivle LD, de Jong K, Dempster TD, Forland TN, Barrett L (2024) Lyd i lakseoppdrett. Rapport fra. havforskningen

Ortega S, Lutfi E, Horn SS, Durland E (2024) Quantification of fat content in the liver of different aquaculture fish species using hyperspectral image analysis. Aquaculture:741924. https://doi.org/10.1016/j.aquaculture.2024.741924

Palaiokostas C (2021) Predicting for disease resistance in aquaculture species using machine learning models. Aquac Rep 20:100660. https://doi.org/10.1016/j.aqrep.2021.100660

Pavlidis M, Samaras A (2020) Mediterranean Fish Welfare: guide to good practices and assessment indicators, p 90

Pavlidis M, Samaras A, Papaharisis L (2024) Welfare assessment study for European sea bass. University of Crete. https://doi.org/10.26248/4V5N-7V19

Petersen C, Jurevicius D (2009) Environmental noise impact of a major transport corridor on a barramundi fish farm. In: Proceedings of ACOUSTICS 2009, 23–25 November 2009, Adelaide, Australia. https://acoustics.asn.au/conference_proceedings/AAS2009/papers/p13.pdf

Petkovski A, Shehu V (2023) Anomaly detection on univariate sensing time series data for smart aquaculture using deep learning. SEEU Rev 18(1):1–16. https://doi.org/10.2478/seeur-2023-0030

Pettersen R, Braa HL, Gawel BA, Letnes PA, Sæther K, Aas LMS (2019) Detection and classification of *Lepeophtheirus salmonis* (Krøyer, 1837) using underwater hyperspectral imaging. Aquac Eng 87:102025. https://doi.org/10.1016/j.aquaeng.2019.102025

Pinkiewicz TH, Purser GJ, Williams RN (2011) A computer vision system to analyse the swimming behaviour of farmed fish in commercial aquaculture facilities: a case study using cage-held Atlantic salmon. Aquac Eng 45(1):20–27. https://doi.org/10.1016/j.aquaeng.2011.05.002

Pitt KA, Haberlin D, Stantic B, Doyle TK (2024) Mitigating and managing the impacts of gelatinous zooplankton on finfish aquaculture. Aquaculture 581:740403. https://doi.org/10.1016/j.aquaculture.2023.740403

Plew DR, Klebert P, Rosten TW, Aspaas S, Birkevold J (2015) Changes to flow and turbulence caused by different concentrations of fish in a circular tank. J Hydraul Res 53(3):364–383. https://doi.org/10.1080/00221686.2015.1029016

Popper AN, Hawkins AD (2018) The importance of particle motion to fishes and invertebrates. J Acoust Soc Am 143(1):470–488. https://doi.org/10.1121/1.5021594

Popper AN, Hawkins AD (2019) An overview of fish bioacoustics and the impacts of anthropogenic sounds on fishes. J Fish Biol 94(5):692–713. https://doi.org/10.1111/jfb.13948

Qiu D, Xu S, Song C, Chi L, Li X, Sun G, Liu B, Liu Y (2015) Effects of spectral composition, photoperiod and light intensity on the gonadal development of Atlantic salmon *Salmo salar* in recirculating aquaculture systems (RAS). Chin J Oceanol Limnol 33(1):45–56. https://doi.org/10.1007/s00343-015-4011-3

Radford C, Slater M (2019) Soundscapes in aquaculture systems. Aquac Environ Interact 11:53–62. https://doi.org/10.3354/aei00293

Ranjan R, Tsukuda S, Good C (2023) MortCam: an Artificial Intelligence-aided fish mortality detection and alert system for recirculating aquaculture. Aquac Eng 102:102341. https://doi.org/10.1016/j.aquaeng.2023.102341

Rasheed A, San O, Kvamsdal T (2020) Digital twin: values, challenges and enablers from a modeling perspective. IEEE Access 8:21980–22012

Rastegari H, Nadi F, Lam SS, Ikhwanuddin M, Kasan NA, Rahmat RF, Mahari WAW (2023) Internet of Things in aquaculture: a review of the challenges and potential solutions based on current and future trends. Smart Agric Technol 4:100187. https://doi.org/10.1016/j.atech.2023.100187

Rather MA, Ahmad I, Shah A, Hajam YA, Amin A, Khursheed S, Ahmad I, Rasool S (2024) Exploring opportunities of Artificial Intelligence in aquaculture to meet increasing food demand. Food Chem X:101309. https://doi.org/10.1016/j.fochx.2024.101309

Rault JL, Bateson M, Boissy A, Forkman B, Grinde B, Gygax L, Harfeld JL, Hintze S, Keeling LJ, Kostal L, Lawrence AB (2025) A consensus on the definition of positive animal welfare. Biol Lett 21(1):20240382. https://doi.org/10.1098/rsbl.2024.0382

Rosten CM, Mathiassen JR, Volent Z (2023) Acoustic environment of aquaculture net-pens varies with feeding status of Atlantic salmon (*Salmo salar*). Aquaculture 563. https://doi.org/10.1016/j.aquaculture.2022.738949

Rountree RA, Juanes F, Bolgan M (2018) Air movement sound production by alewife, white sucker, and four salmonid fishes suggests the phenomenon is widespread among freshwater fishes. PLoS One 13. https://doi.org/10.1371/journal.pone.0204247

RSPCA (2020) RSPCA welfare standards for farmed rainbow trout. Royal Society for the Prevention of Cruelty to Animals, Horsham

RSPCA (2024) RSPCA welfare standards for farmed Atlantic salmon. Royal Society for the Prevention of Cruelty to Animals, Horsham

Saberioon MM, Cisar P (2016). Automated multiple fish tracking in three-dimension using a structured light sensor. Comput Electron Agric 121:215–221. https://doi.org/10.1016/j.compag.2015.12.014

Saberioon MM, Cisar P (2018). Automated within tank fish mass estimation using infrared reflection system. Comput Electron Agric 150:484–492. https://doi.org/10.1016/j.compag.2018.05.025

Saraiva JL, Arechavala-Lopez P, Castanheira MF, Volstorf J, Heinzpeter Studer B (2019) A global assessment of welfare in farmed fishes: the FishEthoBase. Fishes 4(2):30. https://doi.org/10.3390/fishes4020030

Saraiva JL, Volstorf J, Cabrera-Álvarez MJ, Arechavala-Lopez P (2022) Using ethology to improve farmed fish welfare and production. Report Produced for the Aquaculture Advisory Council AAC 67, pp + annexes. https://aac-europe.org/en/publication/using-ethology-to-improve-farmed-fish-welfare-and-production-2/

Sasaki Y, Nishikawa R, Komeyama K (2024) Non-invasive swimming speed estimation method based on tail-beat frequency determined from fish length measurement using stereo-cameras. Fish Sci:1–10. https://doi.org/10.1007/s12562-024-01816-8

NCE Seafood Innovation (2023). Industry Insight report: Data sharing in the Norwegian aquaculture industry. 56 pp. https://seafoodinnovation.no/wp-content/uploads/2023/05/Industry-Insight-Datadeling_FINAL-FULL-VERSION.pdf

Segner H, Sundh H, Buchmann K, Douxfils J, Sundell KS, Mathieu C, Ruane N, Jutfelt F, Toften H, Vaughan L (2012) Health of farmed fish: its relation to fish welfare and its utility as welfare indicator. Fish Physiol Biochem 38:85–105. https://doi.org/10.1007/s10695-011-9517-9

Solstorm D, Oldham T, Solstorm F, Klebert P, Stien LH, Vågseth T, Oppedal F (2018 Feb) Dissolved oxygen variability in a commercial sea-cage exposes farmed Atlantic salmon to growth limiting conditions. Aquaculture. 3(486):122–129. https://doi.org/10.1016/j.aquaculture.2017.12.008

Stehfest KM, Carter CG, McAllister JD, Ross JD, Semmens JM (2017) Response of Atlantic salmon Salmo salar to temperature and dissolved oxygen extremes established using animal-borne environmental sensors. Sci Rep 7:4545. https://doi.org/10.1038/s41598-017-04806-2

Stentiford GD, Bateman IJ, Hinchliffe SJ, Bass D, Hartnell R, Santos EM, Devlin MJ, Feist SW, Taylor NGH, Verner-Jeffreys DW, Van Aerle R (2020) Sustainable aquaculture through the One Health lens. Nat Food 1(8):468–474. https://doi.org/10.1038/s43016-020-0127-5

Sterling P (2012) Allostasis: a model of predictive regulation. Physiol Behav 106(1):5–15. https://doi.org/10.1016/j.physbeh.2011.06.004

Stien LH, Gytre T, Torgersen T, Sagen H, Kristiansen TS (2008) A system for online assessment of fish welfare in aquaculture. ICES

Stien LH, Bracke MB, Folkedal O, Nilsson J, Oppedal F, Torgersen T, Kittilsen S, Midtlyng PJ, Vindas MA, Øverli Ø, Kristiansen TS (2013) Salmon Welfare Index Model (SWIM 1.0): a semantic model for overall welfare assessment of caged Atlantic salmon: review of the selected welfare indicators and model presentation. Rev Aquac 5(1):33–57. https://doi.org/10.1111/j.1753-5131.2012.01083.x

Stien LH, Bracke M, Noble C, Kristiansen TS (2020) Assessing fish welfare in aquaculture. The Welfare of Fish, pp 303–321. https://doi.org/10.1007/978-3-030-41675-1_13

Stockwell CL, Filgueira R, Grant J (2021) Determining the effects of environmental events on cultured Atlantic salmon behaviour using 3-dimensional acoustic telemetry. Front Anim Sci 2:701813. https://doi.org/10.3389/fanim.2021.701813

Stradmeyer L (1994) Survival, growth and feeding of Atlantic salmon, *Salmo salar* L., smolts after transfer to sea water in relation to the failed smolt syndrome. Aquac Res 25:103–112. https://doi.org/10.1111/j.1365-2109.1994.tb00670.x

Su B, Bjørnson FO, Tsarau A, Endresen PC, Ohrem SJ, Føre M, Fagertun JT, Klebert P, Kelasidi E, Bjelland HV (2023) Towards a holistic digital twin solution for real-time monitoring of aquaculture net cage systems. Mar Struct 91:103469. https://doi.org/10.1016/j.marstruc.2023.103469

Sveen L, Timmerhaus G, Johansen LH, Ytteborg E (2021) Deep neural network analysis-a paradigm shift for histological examination of health and welfare of farmed fish. Aquaculture 532:736024. https://doi.org/10.1016/j.aquaculture.2020.736024

Svendsen E, Føre M, Økland F, Gräns A, Hedger RD, Alfredsen JA, Uglem I, Rosten CM, Frank K, Erikson U, Finstad B (2021a) Heart rate and swimming activity as stress indicators for Atlantic salmon (Salmo salar). Aquaculture 531:735804. https://doi.org/10.1016/j.aquaculture.2020.735804

Svendsen E, Volent Z, Schellewald C, Tsarau A, Bjørgan A, Venås B, Bloecher N, Bondø M, Føre M, Jónsdóttir KE, Stefansson S (2021b) Identification of spectral signature for *in situ* real-time monitoring of smoltification. Appl Opt 60(14):4127–4134. https://doi.org/10.1364/AO.420347

Tanaka T, Ikeda R, Yuta Y, Tsurukawa K, Nakamura S, Yamaguchi T, Komeyama K (2019) Annual monitoring of growth of red sea bream by multi-stereo-image measurement. Fish Sci 85:1037–1043. https://doi.org/10.1007/s12562-019-01347-7

Tao F, Qi Q (2019) Make more digital twins. Nature 573(7775):490–491

Tian Q, Huo Y, Yao M, Wang H (2024) A method for detecting dead fish on large water surfaces based on improved YOLOv10. arXiv preprint arXiv:2409.00388

Timmons MB, Guerdat T, Vinci BJ (2018) Water quality. In: Recirculating aquaculture, 4th edn. Ithaca Publishing Company LLC, Ithaca, p 2758

Timms GP, de Souza Jr PA, Reznik L, Smith DV (2011) Automated data quality assessment of marine sensors. Sensors 11(10):9589–9602. https://doi.org/10.3390/s111009589

Tolentino LKS, De Pedro CP, Icamina JD, Navarro JBE, Salvacion LJD, Sobrevilla GCD, Villanueva AA, Amado TM, Padilla MVC, Madrigal GAM (2020) Development of an IoT-based intensive aquaculture monitoring system with automatic water correction. Int J Comput Digit Syst 10:1355–1365

Treasurer JW, Hannah F, Cox D (2003) Impact of a phytoplankton bloom on mortalities and feeding response of farmed Atlantic salmon, *Salmo salar*, in west Scotland. Aquaculture 218(1-4):103–113. https://doi.org/10.1016/S0044-8486(02)00516-1

Tschirren L, Bachmann D, Güler AC, Blaser O, Rhyner N, Seitz A, Zbinden E, Wahli T, Segner H, Refardt D (2021) MyFishCheck: a model to assess fish welfare in aquaculture. Animals 11(1):145. https://doi.org/10.3390/ani11010145

Ubina NA, Cheng SC, Chen HY, Chang CC, Lan HY (2021) A visual aquaculture system using a cloud-based autonomous drones. Drones 5(4):109. https://doi.org/10.3390/drones5040109

Ubina NA, Lan HY, Cheng SC, Chang CC, Lin SS, Zhang KX, Lu HY, Cheng CY, Hsieh YZ (2023) Digital twin-based intelligent fish farming with Artificial Intelligence Internet of Things (AIoT). Smart Agric Technol 5:100285. https://doi.org/10.1016/j.atech.2023.100285

Ulvund JB, Engebretsen S, Alfredsen JA, Kristensen T, Urke HA, Jansen PA (2021) Behavioural response of farmed Atlantic salmon (*Salmo salar* L.) to artificial underwater lights: wavelet analysis of acoustic telemetry data. Aquac Eng 95:102196. https://doi.org/10.1016/j.aquaeng.2021.102196

Vambeseth ØF (2022) Morphometric relationships as computer vision parameters for growth development and sexual maturation status of Atlantic salmon (*Salmo salar*). Master's thesis, The University of Bergen

Veasey JS (2017) In pursuit of peak animal welfare; the need to prioritize the meaningful over the measurable. Zoo Biol 36(6):413–425. https://doi.org/10.1002/zoo.21390

Wang Z, Liu H, Zhang G, Yang X, Wen L, Zhao W (2023) Diseased fish detection in the underwater environment using an improved yolov5 network for intensive aquaculture. Fishes 8(3):169

Warren-Myers F, Hvas M, Vågseth T, Dempster T, Oppedal F (2021) Sentinels in Salmon aquaculture: heart rates across seasons and during crowding events. Front Physiol:12

Weeks R, Bresnan E, Davidson K, Whyte C (2022) FFHF Climate change working group project–SOP. Towards a standardised phytoplankton monitoring operating procedure for the finfish sector. https://pure.uhi.ac.uk/en/publications/ffhf-climate-change-working-group-project-sop-towards-a-standardi

Whitford M, Klimley AP (2019) An overview of behavioral, physiological, and environmental sensors used in animal biotelemetry and biologging studies. Anim Biotelem 7:26. https://doi.org/10.1186/s40317-019-0189-z

Wilkins CL, McGreevy PD, Cosh SM, Henshall C, Jones B, Lykins AD, Billingsley W (2024) Introducing the mellorater – the five domains model in a welfare monitoring app for animal guardians. Animals 14(15). https://doi.org/10.3390/ani14152172

World Health Organisation (2024) © 2024 WHO Available at: https://www.who.int/health-topics/one-health#tab=tab_1. Accessed 1 Dec 2024

World Organisation for Animal Health (2024) Introduction to the recommendations for animal welfare – Chapter 7.1. In: Terrestrial animal health code. World Organisation for Animal Health, Paris. https://doi.org/10.1007/BF01826929

Wu Y, Duan Y, Wei Y, An D, Liu J (2022) Application of intelligent and unmanned equipment in aquaculture: a review. Comput Electron Agric 199:107201. https://doi.org/10.1016/j.compag.2022.107201

Yang L, Liu Y, Yu H, Fang X, Song L, Li D, Chen Y (2021) Computer vision models in intelligent aquaculture with emphasis on fish detection and behavior analysis: a review. Arch Computat Methods Eng 28:2785–2816. https://doi.org/10.1007/s11831-020-09486-2

Yousaf M, Røn Ø, Hagen PP, McGurk C (2022) Monitoring fish welfare using heart rate biologgers in farmed Atlantic salmon (*Salmo salar* L.): an insight into the surgical recovery. Aquaculture 738211. https://doi.org/10.1016/j.aquaculture.2022.738211

Ytrestøyl T, Takle H, Kolarevic J, Calabrese S, Timmerhaus G, Rosseland BO, Teien HC, Nilsen TO, Handeland SO, Stefansson SO, Ebbesson LO (2020) Performance and welfare of Atlantic salmon, *Salmo salar* L. post-smolts in recirculating aquaculture systems: Importance of salinity and water velocity. J World Aquac Soc 51(2):373–392. https://doi.org/10.1111/jwas.12682

Yu G, Zhang J, Chen A, Wan R (2023) Detection and identification of fish skin health status referring to four common diseases based on improved YOLOv4 Model. Fishes 8(4):186

Zahir M, Su Y, Shahzad MI, Ayub G, Rahman SU, Ijaz J (2024) A review on monitoring, forecasting, and early warning of harmful algal bloom. Aquaculture 593:741351. https://doi.org/10.1016/j.aquaculturc.2024.741351

Zeng Y, Yang X, Pan L, Zhu W, Wang D, Zhao Z, Liu J, Sun C, Zhou C (2023) Fish school feeding behavior quantification using acoustic signal and improved Swin Transformer. Comput Electron Agric 204:107580. https://doi.org/10.1016/J.COMPAG.2022.107580

Zena LA, Ekström A, Morgenroth D, Mcarley T, Gräns A, Axelsson M, Johansen IB, Sandblom E (2024) Ischemia-induced alterations in the electrocardiogram of salmonid fish. Aquaculture 581:740482. https://doi.org/10.1016/j.aquaculture.2023.740482

Zhang Q, Su B (2025) A hybrid approach towards real-time monitoring of fish distributions in aquaculture net cage. Aquac Eng *110*:102527. https://doi.org/10.1016/j.aquaeng.2025.102527

Zhang T, Shen T, Yuan K, Xue K, Qian H (2022) A novel autonomous robotics system for aquaculture environment monitoring. arXiv preprint arXiv:2211.00917. https://doi.org/10.48550/arXiv.2211.00917

Zhao J, Bao WJ, Zhang FD, Ye ZY, Liu Y, Shen MW, Zhu SM (2017) Assessing appetite of the swimming fish based on spontaneous collective behaviors in a recirculating aquaculture system. Aquac Eng 78:196–204. https://doi.org/10.1016/j.aquaeng.2017.07.008

Zhao Y, Qin H, Xu L, Yu H, Chen Y (2025) A review of deep learning-based stereo vision techniques for phenotype feature and behavioral analysis of fish in aquaculture. Artif Intell Rev 58(1):1–61. https://doi.org/10.1007/s10462-024-10960-7

Zheng H, Liu R, Zhang R, Hu Y (2014) A method for real-time measurement of respiratory rhythms in medaka (Oryzias latipes) using computer vision for water quality monitoring. Ecotoxicol Environ Saf 100:76–86. https://doi.org/10.1016/j.ecoenv.2013.11.016

Zhou C, Zhang B, Lin K, Xu D, Chen C, Yang X, Sun C (2017) Near-infrared imaging to quantify the feeding behavior of fish in aquaculture. Comput Electron Agric 135:233–241. https://doi.org/10.1016/j.compag.2017.02.013

Zhou C, Xu D, Lin K, Sun C, Yang X (2018) Intelligent feeding control methods in aquaculture with an emphasis on fish: a review. Rev Aquac 10:975–993. https://doi.org/10.1111/raq.12218

Zion B (2012) The use of computer vision technologies in aquaculture – a review. Comput Electron Agric 88:125–132. https://doi.org/10.1016/j.compag.2012.07.010

Zupa W, Alfonso S, Gai F, Gasco L, Spedicato MT, Lembo G, Carbonara P (2021) Calibrating accelerometer tags with oxygen consumption rate of rainbow trout (*Oncorhynchus mykiss*) and their use in aquaculture facility: a case study. Animals 11:1496. https://doi.org/10.3390/ani11061496

Chapter 2
Aquaculture Robotics: Challenges, Opportunities and Future Prospects and the Role of Digital Twin Technology and Autonomous Solutions

Linn D. Evjemo, Sveinung J. Ohrem, Biao Su, Martin Føre, and Eleni Kelasidi

Abstract Aquaculture is a growing industry worldwide and makes up almost half of the world's production of aquatic animals meant for human consumption. Norway is the world's largest exporter of Atlantic salmon, with the sea-based production mainly being placed along the coastline, constantly dealing with challenging weather conditions. Daily operations currently rely heavily on manual labour and experience-based decision-making, with a steady struggle to balance productivity, fish welfare, and the HSE conditions of the workers. Many maritime domains already benefit greatly from the use of underwater robotics to increase efficiency, reduce costs, and ensure safety of operations. Similarly, an increased use of robotic and autonomous solutions in the sea-based aquaculture industry could greatly improve safety of operations. However, the sea-based aquaculture industry poses unique challenges to robotic solutions compared to other parts of the maritime domain because of the dynamically changing environment, the flexible structures, and the living fish. This chapter gives an in-depth overview of the many inspection, maintenance, and repair procedures that need to be performed to maintain safe and sustainable operations in sea-based aquaculture in Norway. The status of robotics, digitalization, and autonomy for these operations is considered, both in applications and the potential for future innovations. Extensive research and development efforts made to enable the use of more robotics and autonomy in daily operations are also presented. Digital twin technology is likely to be a crucial component in developing the new technologies within robotics and autonomy needed in the aquaculture

L. D. Evjemo (✉) · S. J. Ohrem · B. Su · E. Kelasidi
Department of Aquaculture Technology, SINTEF Ocean, Trondheim, Norway
e-mail: linn.evjemo@sintef.no; sveinung.ohrem@sintef.no; biao.su@sintef.no; eleni.kelasidi@sintef.no

M. Føre
Department of Engineering Cybernetics, NTNU, Trondheim, Norway
e-mail: martin.fore@ntnu.no

F. O'Donncha et al. (eds.), *Digital Twin for Marine Fish Farming*, Aquatic Production Systems, https://doi.org/10.1007/978-3-032-10362-8_2

practices of the future. Throughout the chapter, the possible roles of a digital twin in realizing this potential are highlighted.

Keywords Robotics · Aquaculture · Precision fish farming · Control · Autonomy and autonomous operations

2.1 Introduction

Aquaculture covers the practice of farming all plants or animals (biomass) in an aquatic environment. As the water-based equivalent to the term agriculture, it includes both sea-based production open to the environment and land-based closed tank systems. The world's consumption of aquatic food keeps rising and has more than doubled per capita over the past 50 years (Food and Aquaculture Organization 2022). As much as 225 states and territories reported trading activity of aquaculture or fisheries products in 2020 and the total aquaculture and fisheries production reached an all-time high of 214 million tonnes aquatic animals and algae. The proportion comprised of aquatic animals was approximately 178 million tonnes, and aquaculture production counted for 49% of this, an increase from 46% in 2010 and 43% in 2006 (Food and Aquaculture Organization 2022, 2010). Almost 90% of the biomass produced in aquaculture was used for human consumption, while the rest was used for, e.g., fish oil or fish meal. The largest share of the world's aquaculture production is farmed finfish, which made up 57.5 million tonnes of the production in 2020, with a farm gate value of USD 146.1 billion (Food and Aquaculture Organization 2022). Norway is the largest producer of farmed Atlantic salmon (*Salmo salar*) in the world (Food and Aquaculture Organization 2022) and exported approximately 1.2 billion tonnes of farmed Atlantic salmon and sea trout in 2020, with a farm gate value of USD 7.9 billion (NFD 2021).

The sea-based salmon farming industry in Norway is currently facing several challenges. Each net pen typically contains up to 200,000 fish, and this high concentration of fish creates an environment where parasites or pathogens can easily evolve and spread, both among the fish within the net pen and to the surrounding wildlife (Kelasidi and Svendsen 2023). Sea lice are of particular concern, as high densities of this parasite can cause physical damage to the fish such as skin erosion, and increase the susceptibility to additional infections due to compromised immune response or open wounds (Overton et al. 2019). Another challenge is escapes, caused by factors such as holes in the nets or accidents during operations. Escapes represent a financial loss, and escaped farmed fish can negatively impact wild salmon stocks through genetic effects from cross-breeding, by competing for food and habitat, and by the spread of parasites (Andersen et al. 2023; Rydeng 2022; Kelasidi and Svendsen 2023). Mortality in the net pens has also been outlined as an increasing challenge, with 16.7% mortality in 2023 (Sommerset et al. 2023b). This also represents a significant financial loss, as well as a considerable concern for fish welfare. Moreover, Riksrevisjonen, a national governmental audit institution in Norway, recently directed criticism toward the Norwegian authorities for not implementing sufficient measures to reduce disease and poor fish welfare in the Norwegian aquaculture industry (Riksrevisjonen 2022–2023). Fish monitoring is currently conducted either on a group level or by individual sampling, and it is not yet possible to monitor each fish in a way that ensures fish welfare on an individual level or to efficiently identify and remove sick or moribund fish. The law requires farmers to keep the fish in environments that ensures good welfare based on species specific need and that the biomass should be protected as much as possible from harm and unnecessary stress (Nrk.no 2024). This separates fish farming from traditional livestock farming, where the law requires farmers to promptly care for sick or injured animals and to euthanize them if continued life would cause unnecessary suffering (Lovdata 2006). In the same report from Riksrevisjonen, the Norwegian Veterinary Institute suggested that the regulations for aquaculture should move in a direction where fish are regarded more as individuals rather than biomass and stated that the difference between regulations is not based on scientific evidence supporting that there is a distinction in welfare needs of fish and traditional livestock (Riksrevisjonen 2022–2023). This could be an indicator of the possible introduction of stricter regulations in the future.

Sea-based aquaculture operations in Norway are carried out primarily at the surface level along the coast. Even close to shore, strong winds and unpredictable wave and current conditions can present significant challenges for daily operations, impacting both equipment and the human operators involved, especially during the winter months with stronger winds and storms. Because of the growth in the Norwegian salmon farming industry and a shortage of new, suitable areas to expand production, there is also a growing interest in moving salmon farming to more exposed locations (Bjelland et al. 2015). This shift also aims to reduce local negative environmental effects, as the accumulation of faeces and feed spills on the seabed around the net pens, and alter the existing ecosystems (Havbruksutvalget 2024; Kelasidi and Svendsen 2023). Moving production further out from the sheltered coastal areas will entail even more challenging weather conditions, creating a need for robust solutions for both the net pens and the execution of daily operations.

New technologies provide more accessible and efficient methods for monitoring a wide range of parameters, from weather conditions to fish behaviour, enabling us to gather massive amounts of data and to introduce new analytical approaches that can be used, for example, for objective decision support (Føre et al. 2018). This can help increase efficiency in operations and reduce risks to both fish and personnel. Although many variables relevant for fish welfare are beyond our control, monitoring what we can and adjusting daily operations based on an understanding of how these conditions impact the fish can help reduce stress levels and improve overall fish welfare (Bregnballe 2022; Zhang et al. 2024).

Currently, most daily operations rely on workers performing manual labour or executing tasks with assistance from cranes and other equipment directly controlled by experienced human operators (Gumdal 2022; Føre et al. 2018). These operations can be physically demanding and risky for the workers involved (Thorvaldsen et al. 2017, 2023) and need to be performed at different frequencies depending on the season. Operations that should be performed daily or weekly must sometimes be carried out under challenging conditions due to winds, currents, and rough weather. Therefore, there is a constant need to balance the health, safety, and environment (HSE) concerns of the workers, the productivity of the farm, and the welfare of the fish. According to a survey conducted in 2023, in which 1289 employees in the Norwegian aquaculture industry (both sea- and land-based) participated, as much as 69% of the employees working in sea-based aquaculture had experienced near-accidents that could have resulted in bodily harm to human operators in the last year (Thorvaldsen et al. 2023). Close to 17% of the employees had been on work-related sick leave in the past 2 years, with the most common causes being strains and aches in shoulders, neck, and back, as well as fatigue. These figures were in line with a similar survey conducted in 2016, which was answered by 447 employees (Thorvaldsen et al. 2017). Although it is important to note that the vast majority of the employees in this survey felt that HSE was important in their workplace, such numbers are still very high.

The Norwegian sea-based aquaculture industry faces a wide range of challenges that need to be addressed. With the integration of new technologies and data-driven decision-making systems, production could become less vulnerable to human error, reducing the pressure on the workers during stressful operations, which would lead to improved HSE and reduced risk of accidents. This chapter will largely focus on sea-based farming of Atlantic salmon in Norway, exploring the digital and robotic solutions currently being investigated by the research community and utilized in the industry today, and examining the potential for increased use of robotic and autonomous solutions in the future as the industry evolves. Digital twins technology will be a core tool in achieving this, and recent advancements in this field have gained a strong following in many industrial fields (Rasheed et al. 2020). The ongoing digital twin implementation in marine aquaculture (Su et al. 2023; Føre et al. 2024) is thus likely to bring new opportunities for remote and/or autonomous operations of underwater vehicles.

2.2 Robotics in the Maritime Domain

Work in the maritime domain is typically characterized by harsh and challenging operating environments, whether it is work on the sea floor in dark areas and under high pressure or work in the wave zone impacted by rough weather, winds, and currents (Kelasidi and Svendsen 2023). These environments are very challenging for human operators, because of both access and general safety. Divers are often a solution for operations relatively close to the surface. However, the diving depth for humans is limited, and there are always risks involved when diving, especially on greater depths. All diving operations are therefore subject to strict regulations from the government (Arbeidstilsynet 2013; Risberg et al. 2019). To further explore and utilize the underwater domain, there has been a significant increase in the use of unmanned underwater vehicles (UUVs) over the last decades (Zereik et al. 2018; Kelasidi and Svendsen 2023). Using UUVs exclusively for remote-controlled operations will require experienced personnel for operations and support vessels to carry it to location, so increased autonomy is a potential next step. The main motivations for adapting more and more use of robotics systems with increased autonomy level are to reduce risk and to increase efficiency and optimize the operations (Kelasidi and Svendsen 2023).

Robotic solutions have gradually improved our way of handling tasks that are *dull*, *dirty*, *dangerous*, *dear*, and *distant*, and underwater work can often be characterized by all of these terms. There exists a range of different robotic systems developed for applications where the robot is fully or partially submerged, and UUVs are being used in many different applications (Zereik et al. 2018). Completely submerged robotic systems are generally divided into two main categories: remotely operated vehicles (ROVs) and autonomous underwater vehicles (AUVs). Both main branches of UUVs exist in many shapes and sizes, and generally the main criterion that separates the two is whether they are remote controlled by a human operator (ROV) or if they are autonomous (AUV). The topic of autonomy covers energy autonomy, navigation autonomy, and decision autonomy. This means that a fully autonomous robotic system should have a reliable and realizable *power source*, be able to *navigate* safely without human intervention, and be able to *adapt and make decisions* during missions based on its environment. Generally, fully autonomous operations utilizing underwater robotic systems can have major impacts in maritime industries. However, realizing fully autonomous UUVs entails several challenges that need to be addressed through novel enabling technologies (Fig. 2.1).

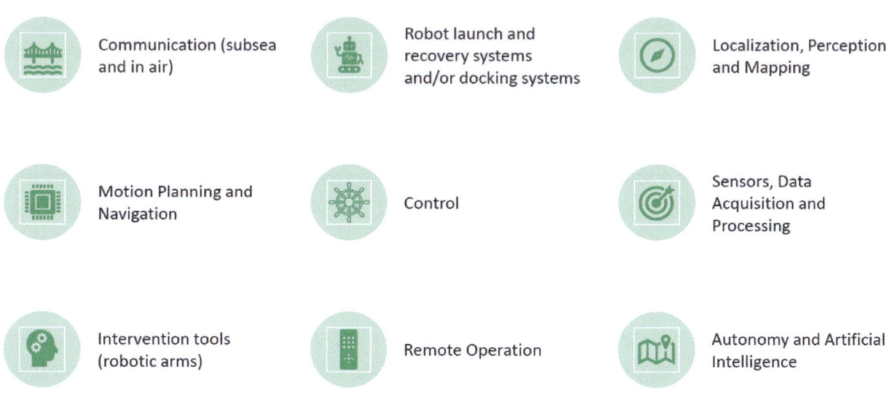

Fig. 2.1 An overview of some of the major enabling technologies for robotics in the maritime domains

2.2.1 Communication

For a UUV to perform complex operations without human assistance while simultaneously being able to compensate for system failures and environmental uncertainties, there is a series of enabling technologies that need to be present. One of the main challenges for underwater robots is communication. Underwater communication has endemic challenges such as low bandwidth and a lack of reliable ubiquitous communication infrastructures such as Global Navigation Satellite Systems (GNSS) or Wi-Fi (Quattrini Li et al. 2023). Radio waves, acoustics, and optical communication are technologies that can be used wirelessly under water. While radio and optical communication can provide high bandwidths, their range in under water is very restricted due to the high attenuation of electromagnetic signals in seawater. Because a certain range is required for in-cage navigation, acoustic communication is therefore often a preferable channel for wireless underwater communication. However, this principle also has its shortcomings, including a limited bandwidth, low data rate, high latency, and general unreliability making it necessary to handle frequent data loss in transmissions (Paull et al. 2013). In cases where conditions allow for a cabled connection from the vehicle to the surface, wired communication through, for example, a tether can provide fast, reliable, and real-time communication. Data transfers can also happen via physical connectors in, e.g., a docking station (Quattrini Li et al. 2023).

2.2.2 Localisation, Perception, and Mapping

The challenge of localization, perception, and mapping is similar to that of communication, as underwater systems lack some of the existing infrastructures used by above surface localization, like GNSS. For underwater robots working in complex environments, such as inside a net pen, robust simultaneous localization and mapping (SLAM) based on sensor inputs from onboard sensors is thus necessary. Having robust and accurate localization systems that estimate the vehicle's own position and pose within the environment is essential for both manned and unmanned operations (Maurelli et al. 2022). Localization sensors commonly used in UUVs include acoustic sensors such as ultrashort baseline (USBL), Doppler velocity logger (DVL) and sonars, and optical sensors such as cameras and lidar. For localization using a USBL, the position of a transponder relative to a transceiver is calculated based on the reception time of acoustic signal transmissions sent by the transponder and received by the transceivers. DVLs measure the position of the vehicle onto which they are attached relative to a fixed surface by emitting acoustic beams. By calculating the distance based on the time-of-flight and using the Doppler shift on the returned beams, they derive the velocity relative to the object reflecting the beams. DVLs are usually placed facing downward to measure the velocity and position of an underwater vehicle relative to the sea bottom. Movement-based sensors are also used in underwater navigation and localization, such as inertial measurement units (IMUs), and pressure sensors, which can determine depth. Several types of sensor data can be utilized in a sensor fusion setup to develop multi-modal robust perception methods and provide inputs to perform online mapping of the explored underwater environment Singh and Alexis (2024); Botta et al. (2024). Such mapping can be utilized for both autonomous navigation and autonomous operations with UUVs. In addition to localization and mapping sensors, UUVs can be equipped with a wide range of **sensors for data acquisition** during missions. Such sensors can gather environmental data, e.g., temperature, oxygen levels, salinity, turbidity, waves, currents, and light, as well as data for monitoring and inspection of the explored environment Kelasidi et al. (2022). The data can either be stored locally on the UUV to be retrieved later at surface level or transmitted through a tether.

2.2.3 Motion Planning and Control

When localization and mapping are in place, so that the vehicle knows its own position and perceives its environment, the vehicle has the basis to be able to perform inspection or intervention tasks autonomously (Kelasidi and Svendsen 2023; Føre et al. 2018). Furthermore, methods for **motion planning and navigation** need to be applied, along with linear and nonlinear **control methods**. Through motion planning and control, the optimal path for the vehicle can be calculated, robust control methods can limit the error of state estimation, and the necessary commands can be sent to the actuators to perform the movement and complete the task. Inspection tasks can be performed using sensors, such as sonars or cameras. Intervention tasks, such as installation assistance, grasping and recovering objects, or valve operations, are often more complex and require dedicated **intervention tools** attached to the UUV, developed for specific tasks. Performing operations via **remote control** is the first step toward full **autonomy** (Su et al. 2023), and remote control can also greatly benefit from control schemes, such as heading- or depth control, to allow assistance.

2.2.4 Mission Execution and Practical Aspects

By equipping underwater robots with sensors for imaging, navigation, and surveying, they can be used in various applications such as archaeological surveys, resource exploration, oceanographic research, and marine biology. Both ROVs and AUVs have distinct advantages and limitations that make them more or less suitable for different tasks and operations. The tether, or umbilical, of an ROV provides both power and communication, allowing an ROV to be in continuous operation for longer periods than an AUV powered by batteries. An ROV can also easily receive real-time commands from the surface and transmit large amounts of data directly to an operating centre (Fossen 2011). Real-time data transfer is essential when control relies on a video feed, as is common in most remote-controlled operations, and the access to easy data transfer eliminates the need for local data storage. However, the umbilical will limit to how far from the operation base the robot can go and restricts the areas and environments the ROV can access. While an AUV can move around relatively freely, an ROV operation requires continuous management of the tether to avoid tangling. ROV operations also require a supply vessel to deploy the ROV, provide power, and, in most cases, serve as a control centre during operations. Although the main distinction between AUVs and ROVs is whether they operate autonomously or are remote controlled, ROVs can also be used in autonomous operations. To make a UUV system fully autonomous, it would also require a **launch and recovery system and/or docking station** for charging (Yazdani et al. 2020; Haugaløkken et al. 2021a). These docking stations can also facilitate data transfer and recharging of batteries for UUVs that do not receive power through a tether.

Underwater robotic systems can work together with partially submerged vehicles, such as unmanned surface vehicles (USVs) or autonomous surface vehicles (ASVs), to perform tasks in a range of industries. Currently, robots are used primarily for inspection tasks, including surveying tasks in the oil and gas industry performed at substantial depths, where underwater robots perform acoustic inspections of pipelines or rig site surveys (Merz et al. 2023). In both search and rescue and archaeology, UUVs can be used in scouting missions, and UUVs are often used in geological surveys, seabed mapping, and coastal mapping. In sea-based aquaculture, UUVs can be used both outside and inside the net pens in combination with other service vessels. In critical situations such as an oil spill, inspection operations can be crucial to get a clear image of the severity of the situation, to monitor how the situation progresses, and to put in place the appropriate countermeasures (Dooly et al. 2016). The level of navigation and decision autonomy varies, often depending on the complexity and scale of the task. For instance, seabed mapping, which is conducted on a large scale with comparatively low demands on precise positioning of the robot, is well suited for an AUV. In contrast, tasks such as inspecting an active oil leak require higher precision in robot positioning, as the risk is greater if something goes wrong. For such scenarios, an ROV operated by an experienced human operator on the surface, supported by automatic control schemes, is generally considered more suitable.

2.2.5 Robotics in Sea-Based Aquaculture

As for many other maritime industries, sea-based aquaculture could greatly benefit from an increased use of robotic and autonomous solutions to replace humans, particularly in risk-prone operations and operations that are difficult to do for human workers. However, the use of underwater robotic systems in sea-based aquaculture introduces a new set of challenges compared to most other parts of the maritime domain, with operations being conducted in the wave zone where currents and waves create a challenging and dynamically changing environment (Bjelland et al. 2015). The structures involved can be soft and flexible, and the net, mooring lines, and other equipment will all move with currents and waves. This means that robotic solutions must not only be able to perform tasks while withstanding impact from waves and current but must also avoid collision with any of the moving structures. The fish itself should not be harmed, complicating things further as a robotic system would ideally have to adjust to the unpredictable movements of up to 200,000 individual fish. Prioritizing welfare also puts limitations on how operations are carried out in order to minimize unnecessary stress on the fish. However, improving the welfare of the fish might also be one of the strongest motivations for adopting robotic and autonomous solutions, as these technologies could make the industry—and its necessary inspection, maintenance, and repair (IMR) procedures—less vulnerable to the elements. Increasing the use of robotic and autonomous solutions in sea-based aquaculture can help improve the level of control we have over operations (Kelasidi and Svendsen 2023). The SINTEF ACE RoboticLab that is associated with the full-scale research infrastructure SINTEF ACE was established to target these challenges, focusing on the new demands for technology developed for aquaculture (Fig. 2.2). Digital twin technology is an essential building block in the ongoing and future development of this laboratory.

Fig. 2.2 Typical Atlantic salmon farm at Tristeinen in central Norway, with ten net pens connected to a feed barge. The feeding tubes can be seen as white lines going from the feed barge to each net pen. (Photo: Courtesy of SINTEF ACE)

2.3 Status and Operations in Sea-Based Aquaculture

As per 2024, a typical sea-based fish farm for Atlantic salmon in Norway consists of 8–16 flexible net pens (Fig. 2.2), each containing a maximum of 200,000 salmon (Holmen 2022; Nrk.no 2024). Net pens can have different flexible designs, such as cylindrical, cone-shaped, or rectangular, the cylindrical type illustrated in Fig. 2.2 being the most common. A typical industrial scale net pen can have a diameter up to 50 metres and be up to 35 metres deep. A top collar consisting of hollow plastic tubes keeps the top of the net pen afloat, while a bottom weight system keeps the net pen containing the fish submerged and extended. The fish are fed automatically pumping feed through feeding tubes (seen as white lines from the feed barge to each net pen in Fig. 2.2) into each cage based on commands sent remotely from a feed barge or from shore. Each farm is usually staffed with one operations manager and 3–6 workers operating from a feed barge that serves as the centre of operations (Holmen 2022). Daily operations include IMR tasks related to fish welfare, such as feeding, removing dead fish or cleaning the net, and operations to maintain the state of the net pen and other parts of the facilities up to standard (Fig. 2.3).

The present desire to expand aquaculture production and make it more cost-effective has led to industrial trends toward the establishment of larger cages and in more exposed locations (Bjelland et al. 2015). More exposed locations will make human presence at and operation of farms more challenging because of the exposure to the elements and will thus increase the need for automated data collection and intervention operations. This section presents the most prominent IMR operations in sea-based salmon farming in Norway, summarized in Tables 2.1 and 2.2. These tables provide an overview of the current status of potential challenges related to safety, productivity, and the current use of robotic solutions and autonomy.

Fig. 2.3 Some of the many inspection, maintenance, and repair operations that are performed at fish farms throughout the year and how they are currently executed

Table 2.1 Inspection operations. Overview of common inspection operations in sea-based aquaculture and the status on how they are currently performed

Operations	Current status	
Net inspection	Inspected via video stream using ROVs remote controlled from the surface by a human operator. The operator inspects the net for holes while simultaneously navigating the ROV along the net manually, aided by systems like depth control and heading control on the ROV. Net inspection can also be performed by divers	 Photo: Artifex
Mooring line inspection	Inspected via video stream using ROV remote controlled from the surface by a human operator. The operator inspects the mooring line for biofouling or weaknesses via video stream while manually navigating the ROV along the mooring line	 Photo: Artifex
Fish monitoring	Fish monitoring and inspection is primarily performed manually by one or more operators. This is done either on a group level using, for example, feeding cameras installed on the net pen to evaluate the fish or through manual inspection via individual sampling	 Photo: Crowdguard
Environmental monitoring	Oxygen levels, temperatures, salinity, currents, etc. are monitored using sensors permanently installed either in the cage, around the net pen, or by a feeding barge close by. Data from these sensors are transmitted and monitored by human operators on the feeding barge	 Photo: SINTEF Ocean AS
Inspection and monitoring of other equipment	Maintenance and cleaning of submerged equipment is performed either at predefined intervals or when a need for additional maintenance is detected via the live camera feed from installed on-site cameras	 Photo: SINTEF Ocean AS

2.3.1 *Common Inspection and Monitoring Operations in Aquaculture*

2.3.1.1 Net Inspections

Net inspections must be performed regularly to check for holes and flaws in the net that, over time, could grow more severe and cause fish to escape. Escapes represent a major challenge for the aquaculture industry. A 2019 report shows that more than 70% of escaped farmed Atlantic salmon and rainbow trout in Norway between

Table 2.2 Intervention operations. Overview of common maintenance and repair operations in sea-based aquaculture and the status on how they are currently performed

Operations	Current status	
Net cleaning	Performed every 2–4 weeks depending on the time of year. Conducted using half-automatic crawling robots or cleaning equipment attached to ROVs or with ROV functionality. Cleaners are deployed using cranes. Farmers need to schedule cleaning operations and book service companies in advance. It requires human workers to be present on the net pens to control and monitor the cleaning operation	 Photo: AKVA group(Akvagroup, 2024b)
Net repair	Tears or holes in the net must be repaired to avoid escapes. While recent research and industry initiatives are investigating robotic net patching solutions, divers are currently conducting all permanent repairs. It might therefore take days from when a hole is discovered until a diving team can perform the repairs	 Photo: Unsplash/Balland
Equipment installation and maintenance	Performed by human workers assisted by cranes and capstan winches, as well as divers for operations like bolt drilling for the mooring lines. This involves a significant use of cranes, manually operated by certified operators	 Photo: SINTEF Ocean AS
Dead fish removal	Dead fish must be removed from each net pen daily, which requires several workers and a service vessel. Dead fish can be pumped from the bottom of the cage over to a service vessel, or a dead fish basket can be installed at the bottom of the cage to gather the fish. This basket is then hoisted up during these dead fish removal using capstans. Dead fish at the surface might be removed manually by a worker using a net	 Photo: AKVA group(Akvagroup, 2024c)
Sea lice counting	Performed every 1–2 weeks depending on water temperature, and done by transferring a few fish from the net pen over to a tank of saltwater and added sedatives aboard a service vessel. Sea lice are then counted manually by human operators before releasing the fish back into the net pen	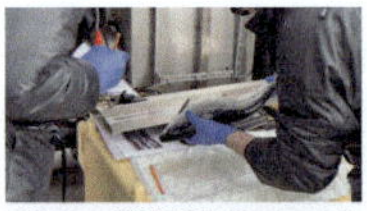 Photo: SINTEF Ocean AS
Delousing	When the number of mature female sea lice reaches the limit, a delousing operation can be performed to lower it. The fish is crowded and pumped over to a service vessel with a treatment system (fresh water, thermal, mechanical). Delousing operations are often time-consuming, and the handling of the fish negatively affects fish welfare and growth rate, which presents a significant cost	 Photo: SINTEF Ocean AS
Crowding operations	Operations performed when the fish must be transferred to a service vessel. The volume of the fish cage is reduced, for example, by lifting parts of the net using cranes or capstan winches and pumping the fish from the net pen. The perceived stress levels of the fish are monitored based on the surface-level activity	 Photo: SINTEF Ocean AS

2010 and 2018 were due to holes in the net (Føre et al. 2019). This averaged to approximately 220,000 escaped fish per year, which poses a significant challenge when considering the potential consequences of escapes. New regulations from 2022 require the farmers to monitor the condition of the net and other structural components before, during, and after operations that include interaction in contact with or near the net structure, such as net cleaning or delousing operations (Nrk.no 2024, §37a). Depending on operations, this inspection may be limited to certain exposed parts of the net rather than a full net inspection, but post-operation inspections must be carried out immediately and cannot be postponed (Directorate of Fisheries 2023). Today, net inspections are most commonly performed using ROVs controlled by skilled human operators, supported by integrated control systems like auto depth and auto heading, which utilize sensors such as altimeters and gyroscopes. Net inspection operations can be performed by a service company, requiring advance scheduling and booking by the farmers, or by ROVs provided by the farmers and operated by their local staff. In either case, this involves several workers and a service vessel to dispatch the ROV. Using ROVs makes the operation more easily accessible for farmers because they do not need licensed divers, thus reducing the risk of injury to diving personnel (Føre et al. 2018). To avoid the tether getting tangled in mooring lines or other equipment, ROV net inspections are performed inside the net pen. Using a real-time video feed, the human operator navigates the robot to inspect the area of the net. This operator is typically also responsible for the inspection operation itself, examining the video feed for signs of holes or damage to the net. During the inspection operation, the ROV is generally kept at a distance of 2–3 metres from the net for effective inspection. However, in cases of poor visibility caused by poor light conditions or high water turbidity, the ROV may be kept as close as 0.5 metres from the net (Solem 2017). Net inspection operations are highly sensitive to external conditions, with strong currents being a particular challenge, as they can make it difficult for the operator to keep the ROV in position.

2.3.1.2 Mooring Line Inspections

The mooring lines anchoring the net pens to the seabed must also be inspected regularly at least every other year, checking for deviations and weaknesses along the mooring line, around the moorings, and at the anchoring (Solem 2017). Similarly to net inspections, this operation is often carried out using an ROV controlled by an experienced human operator, who navigates the ROV along the mooring line from the top seal to the bottom seal. When mooring lines are examined using ROVs, the operator typically controls the ROV remotely while simultaneously monitoring the video feed to inspect the mooring lines, just like for net inspections. Thorough mooring line inspections are performed every 1–2 years, as issues with mooring lines tend to appear more gradually than, for example, holes in the net and do not require immediate attention. Inspection of the mooring lines typically takes less than a day per facility. Due to the low frequency of inspections and the low chance of the need for immediate countermeasures, inspections of the mooring line are not considered a significant cost and are not given much priority when considering the introduction of more autonomous operations (Solem 2017).

2.3.1.3 Fish Monitoring: Nonintrusive

The inspection and monitoring of fish is necessary to ensure the health and welfare of the fish population, including factors such as growth, potential disease or infections, and lice infestations. At the inspection level, the most commonly used tool for monitoring fish behaviour during activities such as feeding operations is the permanently installed feeding cameras within each net pen (Føre et al. 2018). The feeding operation is controlled by a human operator located on a feed barge at the site, not necessarily directly by the specific net pen, so the transmitted video data provides insight into the fish's distribution and their interest in the food. The assessment of the fish behaviour is performed by a human operator who reviews the transmitted video data. Based on this information, they make an experience-based, yet subjective, evaluation of the fish's behaviour and decide when to stop the feeding. Additionally, the video data is used to assess the status of fish health in each net pen, allowing the operator to identify visible ailments such as infections, winter sores, or an increased number of moribund or dead fish. According to Norwegian government regulations, mortality must be checked for every net pen, and any dead fish discovered must be removed at least once a day (Nrk.no 2024, §16). This is to avoid any potential spread of disease or the development of pathogens, with no exceptions from this duty, except in extreme situations such as *force majeure*. Much of the monitoring of fish health is conducted by physical checks and random sampling of fish.

Cameras are also used to monitor fish activity at the surface. This is crucial during crowding operations, where the available volume of the cage is reduced so that the fish gather at higher densities before being transferred to a well boat for transport (Bhadmus 2022). Crowding operations are performed each time the fish need to be moved, e.g., during delousing operations or before slaughter. Human operators evaluate how well a crowding operation is going based on the level of surface activity, the swimming speed and direction of the fish, if the fish are colliding with each other, and whether too many fish are breaking the surface with their dorsal fin or have the white side up because they are on their sides or backs (Noble et al. 2018). If panic develops among the fish during crowding operations, there is a risk of both escapes and mass mortality due to traumatic injury or hypoxia (Overton et al. 2019). Therefore, the expertise of the operators in evaluating the surface activity is crucial.

2.3.1.4 Environmental Monitoring

Monitoring and controlling some of the environmental parameters that affect the welfare of fish in open, sea-based aquaculture facilities can be challenging compared to closed systems, such as land-based aquaculture in closed tanks. The flow and currents of the open water pose challenges such as the transport of sea lice or pathogens between cages and between separate aquaculture facilities. The environmental conditions in and around the net pen must be monitored to ensure good conditions for the fish and to assess how conditions in one net pen or site can impact neighbouring facilities (Solstorm et al. 2018; Klebert et al. 2023; Klebert and Su 2020). Environmental monitoring is done using a range of sensors that measure parameters like temperature, oxygen levels, salinity, or currents. The sensors are usually installed permanently, and the data is transmitted to a central and monitored by human operators. The number of sensors installed on each site varies, and data is sometimes gathered from only a single sensor placed near the feeding barge, rather than from sensors in the individual net pens. As a result, there may be no information available on the environmental differences between the various net pens, although factors like currents and the degree of exposure can affect conditions such as oxygen levels (Ohrem et al. 2022).

2.3.1.5 Equipment Inspection

Equipment installed in and around each net pen, such as environmental sensors, feeding cameras, feeding tubes, collar, and railings, will often require on-site cleaning and other maintenance. For example, cameras permanently installed on winches in the net pen need to be cleaned to maintain functionality, as biofouling on the lens can reduce the image quality or completely remove visibility. Maintenance and cleaning of most equipment is scheduled based on visual inspection by operators working on the cage, either directly or via winched cameras installed in and around the net pen. Alternatively, maintenance may follow a predefined schedule based on experience with, e.g., wear and tear or biofouling growth. The site manager at each aquaculture facility is responsible for ensuring that risk-based supervision is conducted daily for all production equipment, technical instruments and devices, and other installations, with a focus on fish welfare and environmental considerations, provided the weather conditions allow it (Nrk.no 2024, §12). If storms or other bad weather are expected to hit, all installations must be checked to ensure that they are secured sufficiently to withstand the weather. Additional checks must be performed immediately after the weather event to assess any potential damage to the facilities. These inspections are primarily conducted by workers on the site, with assistance from permanently installed cameras or ROVs when, for example, a net inspection is required.

2.3.2 Common Intervention Operations in Aquaculture

2.3.2.1 Cleaning Operations

While a net pen is installed and operational, a wide range of maintenance and repair operations must be performed to ensure the integrity of the facilities and the safety of operations and maintain fish welfare. Cleaning operations are necessary to prevent the development of too much *biofouling* on the net, mooring lines, and other parts of the cage. Biofouling is the unwanted growth of organisms on submerged structures such as nets and other parts of the cage (Bannister et al. 2019). Organisms growing in the net can negatively affect fish welfare and operations in several ways, as summarized by Fitridge et al. (2012). There are three main negative effects to consider (1) restriction of water flow through the cage, leading to lower water quality due to reduced oxygen levels and decreased removal of excess feed and waste, (2) increased risk of disease, as the biofouling serves as growing grounds for pathogenic microorganisms, and (3) cage deformation and extra wear and tear on equipment due to the additional weight from the biofouling causing increased stress on mooring systems. Biofouling has also been found to decrease the effectiveness of cleaner fish since cleaner fish eat sea lice but may also eat biofouling, reducing their effectiveness (Bannister et al. 2019; Fitridge et al. 2012; Ohrem et al. 2022). To reduce these negative effects, the nets need to be cleaned regularly. While antifouling coatings applied to the nets may delay and reduce the accumulation of biofouling organisms, several net cleaning operations must still be performed during the life cycle of most nets (Ohrem et al. 2020; Edwards et al. 2015; Bloecher and Floerl 2020).

In Norwegian sea-based salmon farming, net cleaning operations are typically performed every 2–4 weeks but may be performed even more often during the summer and autumn months, since biofouling growth varies with water temperature and light conditions (Bloecher et al. 2015). Cleaning operations are performed by service providers and must be scheduled and booked by farmers in advance. Net cleaning in cages for Atlantic salmon in Norway is currently performed using a range of different cleaning tools. This includes semi-automatic net cleaning rigs or net crawling robots, which are deployed from a service vessel and moved vertically along the net cage in a controlled manner using a crane (Føre and Lien 2014), as well as cleaners with ROV functionality or cleaners attached to different types of ROVs (Solem 2017). Many of these solutions use water pressure in combination with rotating cleaning discs to flush away the biofouling. While this method is efficient, it can potentially be harmful to the fish because the biofouling is released into the net pen during the pressure washing process. Although the water quality will improve again over time as currents move the particles out of the cage, the period of reduced water quality can still impact the fish through increased stress levels and the potential development of gill disease (Fitridge et al. 2012; Bloecher et al. 2018). Recently, there have also been initiatives to develop new biofouling management systems with permanently installed autonomous grooming robots designed to prevent excessive biofouling growth through more frequent cleaning operations (Skaldebø et al. 2023). Floating collar and sinker tubes are also cleaned with specialized washers using water pressure or brushes, as they cannot be raised above the waterline. Facilities using cleaner fish such as lumpfish or wrasse are often equipped with cleaner fish shelters: artificial seaweed where the cleaner fish can hide and rest. These should be cleaned monthly, depending on the season. This is done by lifting the shelters out of the net pen and onto a service vessel, where they are cleaned using high pressure washers (NorseAqua 2022). Smaller sensors generally need to be taken out of the water for cleaning by workers on the net pen, or they are cleaned by divers. Currently, all available cleaning solutions for different parts of the net pen structure require dedicated personnel for control, deployment, and monitoring of the cleaning process. The need for human operators to be present at the net pens means that these cleaning operations cannot be performed during very rough weather conditions.

2.3.2.2 Net Repair

Normal wear and tear of the net can sometimes cause small holes and tears, with the risk increasing during handling of the net related to net cleaning, crowding operations, or when equipment pulls on the net during storms or strong currents. Operational risks to net integrity include contact with service vessels or ROVs, the installation or moving of equipment that may touch the net, as well as strong waves, currents, debris, or ice in the water (Holmen et al. 2021, 2023). Small holes left unattended may grow larger and eventually allow fish to escape. Holes in nets are currently the leading direct cause of escapes (Holmen et al. 2023). Therefore, it is crucial to both detect and repair holes. Although net inspections are increasingly dominated by underwater robots (Sect. 2.3.1), all *repairs* still need to be performed by a diving team. Divers introduce a significant expense, and due to availability it may take time to schedule and get a diving team on site. It may therefore take days or weeks from when a hole is detected until it is repaired (Solem 2017), or even longer if weather conditions such as strong winds make working conditions too risky for human workers.

2.3.2.3 Equipment Installation and Maintenance

The installation of the net pen, with collar, railings, bottom net weight, and the net itself, is carried out by human operators with the assistance of service vessels equipped with cranes and capstan winches. In addition, divers are needed for operations such as bolt drilling and chain mounting during the installation of the mooring lines (Solem 2017). Installations must follow strict and quality assurance procedures in the specific contractor's manuals, and a complete net inspection is required before releasing fish in the net pen (Nærings- og fiskeridepartementet 2018). The installation of feeding tubes is also performed by human workers assisted by cranes. ROVs are currently not directly involved in the installation of equipment but are used for visual inspection and support. Underwater repairs are performed by divers. Repairs that require the equipment to be removed from the net pen are performed on land or on a service vessel, e.g., changing sections of feeding tubes. Heavy lifts using cranes have the potential to cause significant damage to both equipment and personnel, so this involves its own set of risks. According to interviews with crane operators in aquaculture in 2022, the need to move equipment that may not be suitable for lifts by crane because it does not have satisfactory lifting points introduces an increased risk (Gumdal 2022). Related risk assessments can often be experience based and subjective because there does not exist a best practice or guide for aquaculture crane operations. Therefore, the development of systems and technology that can reduce the number of crane operations was suggested as an important step toward improving HSE and safety in operations (Gumdal 2022). Automating crane operations could also be a solution.

2.3.2.4 Dead Fish Removal

According to regulations from the Norwegian government, the removal of dead fish and monitoring of the mortality must be performed at every net pen each day (Nrk.no 2024). Dead fish removal operations involve direct interaction between a service vessel and the cage collar: Either dead fish are pumped from the bottom of the net pen using compressed air, or a dead fish basket installed at the bottom of the cage can be hoisted up to surface level using a capstan. The dead fish are gathered aboard a service vessel and taken away for immediate destruction. Such operations can involve heavy and potentially straining lifts for the workers involved. As for many other operations, the removal of dead fish requires calm weather, as strong currents or winds pose a significant risk to both personnel and equipment. Longer periods of rough weather could therefore mean that dead fish stay in the cage longer than recommended, posing a threat to the general fish welfare.

2.3.2.5 Fish Monitoring: Sampling

To evaluate the welfare status of fish in a net pen, *welfare indicators* (WIs) are used (Noble et al. 2018), both at the individual level and to evaluate the overall fish population. Some of these indicators can be evaluated by observing the biomass as a whole, both from the surface and using submerged feeding cameras. Examples of these indicators include mortality, appetite or feeding activity, and visible signs of disease and wounds. Sea lice counting is another important WI and a critical inspection task that must be performed regularly to ensure that sea lice levels remain below the accepted limit. If necessary, a delousing operation is initiated. The regulations stipulate that the average number of mature female sea lice per salmon must remain below an upper limit, ranging from 0.2 to 0.5, depending on the season (Lovdata 2023). In the Norwegian farming of Atlantic salmon, sea lice counting is performed every week or every other week, depending on the water temperature (Lovdata 2023; Havbruksutvalget 2024). Sea lice counting is performed through manual sampling, where a few fish are transferred from the net cage to a saltwater tank with sedatives aboard a service vessel (Thorvaldsen et al. 2018). Any lice are then counted manually before the sampled fish are returned to the net pen. If the number of sea lice exceeds the upper limit, a delousing process is scheduled. Both research and industry initiatives are underway to develop automated systems for lice counting. Other WIs, such as growth, disease, and gill conditions, are also addressed through manual sampling (Noble et al. 2018). Some WIs require laboratory analysis, such as blood samples to measure cortisol, glucose or lactate levels to provide information on stress, or analysis of the mucus layer, providing information on the general health and the immune system status. Certain WI tests require the fish to be euthanized, such as laboratory analysis of the intestines to check for parasites or other diseases.

2.3.2.6 Delousing

Delousing operations can be performed using pesticides, but the occurrence of drug resistance in seal lice becoming more common in Norway, there is now a growing shift away from medicinal treatments (Aaen et al. 2015; Overton et al. 2019). The non-medicinal methods used today include fresh water treatments, thermal treatments, or mechanical delousing with water jets or brushes (Bui et al. 2023). All methods require the fish to be pumped from the fish farm and into a service vessel or barge with a treatment system, which means that crowding is necessary (Overton et al. 2019). Thermal delousing was until recently the most common non-medicinal method used in Norwegian salmon farming. This treatment involves transferring the fish through a basin with lukewarm water (up to 34°C) for a period of maximum 30 seconds, with the rapid temperature change effectively killing the sea lice (Sommerset et al. 2022). From 2023, mechanical delousing using brushes or water jets has become the predominant non-medicinal delousing method. Winter sores and complex gill disease, listed as two of the main reasons for reduced fish welfare in the sea phase, are exacerbated by mechanical handling of the fish (Sommerset et al. 2023a; Østevik et al. 2022). Delousing operations as they are currently performed also puts a strain on the human operators involved. According to a survey from 2020 involving fish farmers, operational managers, and employees aboard service vessels, fish farm personnel sometimes work up to 20 hours a day, several days in a row during delousing operations (Thorvaldsen et al. 2020; Holmen et al. 2023). The additional cost of a single delousing operation for one net pen has been calculated to be approximately 400–800,000 NOK (approx. 35–70,000 USD), although it is difficult to accurately calculate the full cost of a delousing operation. The fish are also starved prior to the operation, as they have been shown to handle stress better when starved. Additionally, this has the practical benefits of less faeces when moving the fish to close chambers for delousing (Lusedata 2020). Although the starving procedure may mean that the additional cost of the delousing operation can be reduced somewhat by saving on feed in the days before (Ruggesæter and Haug 2021), starving the fish will lead to reduced growth rate. While compensatory growth after the feed retention period has ended may partly amend for this loss in weight gain, this represents a significant cost, potentially amounting to millions of NOK per delousing operation, depending on the size of the fish (Walde et al. 2022; Ruggesæter and Haug 2021; Iversen et al. 2017).

2.3.2.7 Crowding

Many operations, such as delousing, vaccination, or transport to slaughter, involve crowding operations. Crowding involves gathering the fish into a reduced volume within the cage, typically by lifting parts of the net using cork lines or by lifting the bottom weight system (Holmen 2022), before transferring the fish to a well boat via a pumping system (Bhadmus 2022). This generally involves two service vessels, each with no less than 3–4 operators on board (Iversen et al. 2017). Increasing the density of fish in the net pen restricts their freedom of movement, which can trigger an increased stress response (Noble et al. 2018). Additionally, higher fish densities will lead to lower oxygen levels, while heightened activity caused by stress further raises the fish's oxygen demand. For this reason, a crowding operation carries the risk of mass mortality due to hypoxia if panic breaks out among the fish (Overton et al. 2019), highlighting the importance of carefully monitoring the surface activity during crowding (Noble et al. 2018; Bhadmus 2022). Due to repeated delousing operations and other similar procedures, fish may undergo multiple crowding operations in their life cycle.

2.4 Directions Toward Autonomous Operations

To ensure efficient and safe production, a wide range of IMR operations must be carried out throughout the year, as outlined in Sect. 2.3. Despite a high focus on HSE, many of these operations still pose risks to workers, especially when working under challenging weather conditions (Thorvaldsen et al. 2017, 2023, 2020). The aquaculture industry faces the constant challenge of balancing fish welfare, worker HSE, equipment and facility integrity and durability, and productivity to ensure profits. This delicate balance makes this sector a great candidate for increasing the use of autonomous and robotic solutions. A variety of new UUV systems are being developed to support a range of IMR operations in aquaculture, though most of these technologies are still in the research and development phase. In this section, we present a selection of the IMR operations presented in Sect. 2.3 that have the potential to be addressed through robotics and autonomous solutions and review the current status of the available technological solutions (Table 2.3).

2.4.1 Net Inspection

As detailed in Sect. 2.3.1, net inspections are most often performed by ROVs: A human operator remotely controls the ROV while simultaneously monitoring the live video stream to identify any holes or structural weaknesses. The net inspection operations heavily depend on the operator's skill and experience. These operations are also highly weather-dependent, as controlling an ROV in strong waves and

Table 2.3 Toward autonomous operations. Overview of operations with potential for future robotic solution and automation

Operations	Potential future solutions	
Net inspection	Fully autonomous systems with integrated sensors able to perform 24/7 net-pen inspections, enabling human and weather independent operations. This could leave the operator free to focus more on monitoring the video stream. Computer vision methods have also shown promise in detecting holes from videos, and can potentially be used as support, either live or based on recordings	 Photo: Artifex/ SINTEF Ocean AS
Fish monitoring	Future solutions could aim for static and/or moving sensor platforms to monitor and inspect fish status continuously and provide inputs for objective decision support systems. Cameras and sonars can provide data on fish behaviour on a group level, such as during crowding operations. Using computer vision could help detect wounds to identify fish that should be removed from the net pen	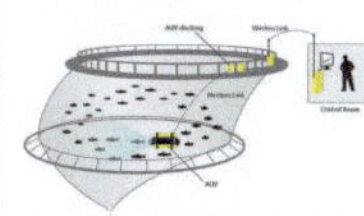 Photo: CageReporter/ SINTEF Ocean AS
Environmental monitoring	Resident UUVs used in for example net inspection could be equipped with sensors to continuously monitor environmental parameters throughout the net pen and provide unique data for each individual net pen	 Photo: SINTEF Ocean AS
Net cleaning	A range of new solutions for dedicated cleaning robots are under development, currently on a testing stage. Some of these solutions function similarly to current cleaning methods using for example water pressure to remove developed biofouling. Others are based on a more continuous grooming procedure that will prevent biofouling from developing at all. Such permanently resident vehicles should have navigation, energy, and decision autonomy	 Photo: NetClean24/7/ Remora Robotics AS
Delousing	Increased focus on fish welfare has led to an increased interest in non-invasive methods. Currently, the Stingray laser delousing system is being tested, aiming to kill the sea lice without stressing or hurting the fish. Combining these types of solutions with resident and autonomous robots could allow for continuous, non-invasive delousing	 Photo: (Stingray AS, 2024)
Equipment repair	Designated repair tools are being developed for ROVs to perform operations such as net patching. However, such tools are currently on a testing stage, and permanent repairs still need to be performed by divers. Robotic systems with dedicated tools and autonomous functionalities able to perform maintenance and repair operations can greatly benefit the industry in the future	 Photo: (Deep Trekker, 2024b)

Fig. 2.4 Remote-controlled ROVs are already being used to perform net inspections, with a human operator performing a visual inspection of the net through video stream while simultaneously navigating the ROV. Autonomous net following could leave the operator free to focus only on the inspection operation (Amundsen et al. 2021). Algorithms for detection of irregularities could improve the net inspection further. (Photo: Courtesy of Artifex-NO256241/SINTEF Ocean AS)

currents is challenging. It is crucial to avoid unintended contact with the net to prevent creating tears or damage to the net that could lead to escapes (Holmen et al. 2021, 2023). According to an ROV pilot interviewed in Solem (2017), it is not uncommon for the ROV to make sufficient contact with the net during net inspection operations to fracture masks, even when supported by depth and heading control. It is important to note that there are no reported cases of ROV net inspections leading to escapes. Considering the repetitive pattern of the net, net inspection operations require intense focus from the operator to identify holes and ensure coverage of all areas of the net during inspection. The operator's attention is also split between manoeuvring the ROV and monitoring the video feed.

Due to the serious consequences of potential escapes, it would be beneficial if net inspections could be performed more often and with fewer resources than what is currently possible. Having daily or even continuous net inspection operations could ensure that tears or holes were detected and repaired sooner. The cages used in most sea-based finfish farming generally have a symmetric and predictable shape, and this can be used as an advantage when developing control schemes for autonomous net inspections (Fig. 2.4) (Amundsen et al. 2021; Kelasidi et al. 2022). The cage itself is typically a flexible structure in a dynamically changing environment, meaning its shape will change somewhat in response to currents and waves. However, net deformation and the conditions of the surrounding environment can be estimated using predictive methods that account for these factors (Su et al. 2019). This could allow for navigation and decision autonomy or semi-autonomy where the operator can interrupt if necessary. Currently, most net inspections are conducted without a digital tracking system for the traversing of the net to support the operator's overview of the progress of the net inspection. By tracking the inspection operation and the position of the ROV, a detailed overview of which areas of the net have been inspected could be generated, improving the precision of the operation (Su et al. 2023; Føre et al. 2024).

By introducing navigation autonomy for the ROV, the operator could focus more on monitoring the video feed. In addition, inspection operations could be supported by computer vision technology to detect irregularities and holes. State-of-the-art computer vision-based methods have demonstrated high efficiency for hole detection in videos recorded in water with good light conditions and low turbidity (Madshaven et al. 2022; Schellewald and Stahl 2022). These methods have also shown promising results under more challenging water conditions, such as uneven lighting, varying water turbidity, and different camera perspectives (Zacheilas et al. 2021; Labra et al. 2023). If this technology could be combined with a robot that traverses the cage and documents the entire area of the net, machine vision methods could be used to detect both fully developed holes and potential problem areas that might evolve into holes over time (Labra et al. 2023). During net inspections in a fully operational net pen, it is likely that a human operator will be required to monitor the operation to ensure that the tether does not get tangled with other equipment. Alternatively, a tether management system (TMS) can be developed for such underwater operations. As long as net inspection operations require live video feed to the surface for monitoring, it will not be possible to switch from ROVs to untethered UUVs. However, if UUVs with integrated sensors could be utilized for continuous inspection of cage infrastructure, this could enable the development of fully autonomous net inspection operations in the future. The information collected could be used to develop maps of the inspected environments, and the vehicle could report only essential information, such as detected irregularities and holes, via wireless communication to the surface. This information and maps of the inspected net pens could also be used for the planning and execution of maintenance and repair operations, in addition to being a component in future autonomous decision support systems for optimal operations in fish farms.

The project CageReporter investigated how an autonomous and tether-less underwater vehicle equipped with vision and environmental sensors for data capturing could be used for real-time monitoring of fish cages, focusing on net inspection, fish monitoring, and environmental monitoring (Kelasidi et al. 2020; Kelasidi 2020). The project concluded that efficient and autonomous inspection of the net and other parts of the cage requires speed control and precise position measurements. Additionally, it identified the need for real-time algorithms for image processing and adequate light conditions and the capability of the AUV to move slowly enough to capture a sufficient number of images for effective image processing. The main challenges identified included strong currents, loss of communication with the AUV, data loss, and sensor failure. An operator should monitor the AUV at all times and have the ability to intervene if necessary.

2.4.2 Fish Monitoring

An increased focus on fish welfare has led to a preference for non-invasive methods, such as camera or sonar monitoring of biomass, over more invasive methods that can stress the fish (Barreto et al. 2022; Svinøy 2023). Sonars have been used to detect behaviour indicative of disease (Måløy 2020) and to study how fish interact with foreign objects introduced into the net pen, providing insight into factors like how the distance maintained to the object varies both with the object's size and the size of the fish itself (Zhang et al. 2024). Additionally, farmed Atlantic salmon have been shown to adjust their preferred depth based on light conditions and water temperature. This causes the vertical position of the main school to vary depending on the time of day or season, a pattern that can be closely monitored using sensors installed within each net pen (Fernö et al. 1995). Just as computer vision has been used for net inspection and hole detection (Madshaven et al. 2022; Schellewald and Stahl 2022), similar methods can be used for the detection of winter wounds. This has shown promise, both using video data recorded in ideal and highly controllable water conditions, such as an indoor pool (Gupta et al. 2022), and using video data filmed at a fully operational fish farm (Labra et al. 2023; Nissen et al. 2024). ROVs working within net pens could collect video data to help with such analyses. Having robotic systems positioned *outside* of the net could prove valuable during operations such as crowding, which currently depend largely on observing the level of surface activity to assess the stress levels and behaviour of the fish (Bhadmus 2022; Noble et al. 2018; Overton et al. 2019). ROVs monitoring the fish under water could provide additional insights into the fish behaviour at deeper levels and increase the understanding of the overall condition of the biomass.

The CageReporter project also explored how an autonomous and tether-less AUV inside the cage could be used for real-time monitoring of fish conditions, including tasks such as fish counting, assessing behavioural indicator for fish welfare, and detection of wounds or physical injuries (Kelasidi et al. 2020; Kelasidi 2020). The project identified high-quality vision data obtained using underwater camera systems and the use of equipment such as stereo cameras and image-based processing algorithms as success criteria, in addition to speed control and precise positioning systems for the AUV. The CHANGE project investigated how fish react to structures of various shape and colour being placed in the net pen (Zhang et al. 2024) and developed intelligent control strategies for collision-free autonomous navigation in dynamically changing environments (Amundsen et al. 2024c,a,b). In conclusion, ensuring fish welfare in the future requires moving away from manual and thus subjective methods for fish monitoring and inspection. Instead, autonomous solutions that use static sensors (e.g., acoustic and/or cameras) and mobile sensor platforms (e.g., robotic systems with integrated sensors) that provide inputs for objective decision support systems should be a core focus area for future aquaculture.

2.4.3 Environmental Monitoring

A significant limitation of current environmental monitoring practices is that sensors are often positioned near the feeding barge, providing less specific information about the specific cages. Alternatively, sensors are placed inside or outside one or a few net pens, which offers more information on that specific cage but lacks a detailed overview of conditions at different parts of the cage or across different cages at the farm (Su et al. 2023). ROVs or other UUVs equipped with sensors would address this information gap by collecting data from various points of interest both inside and outside the cages. As new technological solutions for other IMR operations in the fish pens are introduced—such as resident cleaning robots—such systems could also be issued with sensors for monitoring environmental data. This could provide additional insights on how the environmental parameters vary in different parts of the net pen and how this variation is impacted by seasons, weather, currents, or the placement of the fish during time of day. For example, farmed Atlantic salmon have been shown to react differently to currents and waves depending on the location of the net pen relative to the farm layout (Klebert et al. 2023). Comparing information collected by robotic systems at different cages can then provide information on variations at different parts of a farm.

The CageReporter project explored how an autonomous and tether-less AUV could be used to monitor key parameters such as oxygen, temperature, currents, and waves (Kelasidi et al. 2020; Kelasidi 2020). To assess the environmental conditions based on temperature and oxygen, the AUV should collect data throughout the entire volume of the net pen, identifying deviations from optimal target values. Additionally, current and wave data can be analysed to predict environmental impacts, such as deformations of the flexible net. Precise measurements, alongside the control of the AUV, were identified as the main success criteria.

As part of SINTEF ACE, a digital hub for research on digitalization of aquaculture operations called HAVOPERA has been established at SINTEF Ocean's offices in Trondheim, shown in Fig. 2.5. Sensor data from fish farms can be logged and transmitted to this control room, offering a comprehensive overview of environmental conditions. Such data can be used to inform predictive methods for operation planning. This kind of system could be expanded further, for example, by integrating ROVs capable of collecting data from the full volume of the net pen and geo-referencing this information, providing a deeper understanding of environmental variations and the operational implications.

2.4.4 Net Cleaning

New technological solutions for net cleaning must be robust enough to withstand challenging conditions, especially considering that sea-based fish farming is expected to shift to even more exposed locations in the future (Bjelland et al. 2015).

Fig. 2.5 The HAVOPERA control hub in Trondheim is used for research on digitalization of sea-based aquaculture, gathering sensors data from full-scale fish farms. (Photo: SINTEF Ocean AS)

A range of new solutions are under development, with a strong focus on developing specialized cleaning ROVs. Several companies have developed remote-controlled net cleaning robots (e.g., Akvagroup 2024a; Robotics 2024; Ocein 2024; Probotic 2024; MEOX 2024a). Several new solutions focus on avoiding high-pressure water flow to reduce net wear and tear and avoid the possible harmful effects on fish of dislodged biofouling in the water (Fitridge et al. 2012; Bloecher et al. 2018). One approach has been to develop net groomers—cleaning devices meant to traverse the net so frequently that biofouling is prevented from developing. These groomers are typically brush-based and prevent larval biofouling organisms from settling on the net (Ohrem et al. 2020). Other solutions in development include using one robot inside and one robot outside the net that are interconnected using magnets (Watbots 2024).

To make net grooming through daily traversing of the net financially sustainable, the introduction of navigational autonomy is essential to reduce the need for involvement of human operators. This approach has already been implemented for proactive cleaning of ship hulls (Swain et al. 2022). However, net cleaning in aquaculture net pens presents additional challenges due to the dynamically changing environment with flexible structures that add complexity to the control scheme. A practical solution to reduce the complexity of the task somewhat would be to have the robot stick to the net rather than to operate as a free-swimming unit within the net cage. This approach reduces the number of degrees of freedom (DOF), simplifying navigation, guidance, and control (Skaldebø et al. 2023). The project NetClean 24/7 explored how a tether-less, resident robot could be used for both biofouling prevention and inspection of net integrity in Atlantic salmon farming in Norway (Bloecher 2022; Ohrem et al. 2022). The project used a tether-less design of the cleaning robot, which allows the robot to move more freely around the cage

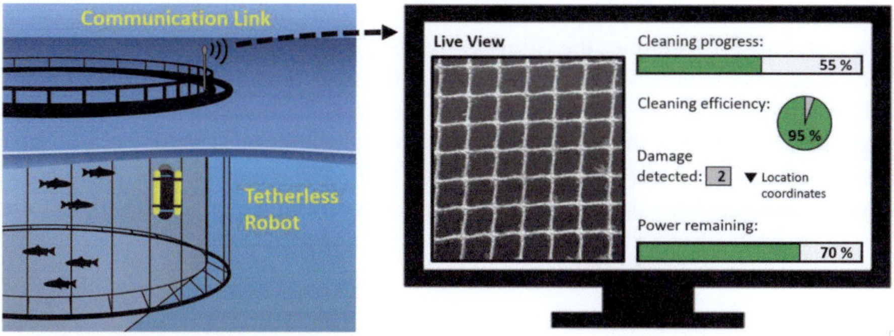

Fig. 2.6 A resident, tether-less cleaning robot could potentially perform daily net grooming to avoid the development of biofouling and simultaneously perform hole and damage detection. (Figure: Courtesy of NetClean 24/7-NO296392)

without the risk of entangling a tether with other equipment (Robotics 2024). A resident cleaning robot would also need a docking station for charging, and sensors to be able to monitor the environment, mapping out potential problem areas in the parts of the net that has been inspected, and the progress of the cleaning operation (Fig. 2.6).

2.4.5 Dead Fish Handling

New tools for UUVs are currently being developed to be used for the removal of dead fish in cages, offering an alternative to the manual methods used today. Among these innovations are specialized ROVs that can closely follow the bottom of the net and collect dead fish while moving (MEOX 2024b) or tools that allow ROVs to push dead fish from one part of the cage bottom toward a centralized collection point with a pump or dead fish lift (Trekker 2024a). These technologies are all currently in a testing phase, and no currently used solutions can collect dead or moribund fish at the surface. However, with the right tools, UUVs could potentially take on this task in the future. Similarly, the removal of fish identified to have severe winter wounds is another challenge that could be addressed using ROV intervention and should be subject to further research.

2.4.6 Delousing

The currently used methods for delousing require that the fish be pumped over to a service vessel to be subjected to delousing methods. Both winter sores and complex gill disease are known to be negatively affected by operations that require direct handling of the fish (Sommerset et al. 2022). New, less invasive methods are under development, for example, the Stingray system, which uses laser rays to kill sea lice directly while attached to the fish within the net pen without this impacting the fish (Moura et al. 2018). The company also includes functionalities such as sea lice counting and fish growth monitoring as part of the continued development of their system (AS 2024). Although trials conducted in 2020 did not find any documented effect of Stingray laser delousing on the sea lice levels (Bui et al. 2020), the system has since been further developed and is currently being tested in production by Norwegian salmon farmers (Ilaks.no 2023; Nærings- og fiskeridepartementet 2022).

2.4.7 *Equipment Repair*

Designated repair tools attached to UUVs offer the potential to perform both temporary and long-term repairs within the net pen more quickly. This could reduce the time between hole detection and hole repairs, reducing the need for divers or other human workers. Currently, prototypes of temporary patching tools for holes are being developed, but they are still at the research stage (e.g., Trekker 2024b; Qysea 2024). More complex repair or maintenance operations could involve grasping with an attached manipulator (Haugaløkken et al. 2018, 2021b). If and when intervention operations using ROVs become more available, divers and autonomous or semi-autonomous ROVs could potentially work together to complete tasks. Divers could focus on the most complex tasks, while ROVs or other UUVs could handle simpler tasks in the vicinity. However, the ability for divers and ROVs to work closely together will depend on the robustness of the ROVs' control algorithms and environmental conditions such as currents and waves, ensuring that the safety of the divers is never compromised.

2.5 Current Enabling Technologies and Demonstrations

A wide range of IMR operations need to be performed during daily operations on a fish farm, and many of these tasks are already supported by or fully performed by ROVs. Ongoing research is exploring what other operations can be supported by UUVs, However, the use of navigational and decision autonomy is still very limited and mostly restricted to the research stage. This section will focus on the enabling technologies required to expand the use of robotics and autonomy in sea-based aquaculture, identifying what areas need to be addressed to make this advancement possible.

2.5.1 Navigation and Localization in Net Pens

Robotic systems with navigation and decision autonomy must be able to self-localize and perceive their surroundings in real time and adapt their motions accordingly. The underwater domain poses unique challenges for state estimation and simultaneous localization and mapping (SLAM) due to potential uncertainties and measurement errors (Xanthidis et al. 2022; Joshi et al. 2019). Traditional above-water navigation techniques are not available because water absorbs and scatters the electromagnetic signals used in communication and localization technologies such as GNSS or Wi-Fi (Quattrini Li et al. 2023). Visibility is also significantly reduced under water compared to in air, with light conditions getting weaker with depth. This is especially challenging during periods of already limited daylight, such as the Norwegian winters. Within net pens, visibility may be further impaired by the presence of fish, feed, faeces, and other particles that increase the water's turbidity. These factors all contribute to reducing the effective range of camera systems in cages to only a few metres.

Acoustic sensors, such as USBL or DVL, have been successfully used for localization in various applications for maritime robotics, including terrain-based navigation and localization around structures (Maurelli et al. 2022). Because these technologies have been developed outside the domain of aquaculture, it was still uncertain whether they could provide accurate positions and velocities when used in an aquaculture setting due to disturbances. This challenge was investigated by Rundtop and Frank (2016). Acoustic communication is more demanding in fish farms compared to other parts of the ocean due to signal scattering and damping caused by the presence of fish (Kelasidi and Svendsen 2023). A DVL is typically used to measure the relative velocity or distance to the seabed and have, for example, been used for approximating bathymetry using a downward facing DVL mounted on an ROV (Dukan and Sørensen 2014). However, in the experiments conducted in Rundtop and Frank (2016), a DVL was mounted facing forwards on a UUV, measuring the distance to objects ahead of the underwater robot. This could potentially also be used for net-relative navigation. The DVL system proved to give reliable information on the net-relative distance, angle, and velocity, despite the net wall having a significantly lower target strength than the seabed. This work was continued in Amundsen et al. (2021), which demonstrated how a forward-facing DVL could be used to autonomously guide an ROV to traverse the net of a net pen. Given the symmetrical and predictable shape and size of a net pen, the net wall can be traversed using net following, maintaining a set distance to the net wall while following a path along the net at a desired speed, keeping the distance (Amundsen et al. 2021). Net following was demonstrated by using the DVL to approximate a local region of the net pen in front of the ROV. In Karlsen et al. (2021), traversing large portions of the net using a forward-facing DVL was demonstrated, and a mission control system for an ROV performing autonomous IMR operations was proposed. In simulations, the ROV successfully inspected the full net cage autonomously, even handling simulated loss of DVL measurements. In practical field trials, managing data loss worked well for the reactive behaviour of the ROV, though the mission planning had some challenges with inaccurate estimations of the net inspection progress, which impacted the quality of the planned pattern for net inspection relative to the ROV's estimated position (Karlsen et al. 2021).

In Bjerkeng et al. (2021), a 3D data camera laser line triangulation method was compared to forward-looking DVL measurements for pose estimation for an ROV within a net pen. This comparison was partially motivated by the significantly lower cost of the laser triangulation equipment compared to the DVL. The method was based on triangulating between two laser lines and one camera. Distance, yaw, and pitch between the ROV and the net cage wall were found by detecting two laser line points in the camera images. However, fish passing between the net cage wall and the ROV, obstructing both the laser lines and the camera view, proved to be significant disturbances. When compared to the DVL data, the laser triangulation method provided very similar distance measurements, though the laser measurements contained more noise before filtering. Both sensors were viable for net-relative navigation, and this method has later been improved to estimate both the ROVs absolute pose and its position within a net pen with a known geometry (Bjerkeng et al. 2021, 2023).

2.5.2 *Motion Planning and Control*

Navigation and decision autonomy for UUVs in aquaculture require that the vehicle can safely navigate to perform a planned mission and adjust its trajectory in response to the dynamically changing environment of deformable structures, such as the net wall, ropes, and equipment within the net pen. To enable robust and resilient autonomous 3D navigation of UUVs in operational net pens, challenges related to control and obstacle avoidance must be addressed. Disturbances caused by visibility and sensing constraints, as well as data loss from USBLs or sonars due to, for example, interference from fish, can result in errors and data loss that require careful handling during state estimation. While safe navigation in dynamic 3D environments using motion planning has been demonstrated under highly controlled conditions, these methods typically do not consider uncertainty or motion errors caused by factors such as poor sensing conditions, waves, or currents (Xanthidis et al. 2023). There is a need for novel real-time solutions that can both maximize safety and achieve desired behaviour from UUVs in aquaculture. To ensure safe and effective operations, future intelligent control strategies must include information on current environmental conditions, with adequate safety guarantees, thus avoiding collisions with equipment or undesirable impacts on the fish (Kelasidi and Svendsen 2023). In the Resifarm project, a general real-time underwater navigation planner called ResiPlan was developed to enable safe autonomy under arbitrary disturbances in proximity to the net pen or other structures. ResiPlan combines path optimization for real-time response with constraints that guarantee continuous collision safety to compute paths with clearance that adapt to the conditions of the environment and the performance of the path-follower (Xanthidis et al. 2023). Inspired by AquaVis (Xanthidis et al. 2021), a framework for active perception in static environment, another framework called ResiVis was developed in the ResiFarm project: enabling active perception of static and dynamic target objects from desired distances to ensure collision-free navigation in dynamic environments while maintaining strong safety guarantees (Xanthidis et al. 2024). A locally optimal motion planning approach for real-time autonomous underwater navigation in 3D cluttered environments called Robust Underwater Motion Planning (RUMP) was also developed through work in the ResiFarm and CHANGE projects: While producing efficient, non-conservative locally optimal paths with safety guarantees, this framework also enables real-time obstacle avoidance from multiple obstacles moving orders of magnitude faster than the UUV (Amundsen et al. 2024b).

2.5.3 *Docking Station for Resident UUV*

If resident UUVs are to autonomously perform tasks such as daily net inspection without unnecessary human intervention, there will be a need for a docking station within the net pen. Tether-less AUVs have limited energy and data storage, making docking stations crucial for their operation. While docking stations for UUVs have already been developed for use in other industries, for instance, on the seafloor for use in oil and gas applications, the aquaculture environment introduces additional challenges because of flexible structure, waves, and currents. A docking station should allow for autonomous docking, charging, and data transfer, such as the transmission of large video files collected by an AUV for off-site analysis. Additionally, a docking station could serve as a more robust depot for the UUV during weather conditions where operation needs to be paused, even for robotic systems. Such a docking station system would need to include launch and recovery methods specific to the vehicle, communication and data transfer methods between the docking station and the vehicle, power transfer options, and autonomous systems to evaluate whether the docking was successful (Haugaløkken et al. 2021a; Matsuda et al. 2019).

2.5.4 *Remote Operations with Digital Twin*

While the preceding sections outline developments that could benefit from using a Digital Twin as a virtual playground for digital prototyping and iterative development, digital twins can also have a direct role in executing and tracking ongoing robotics operations in the field (Su et al. 2023; Føre et al. 2024). A digital twin of an aquaculture net cage (Su et al. 2023) can be used for real-time monitoring

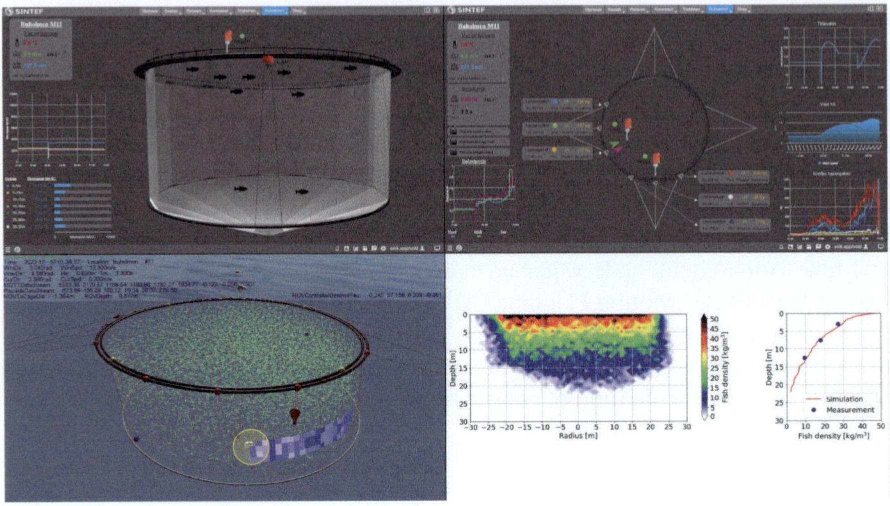

Fig. 2.7 **Digital twin.** Aquaculture net inspection with autonomous ROV navigation based on digital twin (Føre et al. 2024)

of mooring loads, net deformations, and fish distribution within the cage, based on sensor data and a numerical simulation model (Fig. 2.7). By coupling models with real-time monitoring, such systems can provide valuable insights for guidance, navigation, and control of an ROV or UUV throughout the cage, accounting for interactions with the fish. These systems can also be used for tasks such as mapping net inspection operations (Amundsen et al. 2021; Kelasidi et al. 2022).

2.5.5 Interaction Between Fish and Technology

Although a wide range of UUV solutions exist, many are developed for industries such as oil and gas, with few dedicated robotic solutions designed specifically for the aquaculture domain. To ensure both fish welfare and safety of operations, it is crucial that any UUVs operating inside a fish cage can navigate without colliding with the fish and, ideally, without negatively impacting them in any way. The CHANGE project targets the development of methods and technology for UUVs operating within fish cages to better understand how fish are affected by foreign elements introduced to their environment (Fig. 2.8). This involves collecting and analysing data on how farmed Atlantic salmon respond to impact factors such as sound, shape, colour, and size (Zhang et al. 2024), as well as developing intelligent control strategies for the autonomous operation of UUVs within net pens (Amundsen et al. 2024a). With the increasing emphasis on fish welfare in the Norwegian aquaculture industry, there is also a growing need to focus on the interaction between fish and technology. In conclusion, robotic systems operating in fish farms must consider

Fig. 2.8 **Interaction between fish and technology.** It is important to consider how new technological solutions and instruments in the net pen can impact the fish and how best to develop fish-friendly solutions. (Photos: SINTEF Ocean AS)

the interaction between biology and technology to develop fish-friendly solutions, which will be an important foundation for continued and improved fish welfare in the future of fish farming.

2.6 Concluding Remarks

Sea-based aquaculture, such as the farming of Atlantic salmon in Norway, must balance many priorities and challenges: ensuring smooth operations without unnecessary delays to ensure fish welfare and productivity, maintaining the structural integrity of the facilities, and securing the HSE of its human operators, often in unpredictable and challenging weather conditions. As the industry grows, there will be a greater need for increased control over operations and the ability to perform tasks in more challenging environments and at more exposed locations (Føre et al. 2018; Bjclland ct al. 2015; Holmcn ct al. 2023). UUVs will therefore be an essential part of future developments, and remote-controlled ROVs have already been introduced in daily operations. The currents, waves, and complex environment of a fish farm present additional challenges when using UUVs in autonomous inspection and intervention tasks. Extreme weather events, which have become more frequent worldwide, further underline the need for facilities to withstand increased wear and tear from the elements over time and under conditions where it is unsafe for humans to be on-site. The application of robots without autonomous functions will also help reduce risks and increase availability and frequency of inspections, which offers clear benefits. Currently, the primary use-case for robotics in sea-based aquaculture is remote-controlled monitoring and inspection of underwater structures. To expand the role of robotics in daily operations, there needs to be a clear motivation to move away from already established methods, as this will be an investment for both farmers and service providers. With autonomous operations, moving the experienced operators away from performing remote-controlled tasks over to overseeing operations for safety and quality assurance could make this investment more easily justifiable. There are still challenges that need to be addressed related to navigation, decision, and energy autonomy in aquaculture.

A UUV operating autonomously must continuously adapt to avoid obstacles in its environment while performing the tasks efficiently, safely, and according to plan. This introduces even greater demands for robust control methods that feature obstacle avoidance and predictions on the movements of structures in the environment. Ropes and other equipment, in addition to the fish continuously moving throughout the cage in unpredictable swimming patterns, must be detected and avoided. There is a significant potential for multi-purpose robots (Merz et al. 2023), for example, resident AUVs performing daily net cleaning, net inspections, and providing information about environmental parameters such as temperature and oxygen levels (Ohrem et al. 2022). Similarly, robotic solutions performing small repair operations such as temporary patching of holes in nets could also be used to monitor the occurrence of conditions like winter wounds among the fish. Developing methods to increase the efficiency of one operation within sea-based aquaculture might benefit other operations as well.

Increasing the use of technology, robotic solutions, and autonomy in aquaculture can help in the work toward reaching the UN's sustainable development goals by supporting sustainable food production systems and promote safe working environments for human workers. Technological advancements can help achieve higher levels of economic productivity and reduce fuel emissions through more resident solutions and remote operations. This will also be a step on the way to increase both productivity and fish welfare in sea-based aquaculture and to ensure sustainability by enabling production in more exposed locations. However, it is important to evaluate how dedicated robotic solutions for aquaculture should be developed in a way to harmoniously co-exist with the fish, so as to avoid creating new welfare issues while aiming to resolve old ones.

To realize increased use of autonomy and robotics in aquaculture, we must rely on the continued development of methods for underwater communication and navigation and robust control methods for autonomous navigation and decision autonomy for dynamically changing environments. Utilizing digital twins of net pens as tools in this work will be of paramount importance as this provides a virtual arena in which to prototype new solutions with no risk for structures, fish, or humans. A fully fledged digital twin will also enable "what-if" scenarios, which can be used to plan and design robotic operations in aquaculture such that they simultaneously ensure good fish welfare, efficient production, and a safer work environment (Su et al. 2023; Føre et al. 2024).

Acknowledgments This work was funded by the Research Council of Norway (RCN) projects CHANGE (no. 313737) and ResiFarm (no. 327292), the EU project RIMA (ID: 824990), and internal funding from SINTEF Ocean and the RACE initiative.

References

Aaen SM, Helgesen KO, Bakke MJ, Kaur K, Horsberg TE (2015) Drug resistance in sea lice: a threat to salmonid aquaculture. Trends Parasitol 31:72–81

Akvagroup (2024a) AKVA FNC8 - a powerful remote net cleaning rig. Available from https://www.akvagroup.com/net-cleaning/akva-fnc8-category/. Accessed 19 Feb 2024

Akvagroup (2024b) Notvaskrigger. Available from https://www.akvagroup.no/notvasksystemer/notvaskrigger. Accessed 15 April 2024

Akvagroup (2024c) Special nets. Available from https://www.akvagroup.com/sea-based/marineinfrastructure/special-nets/. Accessed 15 April 2024

Amundsen HB, Caharija W, Pettersen KY (2021) Autonomous ROV inspections of aquaculture net pens using DVL. IEEE J Ocean Eng 47:1–19

Amundsen HB, Føre M, Ohrem SJ, Haugaløkken BOA, Kelasidi E (2024a) Three-dimensional obstacle avoidance and path planning for unmanned underwater vehicles using elastic bands. IEEE Trans Field Rob 1:70–92

Amundsen HB, Olsen TF, Xanthidis M, Føre M, Kelasidi E (2024b) Rump: robust underwater motion planning in dynamic environments of fast-moving obstacles. In: 2024 IEEE international conference on robotics and automation (ICRA), IEEE, pp 3743–3750

Amundsen HB, Xanthidis M, Føre M, Kelasidi E (2024c) Spatiotemporal elastic bands for motion planning in highly dynamic environments. IFAC-PapersOnLine 58:27–34

Andersen LB, Grøsvik BE, Karlsen Ø, Kvamme BO, Hansen PK, Husa V, Sandlund N, Stien LH, Solberg MF (2023) Risikorapport norsk fiskeoppdrett 2023—produksjonsdødelighet hos oppdrettsfisk og miljøeffekter av norsk fiskeoppdrett. Rapport fra havforskningen.

Arbeidstilsynet (2013) Forskrift om utførelse av arbeid. Available from https://www.arbeidstilsynet.no/regelverk/forskrifter/forskrift-om-utforelse-av-arbeid/4/26/26-1/. Accessed 17 March 2024

AS SMS (2024) The stingray system. Available from https://www.stingray.no/delousing-with-laser/?lang=en. Accessed 16 May 2024

Bannister J, Sievers M, Bush F, Bloecher N (2019) Biofouling in marine aquaculture: a review of recent research and developments. Biofouling 35:631–648

Barreto MO, Rey Planellas S, Yang Y, Phillips C, Descovich K (2022) Emerging indicators of fish welfare in aquaculture. Rev Aquacult 14:343–361

Bhadmus ZT (2022) Effect of crowding on behaviour of Atlantic salmon (Salmo salar) in aquaculture - Using behaviour as a tool to assess welfare in Atlantic salmon aquaculture production. Master's thesis. UiT The Arctic University of Norway

Bjelland HV, Føre M, Lader P, Kristiansen D, Holmen IM, Fredheim A, Grøtli EI, Fathi DE, Oppedal F, Utne IB et al (2015) Exposed aquaculture in Norway. In: OCEANS 2015-MTS/IEEE Washington, IEEE, pp 1–10

Bjerkeng M, Kirkhus T, Caharija W, Thielemann JT, Amundsen HB, Johan Ohrem S, Ingar Grøtli E (2021) ROV navigation in a fish cage with laser-camera triangulation. J Mar Sci Eng 9:79

Bjerkeng M, Grøtli EI, Kirkhus T, Thielemann JT, Amundsen HB, Su B, Ohrem S (2023) Absolute localization of an ROV in a fish pen using laser triangulation. In: 2023 31st mediterranean conference on control and automation (MED), IEEE, pp 182–188

Bloecher N (2022) Netclean 24/7 tetherless robot for biofouling prevention and inspection in salmon farming. Available from https://sintef.no/en/projects/2019/netclean-247/. Accessed 13 Feb 2024

Bloecher N, Floerl O (2020) Efficacy testing of novel antifouling coatings for pen nets in aquaculture: how good are alternatives to traditional copper coatings? Aquaculture 519:734936

Bloecher N, Floerl O, Sunde LM (2015) Amplified recruitment pressure of biofouling organisms in commercial salmon farms: potential causes and implications for farm management. Biofouling 31:163–172

Bloecher N, Powell M, Hytterød S, Gjessing M, Wiik-Nielsen J, Mohammad SN, Johansen J, Hansen H, Floerl O, Gjevre AG (2018) Effects of cnidarian biofouling on salmon gill health and development of amoebic gill disease. PloS one 13:e0199842

Botta D, Ebner L, Studer A, Reijgwart V, Siegwart R, Kelasidi E (2024) Framework for robust localization of UUVs and mapping of net pens. arXiv preprint arXiv:2409.15475

Bregnballe J (2022) A guide to recirculation aquaculture: an introduction to the new environmentally friendly and highly productive closed fish farming systems. Food & Agriculture Org, Rome

Bui S, Geitung L, Oppedal F, Barrett LT (2020) Salmon lice survive the straight shooter: a commercial scale sea cage trial of laser delousing. Prev Vete Med 181:105063

Bui S, Skern-Mauritzen R, Saito T, Thompson C (2023) Optimising delousing strategies: developing best practice recommendations for maximal efficacy and positive welfare-final report. Rapport fra havforskningen

Directorate of Fisheries (2023) Krav til inspeksjonsfrekvens er knytt til risiko for rømming. Available from https://www.fiskeridir.no/Akvakultur/erfaringsbase-romming/fagstoff1/krav-til-inspeksjonsfrekvens. Accessed 14 April 2024

Dooly G, Omerdic E, Coleman J, Miller L, Kaknjo A, Hayes J, Braga J, Ferreira F, Conlon H, Barry H et al (2016) Unmanned vehicles for maritime spill response case study: exercise cathach. Mar Pollut Bull 110:528–538

Dukan F, Sørensen AJ (2014) Sea floor geometry approximation and altitude control of ROVs. Control Eng Pract 29:135–146

Edwards CD, Pawluk KA, Cross SF (2015) The effectiveness of several commercial antifouling treatments at reducing biofouling on finfish aquaculture cages in British Columbia. Aquac Res 46:2225–2235

Fernö A, Huse I, Juell JE, Bjordal Å (1995) Vertical distribution of Atlantic salmon (*Salmo solar l.*) in net pens: trade-off between surface light avoidance and food attraction. Aquaculture 132:285–296

Fitridge I, Dempster T, Guenther J, De Nys R (2012) The impact and control of biofouling in marine aquaculture: a review. Biofouling 28:649–669

Food and Aquaculture Organization (2010) The state of world fisheries and aquaculture 2010. Available from https://www.fao.org/3/i1820e/i1820e.pdf. Accessed 25 March 2024

Food and Aquaculture Organization (2022) The state of world fisheries and aquaculture 2022: towards blue transformation. Available from https://www.fao.org/documents/card/en?details=cc0461en. Accessed 06 Feb 2024

Føre HM, Lien AM (2014) Utredning og tiltak mot skader på not ved vasking i sjø

Føre M, Frank K, Norton T, Svendsen E, Alfredsen JA, Dempster T, Eguiraun H, Watson W, Stahl A, Sunde LM, Schellewald C, Skøien KR, Alver MO, Berckmans D (2018) Precision fish farming: a new framework to improve production in aquaculture. Biosyst Eng 173:176–193

Føre HM, Thorvaldsen T, Tinmannsvik RK, Okstad EH (2019) Kunnskap og metoder for å forebygge rømming-faglig sluttrapport

Føre M, Alver MO, Alfredsen JA, Rasheed A, Hukkelås T, Bjelland HV, Su B, Ohrem SJ, Kelasidi E, Norton T et al (2024) Digital twins in intensive aquaculture challenges, opportunities and future prospects. Comput Electron Agric 218:108676

Fossen TI (2011) Handbook of marine craft hydrodynamics and motion control. Wiley, Hoboken

Gumdal IS (2022) Sikkerhet i kran-og løfteoperasjoner i havbruksnæringen. Master's thesis. NTNU

Gupta A, Bringsdal E, Knausgård KM, Goodwin M (2022) Accurate wound and lice detection in Atlantic salmon fish using a convolutional neural network. Fishes 7:345

Haugaløkken BOA, Jørgensen EK, Schjølberg I (2018) Experimental validation of end-effector stabilization for underwater vehicle-manipulator systems in subsea operations. Robot Auton Syst 109:112
Haugaløkken BO, Kelasidi E, Mulelid M, Bloecher N (2021a) Docking stations for net-crawling underwater vehicles in aquaculture net pens. In: OCEANS 2021: San Diego–Porto. IEEE, pp 1–10
Haugaløkken BOA, Nissen O, Amundsen HB, Føre M, Kelasidi E (2021b) Modelling and control of a 6 DOF robot manipulator for underwater applications-aquaculture related case studies. In: OCEANS 2021: San Diego. Porto, IEEE, pp 1–10
Havbruksutvalget (2024) NOU 2023:23: Helhetlig forvaltning av akvakultur for bærekraftig verdiskaping. Available from https://www.regjeringen.no/no/dokumenter/nou-2023-23/id2995224/. Accessed 25 Feb 2024
Holmen IM (2022) Safety in exposed aquaculture operations strategies and methods for reducing risk. Doctoral thesis, Norwegian University of Science and Technology
Holmen IM, Utne IB, Haugen S (2021) Identification of safety indicators in aquaculture operations based on fish escape report data. Aquaculture 544:737143
Holmen IM, Thorvaldsen TM, Salomonsen C, Lona E (2023) Sikkerhet og risikostyring i eksponerte havbruksoperasjoner-resultater fra SFI Exposed 2015-2023
Ilaks.no (2023) Nordlaks blar opp: Sikrer seg 50 luselasere. Available from https://ilaks.no/nordlaks-blar-oppsikrer-seg-50-luselasere/. Accessed 26 March 2024
Iversen A, Hermansen Ø, Nystøyl R, Hess EJ (2017) Kostnadsutvikling i lakseoppdrett. med fokus på fôr-og lusekostnader. Nofima rapportserie
Joshi B, Rahman S, Kalaitzakis M, Cain B, Johnson J, Xanthidis M, Karapetyan N, Hernandez A, Li AQ, Vitzilaios N et al (2019) Experimental comparison of open source visual-inertial-based state estimation algorithms in the underwater domain. In: 2019 IEEE/RSJ international conference on intelligent robots and systems (IROS). IEEE. pp 7227–7233
Karlsen HØ, Amundsen HB, Caharija W, Ludvigsen M (2021) Autonomous aquaculture: Implementation of an autonomous mission control system for unmanned underwater vehicle operations. In: OCEANS 2021: San Diego. Porto. IEEE, pp 1–10
Kelasidi E (2020) Seatonomy applied in an operational analysis and overall system design for an autonomous underwater vehicle. Available from https://hdl.handle.net/11250/2647657. Accessed 10 April 2024
Kelasidi E, Svendsen E (2023) Robotics for sea-based fish farming. In: Encyclopedia of smart agriculture technologies. Springer, Berlin, pp 1–20
Kelasidi E, Su B, Thorbjørnsen ES, Moen E, Schellewald C, Yip MH, Remmen BM, Mulelid M (2020) CAGEREPORTER-development of technology for autonomous, bio-interactive and high-quality data acquisition from aquaculture net cages
Kelasidi E, Su B, Caharija W, Føre M, Pedersen MO, Frank K (2022) Autonomous monitoring and inspection operations with UUVs in fish farms. IFAC-PapersOnLine 55:401–408
Klebert P, Su B (2020) Turbulence and flow field alterations inside a fish sea cage and its wake. Appl Ocean Res 98:102113
Klebert P, Su B, Nissen O, Kvæstad B (2023) Response of Atlantic salmon in flexible sea cages to waves and currents: measurements at farm-level and modeling. Appl Ocean Res 138:103599
Labra J, Zuniga MD, Rebolledo J, Ahmed MA, Carvajal R, Jara N, Carvajal G (2023) Robust automatic net damage detection and tracking on real aquaculture environment using computer vision. Aquac Eng 101:102323
Lovdata (2006) Forskrift om velferd for produksjonsdyr. Available from https://lovdata.no/dokument/LTI/forskrift/2006-07-03-885/. Accessed 13 Nov 2024
Lovdata (2023) Forskrift om bekjempelse av lakselus i akvakulturanlegg. Available from https://lovdata.no/dokument/SF/forskrift/2012-12-05-1140/. Accessed 25 Feb 2024
Lusedata (2020) Tiltaksveileder kontroll med lakselus og skottelus. Available from https://lusedata.no/wp-Content/uploads/2022/03/2120.01.20-Tiltaksveileder-Lakselus-og-skottelus.pdf. Accessed 14 April 2024

Madshaven A, Schellewald C, Stahl A (2022) Hole detection in aquaculture net cages from video footage. In: Fourteenth international conference on machine vision (ICMV 2021), SPIE. pp 258–267

Måløy H (2020) EchoBERT: a transformer-based approach for behavior detection in echograms. IEEE Access 8:218372–218385

Matsuda T, Maki T, Masuda K, Sakamaki T (2019) Resident autonomous underwater vehicle: underwater system for prolonged and continuous monitoring based at a seafloor station. Robot Auton Syst 120:103231

Maurelli F, Krupiński S, Xiang X, Petillot Y (2022) AUV localisation: a review of passive and active techniques. Int J Intell Robot Appl 6:246–269

MEOX (2024a) MEOX - cleaning. Available from https://meox.no/en/cleaning/. Accessed 19 Feb 2024

MEOX (2024b) MEOX - Dead fish handling. Available from https://meox.no/en/dead-fish-handling/. Accessed 24 March 2024

Merz M, Transeth AA, Evjemo LD, Kelasidi E (2023) Market study on inspection and maintenance robotics in Norway-suppliers, market needs and challenges

Moura J, Maksan D, Frenzl B, Bron J, Migaud H, Paladini G et al (2018) Stingray: a novel tool for delousing salmon. Fish Vet J 8.12

NFD (2021) Et hav av muligheter . regjeringens havbruksstrategi. Available from https://www.regjeringen.no/contentassets/e430ad7a314e4039a90829fcd84c012a/no/pdfs/et-hav-avmuligheter.pdf. Accessed 06 Feb 2024

Nissen O, Evjemo LD, Ohrem SJ, Haugaløkken BOA, Kelasidi E (2024) Framework for automated wound detection and tracking in industrial scale fish farms. In: 2024 32nd mediterranean conference on control and automation (MED). IEEE. pp 567–572

Noble C, Nilsson J, Stien LH, Iversen MH, Kolarevic J, Gismervik K (2018) Velferdsindikatorer for oppdrettslaks: Hvordan vurdere og dokumentere fiskevelferd. 3. utgave

NorseAqua (2022) Rensefiskskjul - brukerhåndbok. Available from https://norseaqua.no/wpcontent/uploads/2022/09/no-rensefiskskjul-brukerhandbok-1.pdf. Accessed 23 March 2024

Nrk.no (2024) Her blir laksen skoten med laser: kan halvere dødelegheita. Available from https://www.nrk.no/nordland/her-motverkar-selskapet-lakselus-med-laser-1.16786406. Accessed 10 March 2025

Nærings- og fiskeridepartementet (2018) Forskrift om drift av akvakulturanlegg. Available from https://lovdata.no/dokument/SF/forskrift/2008-06-17-822. Accessed 22 March 2024

Nærings- og fiskeridepartementet (2022) Forskrift om krav til teknisk standard for akvakulturanlegg for fisk i sjø, innsjø og vassdrag (NYTEK23). Available from https://lovdata.no/dokument/SF/forskrift/2022-08-22-1484/KAPITTEL_9%C2%A744. Accessed 22 March 2024

Ocein (2024) Net cleaner - stealthcleaner. Available from https://www.ocein.no/en/net-cleaner/. Accessed 19 Feb 2024

Ohrem SJ, Kelasidi E, Bloecher N (2020) Analysis of a novel autonomous underwater robot for biofouling prevention and inspection in fish farms. In: 2020 28th mediterranean conference on control and automation (MED). IEEE, pp 1002–1008

Ohrem SJ, Kelasidi E, Bloecher N (2022) Seatonomy applied in operational analysis of an autonomous net cleaning robot-NetClean 24/7 report for work package H1. 1: Operational analysis and overall system design

Østevik L, Stormoen M, Evensen Ø, Xu C, Lie KI, Nødtvedt A, Rodger H, Skagøy A, Manji F, Alarcón M (2022) Effects of thermal and mechanical delousing on gill health of farmed Atlantic salmon (*Salmo salar l.*). Aquaculture 552:738019

Overton K, Dempster T, Oppedal F, Kristiansen TS, Gismervik K, Stien LH (2019) Salmon lice treatments and salmon mortality in Norwegian aquaculture: a review. Rev Aquac 11:1398–1417

Paull L, Saeedi S, Seto M, Li H (2013) AUV navigation and localization: a review. IEEE J Ocean Eng 39:131–149

Probotic (2024) Probotic - clean nets. Available from https://www.probotic.no/functions/netcleaning. Accessed 19 Feb 2024

Quattrini Li A, Carver CJ, Shao Q, Zhou X, Nelakuditi S (2023) Communication for underwater robots: recent trends. Curr Robot Rep 4:13–22

Qysea (2024) Qysea - Net patch kit. Available from https://store.qysea.com/net-patch-kit.html. Accessed 24 March 2024

Rasheed A, San O, Kvamsdal T (2020) Digital twin: values, challenges and enablers from a modeling perspective. IEEE Access 8:21980–22012

Riksrevisjonen (2022–2023) Myndighetenes arbeid med fiskehelse og fiskevelferd i havbruksnæringen. Riksrevisjonen 3:12

Risberg J, Møllerløkken A, Eftedal OS (2019) Norske dykke- og behandlingstabeller. Available from https://dykketabeller.no/onewebmedia/NDBT%20520utg.pdf/. Accessed 22 March 2024

Robotics R (2024) Remora cleaning robot - a robot that cleans and inspects fish pens while collecting data. Available from https://remorarobotics.no/autonomous-cleaning-robot/. Accessed 19 Feb 2024

Ruggesæter A, Haug KMV (2021) Kostnad ved behandling av lakselus: et litteratursøk og en modelleringstilnærming. Master's thesis

Rundtop P, Frank K (2016) Experimental evaluation of hydroacoustic instruments for ROV navigation along aquaculture net pens. Aquac Eng 74:143–156

Rydeng HE (2022) Regulering av lakselus i oppdrettsnæringa. Med et særlig henblikk på dyrevelferd. Master's thesis. UiT Norges arktiske universitet

Schellewald C, Stahl A (2022) Irregularity detection in net pens exploiting computer vision. IFAC-PapersOnLine 55:415–420

Singh M, Alexis K (2024) Online refractive camera model calibration in visual inertial odometry. In: 2024 IEEE/RSJ international conference on intelligent robots and systems (IROS). IEEE, pp 12609–12616

Skaldebø M, Ohrem SJ, Amundsen HB, Kelasidi E, Bloecher N (2023) Framework for autonomous navigation for a permanent resident aquaculture net grooming robot. In: 2023 31st mediterranean conference on control and automation (MED). IEEE, pp 356–363

Solem AJL (2017) Analysis of current ROV operations in the Norwegian aquaculture-reducing risk in exposed aquaculture operations. Master's thesis. NTNU

Solstorm D, Oldham T, Solstorm F, Klebert P, Stien LH, Vågseth T, Oppedal F (2018) Dissolved oxygen variability in a commercial sea-cage exposes farmed Atlantic salmon to growth limiting conditions. Aquaculture 486:122–129

Sommerset I, Wiik-Nielsen J, De Olivera VH, Moldal T, Bornø G, Haukaas A, Brun E (2022) Fiskehelserapporten 2022. Fiskehelserapporten 2022

Sommerset I, Wiik-Nielsen J, Moldal T, De Olivera VH, Svendsen J, Haukaas A, Brun E (2023a) Fiskehelserapporten 2023. Fiskehelserapporten 2023

Sommerset I, Wiik-Nielsen J, Moldal T, Oliveira VHS, Svendsen JC, Haukaas A, Brun E (2023b) Fiskehelserapporten 2015. Veterinærinstituttets rapportserie 8

Su B, Reite KJ, Føre M, Aarsæther KG, Alver MO, Endresen PC, Kristiansen D, Haugen J, Caharija W, Tsarau A (2019) A multipurpose framework for modelling and simulation of marine aquaculture systems. In: International conference on offshore mechanics and arctic engineering. American Society of Mechanical Engineers, p V006T05A002

Su B, Bjørnson FO, Tsarau A, Endresen PC, Ohrem SJ, Føre M, Fagertun JT, Klebert P, Kelasidi E, Bjelland HV (2023) Towards a holistic digital twin solution for real-time monitoring of aquaculture net cage systems. Mar Struct 91:103469

Svinøy M (2023) Challenges and opportunities of underwater image analysis for lice counting and welfare assessments of Atlantic salmon (Salmo salar) during the sea-phase. Master's thesis. NTNU

Swain G, Erdogan C, Foy L, Gardner H, Harper M, Hearin J, Hunsucker KZ, Hunsucker JT, Lieberman K, Nanney M et al (2022) Proactive in-water ship hull grooming as a method to reduce the environmental footprint of ships. Front Mar Sci 8:808549

Thorvaldsen T, Holmen I, Kongsvik T (2017) HMS-undersøkelsen i havbruk 2016. (Report No. OC2017 A-113). SINTEF Ocean AS. https://hdl.handle.net/11250/2443489/

Thorvaldsen T, Frank K, Sunde LM (2018) Lusetellingsmetoder i lakseoppdrett - en beskrivelse av dagens status. (Report No. 2018:00483). SINTEF Ocean AS. https://hdl.handle.net/11250/2496606/

Thorvaldsen T, Kongsvik T, Holmen IM, Størkersen K, Salomonsen C, Sandsund M, Bjelland HV (2020) Occupational health, safety and work environments in Norwegian fish farming-employee perspective. Aquaculture 524:735238

Thorvaldsen T, Kongsvik TØ, Holmen IM, Størkersen KV, Salomonsen C (2023) undersøkelsen i havbruk. (Report No. 2023:01362). SINTEF Ocean AS. https://hdl.handle.net/11250/3104876/

Trekker D (2024a) Deep trekker - mort pusher. Available from https://www.deeptrekker.com/shop/products/mortpusher-kit-dtg3. Accessed 24 March 2024

Trekker D (2024b) Deep Trekker - net patch kit. Available from https://www.deeptrekker.com/no/shop/products/net-patch-kit-dtg3. Accessed 24 March 2024

Walde CS, Stormoen M, Pettersen JM, Persson D, Røsæg MV, Jensen BB (2022) How delousing affects the short-term growth of Atlantic salmon (*Salmo salar*). Aquaculture 561:738720

Watbots (2024) Autonom vaskerobot med notinspeksjon 24/7. Available from https://watbots.com/. Accessed 19 Feb 2024

Xanthidis M, Joshi B, Roznere M, Wang W, Burgdorfer N, Li AQ, Mordohai P, Nelakuditi S, Rekleitis I (2022) Towards mapping of underwater structures by a team of autonomous underwater vehicles. In: The international symposium of robotics research. Springer, pp 170–185

Xanthidis M, Kalaitzakis M, Karapetyan N, Johnson J, Vitzilaios N, O'Kane JM, Rekleitis I (2021) AquaVis: a perception-aware autonomous navigation framework for underwater vehicles,. In: 2021 IEEE/RSJ international conference on intelligent robots and systems (IROS). IEEE, pp 5410–5417

Xanthidis M, Kelasidi E, Alexis K (2023) ResiPlan: closing the planning-acting loop for safe underwater navigation. In: 2023 IEEE international conference on robotics and automation (ICRA), pp 3138–3145

Xanthidis M, Skaldebø M, Haugaløkken B, Evjemo L, Alexis K, Kelasidi E (2024) ResiVis: a holistic underwater motion planning approach for robust active perception under uncertainties. IEEE Robot Autom Lett 9:9391–9398. https://doi.org/10.1109/LRA.2024.3455893

Yazdani AM, Sammut K, Yakimenko O, Lammas A (2020) A survey of underwater docking guidance systems. Robot Auton Syst 124:103382

Zacheilas T, Moirogiorgou K, Papandroulakis N, Sotiriades E, Zervakis M, Dollas A (2021) An FPGA-based system for video processing to detect holes in aquaculture nets. In: 2021 IEEE 21st international conference on bioinformatics and bioengineering (BIBE). IEEE, pp 1–6

Zereik E, Bibuli M, Mišković N, Ridao P, Pascoal A (2018) Challenges and future trends in marine robotics. Annu Rev Control 46:350–368

Zhang Q, Bloecher N, Evjemo LD, Føre M, Su B, Eilertsen E, Mulelid MA, Kelasidi E (2024) Farmed Atlantic salmon (*Salmo salar l.*) avoid intrusive objects in cages: the influence of object shape, size and colour, and fish length. Aquaculture 581:740429

Part II
In the Cloud

Chapter 3
Machine Learning in Fish Farming

Fearghal O'Donncha, Nikos Papandroulakis, Jennie Korus, Abigail Langbridge, Alexander Timms, Konstantinos Topouzelis, Abdul Baseer Khan, Shree Rama Kamal Kumar Vegu, Mahtab Sarvmaili, Ryan Mowat, Rhanna Turberville, Tyler Sclodnick, and Christopher Whidden

Abstract This chapter explores how machine learning (ML) is transforming aquaculture, with a particular focus on enhancing decision-making processes and improving operational efficiency. The chapter is structured to first introduce the challenges in aquaculture and the role of AI and then provide an overview of

F. O'Donncha (✉)
IBM Research Europe, Dublin, Ireland
e-mail: feardonn@ie.ibm.com

N. Papandroulakis
Hellenic Centre for Marine Research, Crete, Greece
e-mail: npap@hcmr.gr

J. Korus · T. Sclodnick
Innovasea Marine Systems Canada, Bedford, NS, Canada
e-mail: jennie.korus@innovasea.com; tyler.sclodnick@innovasea.com

A. Langbridge · A. Timms
Imperial College London, London, UK
e-mail: abigail.langbridge18@imperial.ac.uk; alexander.timms21@imperial.ac.uk

K. Topouzelis
University of the Aegean, Mitilini, Greece
e-mail: topouzelis@marine.aegean.gr

A. B. Khan · S. R. K. K. Vegu · M. Sarvmaili
Dalhousie University, Halifax, NS, Canada
e-mail: ab916813@dal.ca; kamal@dal.ca; mahtab.sarvmaili@dal.ca

R. Mowat
RS Aqua, Portsmouth, UK
e-mail: r.mowat@rsaqua.co.uk

R. Turberville
Scottish Sea Farms, Orkney, UK
e-mail: rhanna.turberville@scottishseafarms.com

C. Whidden
Dalhousie University and DeepSense, Halifax, NS, Canada
e-mail: cwhidden@dal.ca

F. O'Donncha et al. (eds.), *Digital Twin for Marine Fish Farming*,
Aquatic Production Systems, https://doi.org/10.1007/978-3-032-10362-8_3

ML techniques in the context of aquaculture, followed by applications, emerging trends, future directions, and case studies. The focus is on real-world applications of ML techniques, including Random Forest, Convolutional Neural Networks (CNNs), and Recurrent Neural Networks (RNNs), as well as emerging technologies such as Graph Neural Networks (GNNs) and large language models (LLMs). Key applications include biomass estimation, species recognition, behavioural analysis, and environmental forecasting. The chapter also highlights the synergy between ML and the Internet of Things (IoT) for real-time monitoring and decision support. Ultimately, ML-driven innovations have the potential to revolutionise fish farming, leading to more efficient, sustainable, and productive practices in the aquaculture industry.

Keywords Machine learning · Artificial intelligence · Aquaculture · Large language models · Foundation models · Digital twin

3.1 Introduction

Aquaculture has emerged as a critical component of global food security, supplying nearly 50% of the world's fish consumption. It is one of the fastest-growing food production sectors, contributing significantly to the supply of high-quality protein, particularly in regions where access to such food sources is limited (Naylor et al. 2021; Béné et al. 2016). As global demand for seafood continues to rise, sustainable expansion of aquaculture becomes essential. However, the industry faces multiple challenges that threaten its long-term viability.

3.1.1 Challenges in Aquaculture

Despite its rapid growth, aquaculture is constrained by several critical challenges that affect sustainability, efficiency, and profitability.

- **Environmental Impact:** Intensive aquaculture can degrade habitats, increase nutrient loads, and trigger issues such as algal blooms and oxygen depletion (Godfray et al. 2010).
- **Disease Risk:** Infectious outbreaks remain a major source of loss, requiring better early detection and intervention (Mandal and Ghosh 2023).
- **Feeding Inefficiency:** Feed is the largest cost driver, and poor conversion rates increase waste and environmental burden (Cooney et al. 2021).
- **Workforce and Logistics:** COVID-19 exposed labour and supply chain fragilities that continue to affect operational resilience (Love et al. 2021).

These challenges underscore the need for greater automation and data-driven approaches to ensure efficient, sustainable, and resilient aquaculture operations (Klinger and Naylor 2012).

3.1.2 The Role of Artificial Intelligence in Aquaculture

Artificial intelligence (AI) is revolutionising aquaculture by providing advanced analytics, automation, and predictive capabilities. Machine learning (ML), a subset of AI, enables data-driven decision-making and enhances key aspects of aquaculture management, including:

- **Optimised Feeding Strategies:** AI-driven analysis of fish behaviour and biomass estimation enables more precise feeding schedules, minimising waste, and improving growth rates.
- **Early Disease Detection:** ML models analyse behavioural patterns, water quality data, and historical trends to detect early signs of disease outbreaks, allowing for timely interventions.
- **Environmental Monitoring and Prediction:** Predictive models improve water quality management by forecasting dissolved oxygen levels, temperature fluctuations, and harmful algal blooms (O'Donncha and Grant 2019).
- **Automation of Farm Operations:** AI-driven IoT systems enable automated monitoring of fish health, water conditions, and infrastructure maintenance, reducing reliance on manual labour (Mandal and Ghosh 2023).

The integration of AI into aquaculture is not merely a technological advancement but a necessary step towards increasing efficiency and ensuring long-term sustainability (Maksimovic 2018).

3.1.3 Scope of This Chapter

This chapter explores the role of machine learning in aquaculture, focusing on its applications in farm management, decision-making, and operational efficiency. Given the complexity of aquaculture, which intersects with fields such as oceanography, environmental science, industrial automation, and logistics, the discussion is centred on how AI-driven approaches can enhance on-farm operations. The key question addressed is *How can the combination of human expertise and AI-based insights lead to more effective and sustainable aquaculture practices?*

The chapter is structured as follows:

- **Section 3.2** provides an overview of machine learning techniques relevant to aquaculture.

- **Section 3.3** discusses key applications, including computer vision for fish monitoring, IoT-based environmental tracking, and predictive modelling for disease detection and feeding efficiency.
- **Section 3.4** examines emerging trends, such as Graph Neural Networks (GNNs), large language models (LLMs), and AI-centred digital twins for aquaculture.
- **Section 3.5** presents case studies illustrating real-world AI applications in aquaculture.
- **Section 3.6** provides conclusions and outlines directions for future research.

This structured approach ensures a comprehensive exploration of machine learning in aquaculture, highlighting its potential to drive innovation, efficiency, and sustainability in the industry.

3.2 Machine Learning

Machine learning (ML), a data-driven branch of artificial intelligence (AI), enables models to learn relationships directly from data rather than relying on handcrafted rules or predefined equations. In aquaculture, this allows algorithms to process raw sensor streams and video footage to forecast dissolved oxygen levels, estimate biomass, detect anomalous behaviours, or identify sea-lice infestations. These are tasks that would be labour-intensive or impractical to perform manually at scale.

Three forces have converged over the past decade to make ML practical for aquaculture:

- The mass deployment of IoT sensors, cameras, and telemetry systems across farm infrastructure
- The wide availability of computational power via GPUs, TPUs, and cloud services, reducing training time from weeks to hours
- Algorithmic innovation such as ensemble trees for structured data, convolutional networks for images, and transformers for sequences and multi-modal fusion

These advances enable ML models to capture complex non-linear phenomena, such as oxygen fluctuations driven by biomass, weather, and depth, without requiring explicit mechanistic equations. While physics-based models rely on known processes and can struggle with noise, heterogeneity, or unmeasured processes or influences, ML offers an empirical alternative that adapts to real-world data and uncovers patterns that might otherwise go unnoticed.

High-quality data is the foundation of machine learning. No matter how advanced the algorithm, the performance of any ML system is fundamentally limited by the quality of its training data. Before any model training begins, practitioners must prioritise careful sensor calibration, ensure data is representative across sites, seasons, and operational contexts, and maintain rigorous metadata documentation and version control. While fine-tuning hyperparameters can deliver marginal improvements, it is high-quality, structured data that ultimately determines the robustness and reliability of any ML application.

3.2.1 Learning Paradigms and Typical Tasks

Machine learning problems in aquaculture span several learning paradigms, depending on the structure of the task and the availability of labelled data. Table 3.1 summarises the four major paradigms and their representative applications in aquaculture.

Tasks within these paradigms can be further classified by output type including *classification* (e.g. lice vs. no-lice), *regression* (e.g. predicting oxygen concentration), *clustering* (e.g. grouping cage performance profiles), and *dimensionality reduction* (e.g. summarising hyperspectral or multivariate sensor data).

3.2.2 Machine Learning Algorithms

Different machine learning algorithms are suited to different types of data and problem structures. In aquaculture, model selection balances predictive performance, interpretability, computational efficiency, and data availability.

Tree-based ensembles, such as Random Forest and XGBoost, are widely used for tabular and time series data generated by sensors. These models are fast to train, robust to missing values, and offer interpretability via built-in feature importance measures. Their reliability and simplicity make them a common first choice for tasks like environmental forecasting or estimating biomass from probe readings.

Kernel-based methods, including support vector machines (SVMs) and Gaussian processes, perform well on small, high-dimensional datasets where relationships are complex but data is scarce. They are effective for anomaly detection and

Table 3.1 Learning paradigms in aquaculture and representative applications

Paradigm	Data type	Typical applications in aquaculture
Supervised learning	Labelled input–output pairs	Forecasting dissolved oxygen levels; grading sea-lice severity from images
Unsupervised learning	Unlabelled data	Clustering feeding-rate signals; anomaly detection in behavioural or sensor data
Semi-supervised learning	Sparse labels + unlabelled data	Fish tracking or behaviour analysis in video with minimal annotation
Reinforcement learning	Interaction + reward feedback	Adaptive feeding control to balance growth and feed efficiency

early warning applications, although their scalability becomes limiting beyond tens of thousands of training samples.

Neural networks offer a broader and more flexible modelling class, with architectures that can be tuned for specific data modalities:

- *Convolutional Neural Networks (CNNs)* are highly effective at extracting spatial features from image or video frames, making them ideal for fish detection, biomass estimation, and sea-lice classification.
- *Recurrent Neural Networks (RNNs)* and *Long Short-Term Memory (LSTM)* networks are designed for sequential data and are well suited to tasks involving temporal dynamics, such as forecasting oxygen fluctuations or tracking behavioural changes over time.
- *Transformers* employ self-attention mechanisms rather than recurrence and are increasingly used for modelling long-range dependencies in sequential or multimodal data. They underpin modern large language models and are now being adapted to tasks such as environmental scenario modelling and cross-sensor data fusion.

In practice, model choice is often governed by the volume and structure of available data, latency requirements, and the need for interpretability. Following the principle of Occam's Razor, which favours the simplest solution that adequately explains the data, it is often advisable to begin with well-understood, lightweight models such as tree ensembles. More complex architectures, including deep neural networks, should be considered only when the data scale, task complexity, or performance demands clearly warrant the added computational and operational overhead.

3.2.3 Feature Engineering and Model Interpretability

In classical machine learning, much of a model's performance hinges on the quality of its input features. Carefully designed features, such as tidal phase indices, rolling-window statistics of oxygen levels, or colour histograms that compensate for lighting variation in underwater images, can substantially improve predictive accuracy. These transformations embed domain knowledge directly into the model's input, guiding learning in data-scarce or noisy settings.

While modern ML pipelines increasingly adopt automated approaches such as *feature hashing*, *principal component analysis*, or *learned embeddings*, these techniques often require substantial data and still benefit from domain-informed pre-processing. Automated methods reduce manual effort but do not replace the value of expert insight, particularly when interpretability and operational trust are priorities.

Interpretability remains essential in aquaculture settings where decisions impact animal welfare, compliance, and costs. Model explanation tools such as SHAP (Shapley Additive Explanations) for tree-based models or Grad-CAM (Gradient-weighted Class Activation Mapping) for convolutional neural networks help translate opaque predictions into biologically meaningful insights (Selvaraju et al. 2017). These techniques support validation by domain experts and help ensure that automated decisions align with observed farm conditions and regulatory frameworks.

3.3 Applications of Machine Learning in Aquaculture

The multifaceted nature of aquaculture, encompassing aspects such as environmental monitoring, disease management, feeding efficiency, and yield optimisation, makes it an ideal candidate for the integration of ML techniques. The incorporation of robust ML methodologies enhances the accuracy and efficiency of operations while supporting more sustainable aquaculture practices. The adaptability of ML to the diverse datasets generated in aquaculture, from IoT sensor data to satellite imagery, underscores its potential as a key driver of innovation and productivity in the sector.

In a recent review by Zhao et al. (2021), the authors discussed several applications of machine learning in aquaculture:

1. **Biomass Detection of Fish:** ML supports the estimation of the size, weight, and count of fish, which is crucial for managing fish populations and predicting yields.
2. **Recognition and Classification of Fish:** ML techniques are used to identify fish species, age, and sex, using machine learning algorithms. This is important for species management and scientific research.
3. **Behavioural Analysis:** Machine learning aids in quantifying fish behaviours such as feeding and group dynamics, which is vital for welfare and environmental assessments.
4. **Water Quality Parameter Prediction:** Machine learning algorithms are used to predict and monitor various water quality parameters, ensuring optimal conditions for aquaculture.

These examples illustrate ML's growing role in driving more efficient, sustainable, and productive aquaculture practices. The subsequent sections of this chapter will delve further into the diverse applications of machine learning in fish farming, exploring its various implementations and impacts in greater detail.

3.3.1 Computer Vision in Aquaculture

A significant share of research focuses on using computer vision technology to improve observation, detection, and management of caged fish. Traditional methods using manual sampling often harm fish and rely on human judgement, making them unsuitable for modern aquaculture's demands (Jeong and Revie 2020). The integration of machine learning and machine vision presents a new, intelligent approach for estimating fish size, weight, numbers, and other biological information more accurately.

The inherent nature of farming fish underwater makes computer vision applications in aquaculture challenging. Yang et al. (2021) summarised some of these challenges, including:

- **Turbidity:** Changes in water clarity, due to factors like plankton, can lead to false positives in fish movement tracking.
- **Occlusion and Disappearance:** Fish may get obscured or disappear temporarily during movement, resulting in tracking failures and loss of continuity across frames.
- **Scale Change:** Variations in distance between camera and fish affect apparent size in images, influencing detection and counting accuracy. Smaller fish scales include more background, causing errors, while larger scales may miss multiple targets in a single image, leading to tracking and counting issues. Scale adaptation is essential.

Considering the intricate nature of computer vision applications in aquaculture, considerable research efforts have been dedicated to creating algorithms designed to effectively address these specific challenges. The following subsections summarise key computer vision tasks and representative approaches.

3.3.1.1 Biomass Detection of Fish

In aquaculture, accurately estimating fish biomass is crucial for effective management and sustainable production. For size estimation, researchers have used models trained on datasets like ImageNet and the Atlantic fish dataset. Monkman et al. (2019) developed an R-CNN model for estimating European bass length, achieving an average deviation of 2.2%. However, this model's applicability is limited, especially for overlapping fish. Li et al. (2020) used a CNN model based on transfer learning for small sample pond fish data, reaching an accuracy of 93.93%. Voskakis et al. (2021) developed and deployed a CNN model to estimate the length of European seabass and gilthead seabream in cages with relative error of 3.1% and 7.4%, respectively

Accurate weight estimates are crucial for farmers since the weight of fish determines the price received at harvest time, impacting the economic viability of aquaculture operations. Therefore, precise weight estimation is essential for effective planning, resource allocation, and maximising the returns from the harvest. Typically, researchers focus on predicting quality based on fish body characteristics, using image processing to extract size and shape features. Fernandes et al. (2020) combined linear regression and CNN for predicting weight from body area, while (Saberioon and Císař 2018) used infrared light for image collection and RF and SVM models for back geometry prediction, mitigating lighting constraints.

3.3.1.2 Recognition and Classification of Fish

Fish farmers need comprehensive information on each fish's attributes such as size, weight, gender, maturity stage, and skin colour throughout their growth phases to efficiently track their development and manage stocks (Saberioon et al. 2017). Machine learning facilitates automated fish identification, extracting their features, and classifying them through conventional computer vision methods. For instance, the Stingray system employs computer vision to detect and eliminate individual lice on salmon, addressing parasitic infections in aquaculture (Bui et al. 2020).

The task of fish target detection in aquaculture has seen significant advancements with the application of deep learning models like YOLO (You Only Look Once) and CNNs (Convolutional Neural Networks). YOLO-based methods have been employed for tasks such as fish count (Armalivia et al. 2021) and trajectories (Mohamed et al. 2020). These methods have been further refined, for instance, by Cai et al. (2020), who replaced the Darknet-53 network in YOLOv3 with MobileNet, enhancing feature extraction and overall accuracy. Additionally, CNN models have been increasingly favoured for their efficacy in fish detection. Villon et al. (2018) used the GoogLeNet architecture combined with soft-max classification, achieving an accuracy rate of 94.9%. Extensions of CNN models have further improved accuracy, with (Rauf et al. 2019) deepening convolutional layers to reach a maximum accuracy of 96.63%.

Beyond basic CNN architectures, advanced models like R-CNN and LSTM have also been applied to fish detection. Li et al. (2016) utilised the Faster R-CNN model, initialised with the pre-trained Zeiler and Fergus model, to enhance detection speed and accuracy. Techniques like background subtraction have been employed to distinguish fish from other objects, with varying degrees of success (Shevchenko et al. 2018; Gaude and Borkar 2019). Transfer learning methods, as used by Tamou et al. (2018), have achieved high accuracy in challenging underwater environments. Additionally, innovative methods like 360-panoramic systems and binocular vision-based image processing have been developed, further expanding the scope of aquatic species recognition, including live crabs and shrimp, with high accuracy rates reported (Cao et al. 2020; Hu et al. 2020). These diverse approaches demonstrate the evolving landscape of target detection in aquaculture, offering valuable insights for future research in low-resolution image recognition and underwater species identification.

3.3.1.3 Behavioural Analysis

Accurate recognition of feeding behaviour and activity patterns is essential for optimising feeding strategies, reducing waste, and improving fish welfare (Lopes et al. 2017). Intelligent systems using computer vision and machine learning now offer scalable alternatives to manual observation.

Several approaches have been proposed for feeding behaviour analysis. Zhou et al. (2019) used near-infrared imaging combined with SVMs and a grey gradient symbiotic matrix to quantify feeding intensity, achieving strong agreement (correlation 0.945) with expert scores. Deep learning models have also shown promise: Måløy et al. (2019) applied a CNN–LSTM pipeline to classify feeding and non-feeding states in salmon with 80% accuracy, while Adegboye et al. (2020) used Fourier descriptors to reach near-perfect accuracy on benchmark datasets. Real-time systems such as that by Georgopoulou et al. (2024), integrating YOLO and Deep SORT, demonstrate the viability of continuous monitoring in production settings.

Fish group behaviour offers additional insights into environmental and social dynamics. Local individual responses can lead to emergent patterns at the group level (Giardina 2008; O'Donncha et al. 2021). Han et al. (2020) used CNNs with spatiotemporal fusion to classify six distinct group behaviours with 82.5% accuracy.

Beyond routine behaviour, monitoring anomalies is key to non-invasive welfare assessment. Methods based on RNNs (Zhao et al. 2018) and machine vision (Zhao et al. 2016; Morimoto et al. 2018) have been developed to detect abnormal movement patterns indicative of disease or oxygen stress. These efforts highlight the growing role of behavioural analytics in precision aquaculture.

3.3.2 AI to Forecast Environmental Conditions

Artificial intelligence (AI), when combined with high-performance computing and digital twin technologies, is transforming how we model, understand, and manage the environment. A leading example is the European Union's Destination Earth initiative (DestinE 2024), a flagship of the European Green Deal. Its goal is to build a high-fidelity digital replica of Earth's systems by 2030 to support climate adaptation, environmental stewardship, and sustainability policymaking (Bauer et al. 2021).

At the centre of this initiative is the Digital Twin of the Ocean (DTO) (EDITO 2024), which aims to simulate ocean processes at high spatial and temporal resolution. The DTO provides actionable insights into ocean health, supports assessments of climate change impacts, and enables scenario-based exploration of marine policy interventions. By modelling conditions such as ocean temperature, current dynamics, and biogeochemical variables, the DTO facilitates informed decision-making for goals like sustainable fisheries management, pollution reduction, and marine biodiversity conservation (Tzilivakis 2022).

Fish farming plays a dual role in this ecosystem, serving both as a key generator of high-resolution environmental data and as a testbed for the deployment of AI-driven forecasting models. A growing body of research demonstrates the utility of machine learning in predicting water quality and environmental parameters such as ocean currents, wave height, temperature, and dissolved oxygen (O'Donncha et al. 2024; Yang et al. 2021). While significant progress has been made in freshwater applications (Oruche et al. 2021; Oruche and O'Donncha 2023), operational adoption in marine aquaculture remains limited (Heimbach et al. 2024). Challenges include fragmented infrastructure, lack of standardised data protocols, and limited integration of AI models into real-time decision systems.

Nevertheless, projects such as the DTO highlight the potential for AI-enabled digital twins to serve as shared platforms for both policy and operations, bringing together climate models, remote sensing data, and aquaculture-specific dynamics within a unified predictive framework.

3.3.2.1 AI Models for Environmental Forecasting in Aquatic Systems

Recent advances in ML have enabled more accurate and computationally efficient forecasting of environmental conditions critical to aquaculture, such as dissolved oxygen, temperature, salinity, and wave dynamics. These methods increasingly complement physics-based models, which often require significant computational resources and may struggle to capture fine-scale variability in complex ocean systems (de Bezenac et al. 2017; Wiewel et al. 2018).

Classical time series forecasting methods, such as multilayer perceptrons (MLPs), can capture short-term dependencies but often struggle to represent longer-term temporal structure. Recurrent models like LSTM and BiLSTM networks offer improved handling of such dependencies and have been used to forecast DO levels and water temperature based on in situ sensor networks (O'Donncha et al. 2022). CNNs can also be applied to time series by treating them as structured inputs and have the added advantage of automatic feature extraction. Hybrid CNN–LSTM models have been used for sequence modelling of ocean surface conditions, combining spatial pattern recognition with temporal forecasting capabilities (Yang et al. 2017; Xingjian et al. 2015).

To reduce reliance on computationally expensive physical simulations, surrogate models trained via ML offer fast approximations. These have been applied in wave forecasting (James et al. 2018; O'Donncha et al. 2018), pollution monitoring (Hähnel et al. 2020), and satellite imagery downscaling (Ducournau and Fablet 2016). Such approaches hold promise for near real-time applications on farms, particularly when deployed on edge devices.

An emerging direction involves embedding physical constraints or partial differential equations (PDEs) directly into neural network training. These physics-informed networks preserve consistency with known laws while learning from empirical data (Rao et al. 2020). They are especially attractive in domains where data is sparse but physical theory is well understood.

3.4 AI and Emerging Trends in Aquaculture

AI applications in aquaculture remain in an early phase. Technologies such as computer vision, IoT, and deep neural networks offer strong potential, but large-scale operational deployments are still limited.

The weather forecasting community offers a compelling example of how AI can transform scientific modelling. A recent surge in AI-driven forecasting has led to systems that now outperform traditional numerical weather prediction models in both speed and accuracy (Maskell 2023). These developments signal the kinds of AI-enabled infrastructure and digital twin technologies that could soon shape the future of aquaculture (Alvarez Fanjul et al. 2024).

The European Centre for Medium-Range Weather Forecasts (ECMWF), one of the world's leading weather agencies, has recently adopted experimental AI-based operational forecasts. Three different architectures have been deployed:

- **FourCastNetv2-small:** an updated version of Nvidia's FourCastNet, employing Spherical Fourier Neural Operators to model spatial dependencies. It is trained on ERA5 reanalysis data and operates at 0.25° resolution (Bonev et al. 2023).
- **GraphCast:** Developed by Google DeepMind, this system uses a graph neural network (GNN) with an encoder–processor–decoder structure to model geospatial relationships. It is trained on ERA5 data and refined using ECMWF's HRES forecasts (Lam et al. 2022).
- **Pangu-Weather:** Created by Huawei, this model uses a 3D transformer architecture and includes separate modules for different forecast horizons. It was also trained on ERA5 data and operates at the same resolution (Bi et al. 2023).

These advances in model architecture and training pipelines illustrate how GNNs and transformer-based systems can handle complex, spatiotemporal environmental data.

3.4.1 Graph Neural Networks

Graph Neural Networks (GNNs) are designed for structured relational data, making them well suited for aquaculture systems governed by spatial, biological, and environmental interactions. In net-pen farming, where factors such as water flow, pathogen spread, and cage proximity influence outcomes, GNNs offer a powerful tool for modelling interdependencies.

A key example is the transmission of sea lice across cages or farms. GNN-based models can represent these spatial and biological dynamics as evolving networks, enabling predictive analytics for early intervention and reduced chemical treatment use (Langbridge et al. 2023).

Beyond disease spread, GNNs have shown promise for forecasting environmental conditions such as dissolved oxygen and temperature, by integrating sensor data across sites. They can also support behavioural analysis and feed optimisation, by modelling fish movements and stress responses as relational patterns within and across cages.

Effective graph construction is critical. Approaches that encode similarity between cages, farms, or regions, based on geography, ocean currents, or historical performance, can significantly improve predictive performance (Azmat et al. 2023; Ba et al. 2023). As aquaculture systems become more sensor-rich and interconnected, GNNs are likely to play an increasingly central role in modelling complex farm dynamics.

3.4.2 Large Language Models

Large language models (LLMs), also known as foundation or generative AI, have transformed the AI landscape through their scalability, generality, and ability to transfer across tasks. Trained on massive text corpora, these models (e.g. GPT-4 (Achiam et al. 2023) and BERT (Jawahar et al. 2019)) can be fine-tuned with minimal task-specific data to support a wide array of downstream applications, from translation and summarisation to code generation and scientific writing (Bommasani et al. 2021).

This rapid advancement stems from three core innovations (Tunstall et al. 2022):

- **Transformer architecture:** Introduced in "Attention Is All You Need" (Vaswani et al. 2017), transformers allow efficient parallel processing of sequences through self-attention, enabling context-aware reasoning over long inputs.
- **Pre-trained foundation models:** Foundation models are trained once on broad, general-purpose data and then reused across many tasks and domains. This "train once, deploy many" paradigm dramatically reduces the need for large, task-specific datasets. Through fine-tuning or even simple prompting, these models can be adapted to diverse downstream applications with minimal additional supervision (Bommasani et al. 2021).
- **Accessible model hubs:** Platforms like Hugging Face have accelerated adoption by standardising access, fine-tuning, and deployment pipelines for state-of-the-art models (Wolf et al. 2025).

Although best known for natural language tasks, transformers now underpin applications in code synthesis (Jimenez et al. 2023), vision, time series forecasting (Liu et al. 2025), and geospatial analysis (Mukkavilli et al. 2023). Their ability to integrate diverse modalities makes them particularly well suited to data-rich environments, an increasingly relevant feature for aquaculture.

In aquaculture, LLMs are not yet widely deployed, but their potential is increasingly recognised. As digital twins, autonomous agents, and real-time decision-support systems evolve, foundation models may support multiple integrative tasks, including translating between heterogeneous data sources, responding to user queries, and synthesising operational insights. Their role is likely to complement, rather than replace, domain-specific predictive models, particularly in data-sparse settings. Their practical impact in aquaculture will depend on how effectively they can be integrated with existing models, data pipelines, and decision-support workflows (Fig. 3.1).

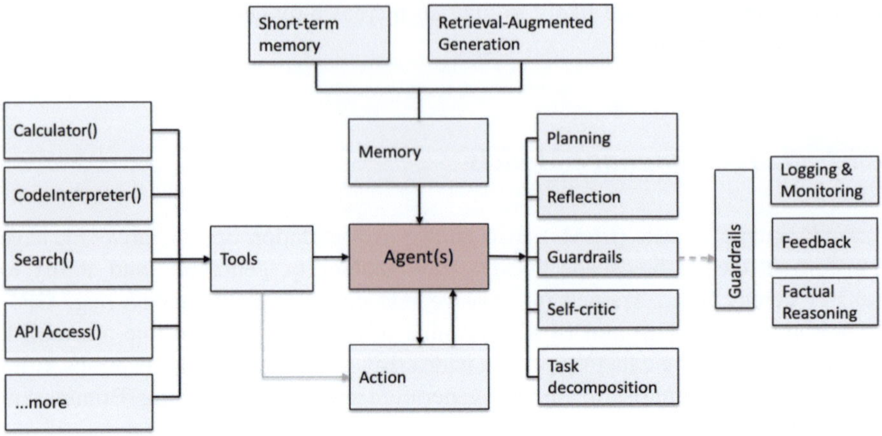

Fig. 3.1 This diagram illustrates the key components and data flows in a large language model (LLM)-driven agent architecture. The central agent coordinates interactions between tools (e.g. calculators, code interpreters, APIs), memory systems (short-term and retrieval-augmented), and action modules. Cognitive processes such as planning, task decomposition, self-critique, and reflection support goal-oriented behaviour and adaptive reasoning. Guardrails ensure system robustness through safety, factual reasoning, feedback, and monitoring. The architecture enables scalable, flexible execution of complex tasks in dynamic environments

3.4.3 LLM-Powered Autonomous Agents

LLM-powered autonomous agents are emerging as transformative tools for data-rich, dynamic domains. By leveraging natural language capabilities, these agents can interpret, reason, and act on complex queries, making them valuable for real-time monitoring, resource optimisation, and decision support.

In an LLM-agent system, the model serves as the agent's "brain," supported by three key components (Weng 2023):

- **Planning:** Decomposes complex tasks into smaller, sequential goals.
- **Memory:** Utilises in-context learning and vector stores to retain short- and long-term information.
- **Tool use:** Refers to calls to external APIs, such as search engines, code interpreters, or databases, to extend capabilities beyond static knowledge.

A key advantage of such systems is tool calling, which enables agents to autonomously interact with external systems. In aquaculture, agents can:

- **Query databases:** Retrieve historical records on fish health, water quality, and feeding patterns.
- **Search the web:** Access literature, regulations, and industry updates.
- **Use weather APIs:** Incorporate forecasts to anticipate temperature shifts or storm events.

- **Monitor sensors:** Track dissolved oxygen, salinity, and current patterns via live environmental feeds.

The ability to synthesise information from diverse data streams makes LLM agents especially valuable in aquaculture, where multiple interdependent factors affect outcomes.

Because they understand natural language, these agents also act as intuitive interfaces. Non-technical users can interact with them using everyday questions such as, "What's the expected water temperature this week?" or "Summarise dissolved oxygen trends over the past 24 hours." The agent interprets the query, retrieves relevant information, and returns concise, actionable responses.

More advanced interactions are also possible, such as generating feeding efficiency reports or identifying anomalies in fish behaviour over time. This accessibility encourages broader adoption and supports faster, data-informed decisions.

While still in early development for aquaculture, LLM agents show strong potential. Future enhancements may include integration with remote sensing, predictive models for fish health, and fully autonomous decision-making frameworks. As these systems evolve, they may shift from passive assistants to proactive managers which helps to optimise operations with minimal oversight and promoting resilience and sustainability across aquaculture systems.

Taken together, GNNs, transformer-based LLMs, and autonomous agents form a complementary stack: GNNs model farm dynamics, LLMs translate data into insight, and agents close the loop by automating action. Their convergence will be the engine of the next wave of precision aquaculture.

3.5 Case Studies

While AI and machine learning applications in aquaculture are rapidly expanding, much of the current work remains exploratory or limited to isolated deployments. Unlocking the full value of these technologies depends on building robust data infrastructures, scaling models effectively, and integrating AI seamlessly into existing operational workflows. Advances in areas such as unmanned monitoring platforms, remote sensing, edge computing, and simulation offer exciting possibilities for accelerating the adoption of AI-driven solutions. Notably, the convergence of machine learning with digital twins and real-time environmental modelling stands to transform aquaculture from reactive management to proactive, predictive control.

The following case studies illustrate how some of these emerging technologies are already being applied in practice. They highlight different strategies to improve environmental forecasting and operational decision-making in marine and aquaculture settings. Specifically, the case studies demonstrate:

- The deployment of real-time sensor networks and machine learning models to forecast dissolved oxygen on Atlantic salmon farms in Scotland;

- The use of satellite remote sensing combined with machine learning for spatial estimation of water quality variables in aquaculture zones;
- A novel LLM-based agentic architecture applied to digital twin workflows for shipping, showcasing how agentic AI can support complex operational decision-making.

Together, these examples provide a snapshot of how AI and digital twin technologies are beginning to reshape the aquaculture and marine sectors.

3.5.1 Case Study 1: Application of a Real-Time Sensor Network in Developing Machine Learning Models to Forecast Dissolved Oxygen on Atlantic Salmon Farms in Scotland

3.5.1.1 Introduction

Oxygen is a critical and dynamic parameter that affects fish metabolism and stress and ultimately impacts growth efficiency (Remen et al. 2012). Widescale environmental monitoring has been adopted by the industry to monitor the complex dynamics of dissolved oxygen at the pen level and is used to inform feeding decisions (Burke et al. 2021). This adoption has created long-term environmental datasets that can be used in model development to forecast oxygen in open net pens. Oxygen forecasting can help to improve decision-making to maximise growth efficiency. During feeding, oxygen levels must be above a given threshold (which varies depending on temperature but generally, 70% saturation is used) and there must be reasonable confidence that oxygen will remain above this threshold hours later, during the digestive process that uses up additional oxygen (Remen et al. 2016). If oxygen levels are too low, fish become stressed and energy is allocated to other more critical life functions instead of growth, and in severe cases fish will regurgitate feed resulting in feed waste and reduced feed conversion ratios (FCRs). Providing confidence in near-term oxygen levels using machine learning predictions can assist with real-time decision-making and helps to maximise growth and enhance welfare by feeding only under optimal environmental conditions.

Machine learning models have been proposed as an alternative to traditional ocean models to capture the fine-scale processes that a pen-level forecast necessitates and to account for the effect of the biomass on dissolved oxygen levels in the pen. This project scrutinises which machine learning models best capture oxygen dynamics at the pen level and examine the robustness of a model as it is scaled up and tested on different pens within the same farm, on pens at different farms within the same geographic region and on pens at farms in different regions. The model was first trained on a farm located in the Shetland Islands, and then tested on a farm on the west coast of mainland Scotland and finally performance was evaluated on a farm located on the west coast of Canada. This work is ongoing, but some preliminary results are presented below.

3.5.1.2 Methodology

Pre-processing of operational data is a critical step as datasets often contain many imperfections such as erroneous measurements, anomalies, and missing data. The data consisted of historical observations of oxygen concentration, temperature, and salinity for individual cages on various farms. During preliminary analysis, temperature and salinity were identified as key features for normalisation and time-based variables such as day of week, hour of day, and month of year were also incorporated (Linardatos et al. 2020). These variables are crucial for capturing trends and seasonality. A circular numeric encoding strategy was adopted for time-based variables to maintain their cyclical nature, enhancing the model's ability to discern patterns in oxygen levels. Many different models were tested and compared including linear regression, random forest, and Long Short-Term Memory (LSTM). However, the Temporal Fusion Transformer (TFT) model reported best performance due to its effectiveness in handling complex time series data. The adoption of this architecture was informed by several standout features:

- Multi-headed Self-attention Mechanism: This feature allows the model to assign differential significance to various temporal intervals, a critical function for decoding complex time-related patterns in data, such as seasonal fluctuations and the influence of specific events on oxygen concentration. By prioritising relevant periods over others, the model achieves superior forecasting precision.
- Dual-Stage Architecture: The model employs an encoder–decoder framework that combines historical data analysis with future event prediction. The encoding phase processes past data, whereas the decoding phase is dedicated to forecasting the target time series. This method increases the model's understanding of time patterns, improving its prediction accuracy and reliability for different forecasting horizons.
- Quantile Forecasting: This enables the prediction of a spectrum of outcomes at varying probability thresholds. By forecasting seven quantiles of future oxygen concentrations, the model provides a comprehensive overview of potential scenarios, facilitating enhanced decision-making capabilities. Seven is the standard number of quantiles for this model and uses an uneven distribution of thresholds for the output (0.02, 0.1, 0.25, 0.5, 0.75, 0.9, 0.98).

The refinement of the model involved a detailed evaluation of its forecasting per formance, at the individual pen level and across entire farms, through comparative analysis with baseline models, utilising its dual-stage and multi-horizon forecasting functionalities. The models were trained using a structured cross-validation method to verify its efficiency across assorted data segments, including different cages, farms, and geographical regions. This approach, coupled with strategic parameter selection, demonstrated the model's adaptability and applicability across different farms. To assess the model's predictive performance, particularly its ability to forecast extreme changes in oxygen levels, a set of evaluation metrics was carefully chosen. These included mean absolute error (MAE) and quantile-based metrics to evaluate accuracy within specific ranges (Fig. 3.2).

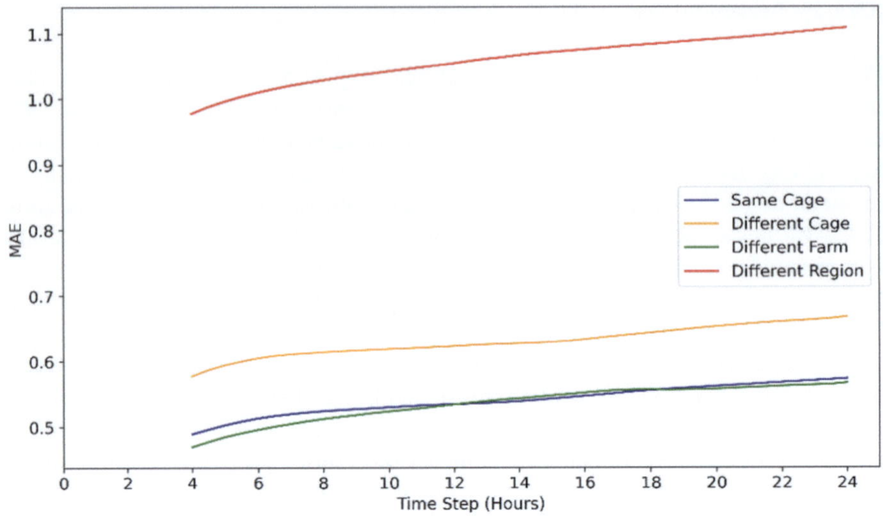

Fig. 3.2 Mean average error (MAE) of 24-hour prediction for various tests of the model on the same cage it was trained on, a different cage from the same farm, a cage from a different farm within the same region, and a cage on a farm in a different region

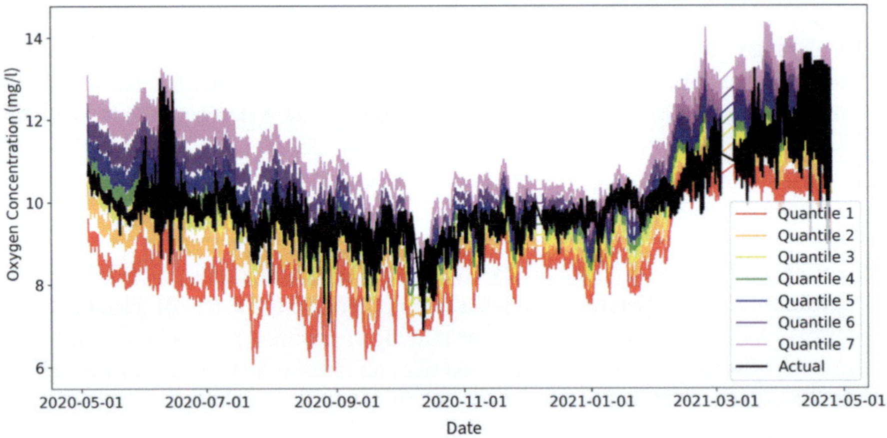

Fig. 3.3 Actual raw oxygen concentration and model output of a 24-hour prediction using seven quantiles that indicate a confidence interval for the forecasted values. Data presented shows the output of the model on a single cage that was trained on the entire farm dataset

Regarding the quantile confidence interval accuracy, the model analyses data at time t and compares it with the quantiles for time t + 24 hours (ranging from the lowest quantile, Q1, to the highest, Q7), measuring how often the actual oxygen levels fall within these predicted ranges, providing a clear percentage of coverage (Fig. 3.3) and allowing operators to derive a range of model outputs based on their risk tolerance. The practical deployment of the model will highlight its utility in generating real-time forecasts. By analysing data from the preceding 48 hours, it can project oxygen levels for the upcoming 24-hour period across seven quantiles, delivering insights into potential variances.

3.5.1.3 Conclusions

The model was capable of outputting 24-hour predictions for various farms within the same region, with an MAE less than 0.7 mg/l. These findings demonstrate that reasonably accurate pen-level forecasting is achievable, validating ML as a viable alternative to traditional environmental models. The MAE increased when the model was tested on farms in different geographic regions, which suggests that models should be trained on data within a reasonable geographic area to achieve the highest possible accuracy. Operational forecasting is a growing area of research in precision fish farming, aimed at equipping farmers with data-driven tools for more proactive decision-making. Oxygen concentration is a dynamic and complex measurement that is impacted by many distinct factors, including tidal and water exchange, biomass, farm orientation, and plankton presence, and this research shows that machine learning models can accurately capture these complexities, and there is potential for these insights to be used on farms to help with feeding decisions under challenging conditions.

This chapter highlights how machine learning can significantly enhance pen aquaculture forecasting by overcoming the limitations of traditional numerical models, which are bound by their grid resolution and sub-grid process parametrization. Unlike these conventional methods, machine learning models adeptly adjust to the scale of observational data. Coupled with the influx of high-resolution data from Internet of Things (IoT) devices, machine learning enables the development of more detailed and adaptable forecasting frameworks, paving the way for improvements in aquaculture efficiency and sustainability. Further details on this work can be found in Khan et al. (2024).

3.5.2 Case Study 2: Dissolved Oxygen Estimation in Aquaculture Sites Using Remote Sensing and Machine Learning

3.5.2.1 Introduction

Dissolved oxygen (DO), together with temperature, is one of the most critical environmental parameters affecting the physiology and performance of marine organisms. Consequently, accurate monitoring is of great importance for finfish aquaculture as it directly affects the survival, growth, food intake, and health of the farmed fish (Oldham et al. 2019; Burt et al. 2012). For net-pen aquaculture in particular, the continuous monitoring of DO is a necessity as environmental conditions at the rearing sites are outside human control and are therefore prone to fluctuations that may threaten production.

DO monitoring has utilised manual, semi-automated, and fully automated in situ sensors (Parra et al. 2018). While such modes of DO monitoring are simple and have been effective in the past, the rapid expansion of aquaculture towards precision farming methods necessitates advanced systems that contribute towards real-time monitoring and forecasting (O'Donncha and Grant 2019; Raju and Varma 2017). Accurate forecasting of dissolved oxygen depends heavily on capturing the unique spatial and environmental characteristics of each aquaculture site. While in situ sensors are promising, they do not provide complete coverage at the pen-, farm-, and bay-scale level. Furthermore, organised zones of aquaculture activity with several farms installed have similar needs, and large-scale monitoring is essential for forecasting. Remote sensing is a powerful tool to enhance farm monitoring and decision-making.

Although remote sensing is widely used to detect variables that affect the optical properties of water at specific wavelengths such as chlorophyll-a (Chl-a), total suspended matter (TSM), and temperature, it is unlikely to record the concentration of DO directly from the reflection values of the satellite sensors. Instead, it can be estimated indirectly because of its correlation with other parameters, such as temperature and Chl-a (Guo et al. 2021; Kim et al. 2020).

DO generally exhibits a negative correlation with sea temperature (Matear and Hirst 2003; Mavropoulou et al. 2020). Primary production also affects DO levels, which is often measured by Chl-a, an indicator of phytoplankton abundance in surface waters (Guy et al. 1993; Lewis et al. 2016).

Based on the above theoretical mechanisms, several researchers have tried to develop models for the correlation of DO with environmental parameters, such as sea surface temperature (SST) and Chl-a (Karakaya and Evrendilek 2011; Batur and Maktav 2018; Kim et al. 2020; Guo et al. 2021). This aims to provide the potential to monitor DO from satellite sensors. Different approaches such as Response Surface Regression (RSR), Multiple Linear Regression (MLR), Artificial Neural Network (ANN), Support Vector Machines (SVMs), and Support Vector Regression (SVR) have been scrutinised. While previous research has been successful in estimating DO levels, data sparsity and the demand for high-resolution estimation in coastal zones have limited the value of the approach for aquaculture. In our study, we developed a methodology for the estimation of DO in coastal areas with aquaculture facilities, based on daily CMEMS data and machine learning techniques.

3.5.2.2 Methods and Data

The satellite data used was from Copernicus Marine Environment Monitoring Service (CMEMS) that provides regular and systematic biogeochemical and physical information on the marine environment. In our study, we exploited level-4, daily, gap-free satellite observations for Chl-a concentration and SST from multi-platform observations. Our in situ dataset includes field measures of DO and SST from a fish farm in Lesvos Island, Greece. Dissolved oxygen was measured daily between 8:00 am and 10:00 am using an OxyGuard sensor with a scavenging packet. 80% of the dataset served as the training sample, while the remaining 20% was utilised as the test sample.

The farm is located at Lesvos Island, in North Aegean Sea, an area which is generally characterised as an oligotrophic sea (Krom et al. 1991). The capacity of the farm is 380T per year, and the main fish farmed are the European sea bass (*Dicentrarchus labrax*) and the gilthead sea bream (*Sparus aurata*). Water temperature in the surrounding area ranges between 14 and 24 °C, and the salinity is constant at 39 psu. The average depth in the farm is approximately 50 m.

For the study, we developed an approach for DO estimation based on a Support Vector Regression machine learning model, using CMEMS data and in situ observations. For each spatial coordinate, an input array was constructed incorporating chlorophyll-a (Chl-a) and sea surface temperature (SST) values across a 3-day lagged window prior to sampling.

Support Vector Regression (SVR) has been demonstrated to effectively tackle both linear and non-linear problems, showing strong performance even with small datasets (Guo et al. 2021). SVR, a variant of Support Vector Machines, is a non-linear, non-parametric machine learning model that adheres to Vapnik's principle of structural risk minimisation for supervised learning (Vapnik 2013). Unlike linear regression, which seeks to minimise squared error sums, SVR aims to manage errors within a predefined tolerance, enabling the selection of error margins (ϵ) and the determination of the optimal hyperplane for decision-making. This approach provides flexibility in model accuracy by adjusting (ϵ) and incorporating slack variables (ξ) to tolerate training errors, enhancing the model's resilience to noise and outliers. The model's formulation involves an objective function and constraints, detailed in Eqs. 3.1 and 3.2, which guide the optimisation process:

$$\min \frac{1}{2}\|w\|^2 + C\sum_{i=1}^{n}|\xi_i| \tag{3.1}$$

$$|y_i - w_i x_i| \leq \epsilon + |\xi_i| \tag{3.2}$$

where y_i is the target, w_i is the coefficient, and x_i is the predictor (feature). C corresponds to a user-selected parameter to control the complexity of the model, acting as a trade-off parameter between non-linearity and number of training errors.

To capture more intricate hyperplane shapes beyond linear approaches, one can utilise kernel functions like polynomial, radial basis function (RBF), and sigmoid kernels. In the current study, the RBF kernel was chosen for its promising capabilities (Guo et al. 2021). SVR, with its simple structure, is supported by various programming languages. Model development was performed in Python 3.4 using the scikit-learn library (v0.18.1), which provides efficient support for SVR and related regression techniques. Model evaluation was conducted using residual analysis alongside mean absolute error (MAE) and root mean square error

(RMSE) as standard performance metrics. MAE calculates the precision expressing the average distance between the estimated and the real values and is described by Eq. 3.3. RMSE describes the standard deviation of the residuals (Eq. 3.4).

$$\text{MAE} = \frac{\sum_{i=1}^{n} |y_i - x_i|}{n} \tag{3.3}$$

$$\text{RMSE} = \sqrt{\frac{\sum_{i=1}^{n} (y_i - x_i)^2}{n}} \tag{3.4}$$

where y_i is the predicted values and x_i the measured values. The initial dataset included dates, in situ DO measurements, latitude, and longitude together with the values of Chl-a and SST from CMEMS for each date and the previous 3 days. In Fig. 3.4, a graphical representation of the approach is presented.

3.5.2.3 Results

The data collected from CMEMS (Chl-a, SST) and the estimated DO were used to create daily maps for examining their spatial distribution with a resolution of 1 km. Figure 3.5 presents snapshots of SST, Chl-a, and DO in the surrounding sea area of Lesvos.

Figure 3.5c presents modelled estimates of DO. Naturally, the spatial distribution follows similar patterns to those of SST and Chl-a. Higher concentrations are observed in the north-eastern waters, near the coastline, presenting a clear negative correlation with SST and a weaker positive correlation with Chl-a. In general, DO concentration varied from 7 to 7.8 ppm across the year with four pronounced drops below 6.5 ppm. The highest values were detected in the winter and spring months with the lowest values reported during summer and fall.

The accuracy of our model was tested using the value of mean absolute error (MAE) and root mean squared error (RMSE), as well as plots of the residuals. These are the most widely used metrics for evaluating a regression model. Both metrics quantify the distance between the actual and the predicted values and are measured in the same units as the output variable, which contributes towards an interpretable performance metric. Our model performed well with MAE of 0.11 and RMSE of 0.13, and the residuals were evenly distributed around 0. These indicate accurate model performance with no obvious bias present.

We have chosen to use MAE and RMSE indices to validate our model as they give a clearer view of the error and distribution of the residuals. On the contrary, R^2 was poor (0.32) compared to other studies. R^2 is generally used in similar studies to examine the fitness of the model instead of the performance, and it does not adequately consider the magnitude and direction of error. In this context, we do not consider the poor value of R^2 to be informative for the practical applicability of the method.

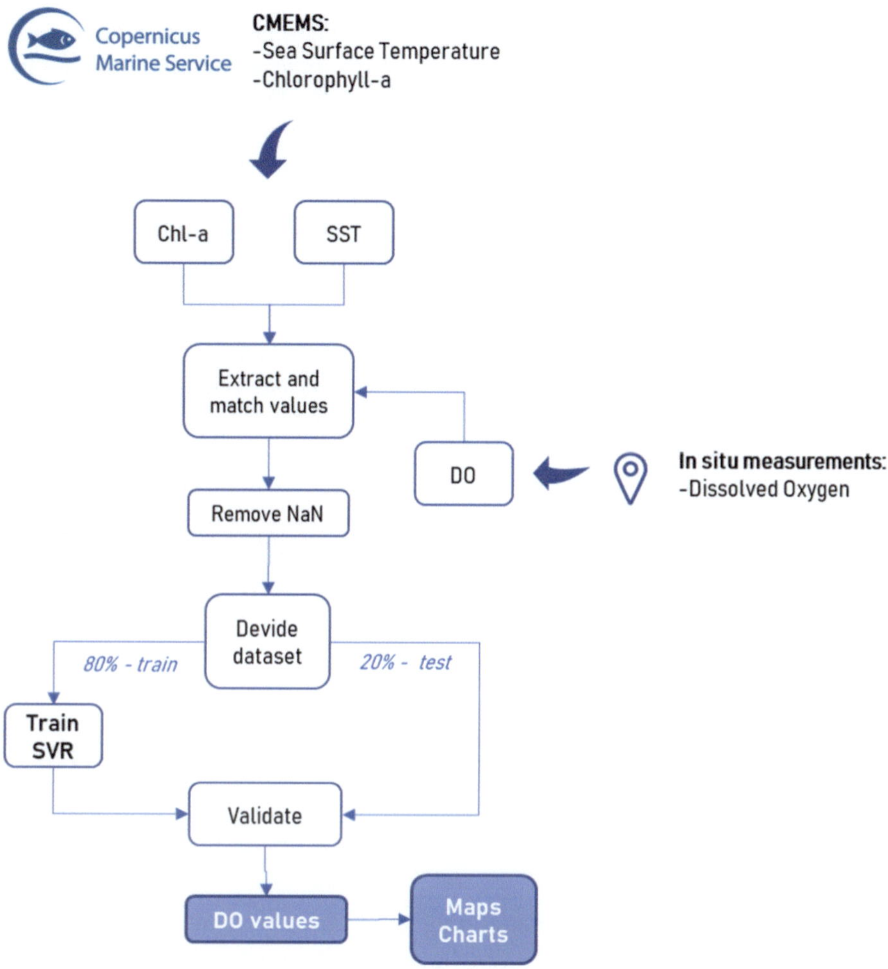

Fig. 3.4 Flowchart of the implemented methodology. (From Chatziantoniou et al. 2022)

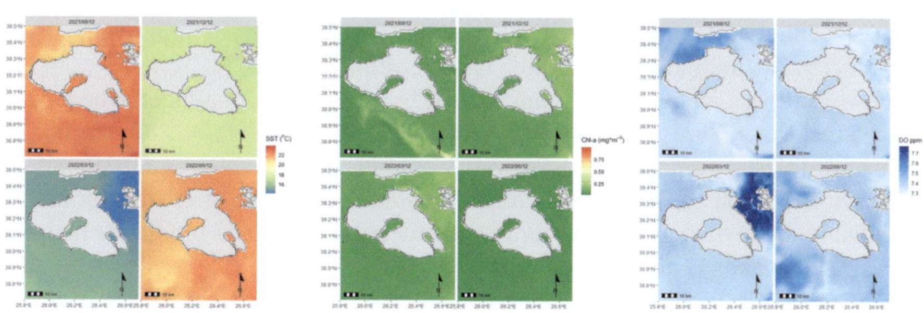

(a) (b) (c)

Fig. 3.5 The spatial distribution of (**a**) SST, (**b**) Chl-a, and (**c**) DO around Lesvos Island for four representative dates (fall, winter, spring, and summer). (From Chatziantoniou et al. 2022)

Our results show a promising approach to estimate DO at aquaculture sites, which paves the way for the development of real-time monitoring systems for aquaculture. The methodology successfully correlates DO with SST and Chl-a, reporting high agreement with field data. The model replicates not only shifts in the seasonal patterns of DO but also sudden drops in concentration in at least four occasions during the summer. While the drops in the analysed time series are of minor practical interest because they are relatively small and thus do not constitute a concern for fish, the capability of the method to realistically capture them is reflective of its potential as a monitoring tool that can accurately estimate high-resolution dynamics.

Another advantage of the proposed approach is that it increases the spatial scale at which monitoring can occur. While in situ sensors are paramount for precise measurements within a single cage of a farm, a remote sensing approach offers a better representation of the oxygen conditions, and thus, potential dangers, at a larger area. Considering that aquaculture farms are often aggregated within the limited coastal space of suitable farming areas, either as single operations or as an organised industrial park, it is imperative for the aquaculture producers to monitor the DO at a higher spatial scale. The same applies to administrative authorities that may be interested in the supervision of zones of organised aquaculture activity at a regional level.

3.5.3 Case Study 3: Agentic AI for Digital Twin Applications in Shipping

3.5.3.1 Introduction

The global shipping industry, comprising over 100,000 vessels, facilitates the transport of 90% of the world's goods, yet it also contributes 3% of global greenhouse gas emissions, which have risen by 20% in the past decade (United Nations Conference on Trade and Development 2023; Stuchtey et al. 2023). To meet its environmental targets, reducing emissions by 20% by 2030 and 70% by 2040, the International Maritime Organization (IMO) prioritises improvements in ship efficiency (International Maritime Organisation 2023).

However, these ambitions are complicated by an ageing global fleet, where over half of the vessels are older than 15 years, making them difficult to retrofit yet premature for scrapping (United Nations Conference on Trade and Development 2023). In the midst of these constraints, digitalisation and AI offer transformative potential in areas such as route optimisation, fuel efficiency, and predictive maintenance (Katsoulakos et al. 2024).

Key priorities include seamless data integration across fragmented systems, real-time monitoring for predictive maintenance, scalability for diverse fleets, and transparent adherence to regulatory compliance across global frameworks (Katsoulakos et al. 2024). Additionally, solutions must provide automation and configurability to address the varied needs of industry personas without relying on specialised expertise (O'Donncha et al. 2024).

3.5.3.2 AI Agents

Artificial intelligence (AI) agents are systems designed to autonomously perform tasks by perceiving their environment, making decisions, and executing actions to achieve specific goals. Recent advances in large language models (LLMs) have significantly enhanced the capabilities of AI agents, enabling them to handle more complex and dynamic tasks.

LLM-based AI agents can now plan and act by analysing tasks, formulating strategies, and adjusting their actions based on feedback. This mirrors human problem-solving processes, allowing agents to refine their approaches through self-critique and iterative improvements. For example, these agents can autonomously search the web, solve mathematical problems, and even review and correct their work, thereby increasing their effectiveness in various applications.

To manage these complex tasks, engineers are enhancing LLMs with additional modules that improve memory, planning, and tool-utilisation capabilities. This development marks a shift from traditional, monolithic AI models to more sophisticated, multicomponent systems. Such systems integrate various functionalities, enabling AI agents to perform a broader range of tasks with greater autonomy and adaptability.

3.5.3.3 Digital Twin for Green Shipping AI Agents

This case study introduces a digital twin for green shipping (DT4GS) framework that uses large language model (LLM) agents to enhance automation and efficiency in the shipping industry (Timms et al. 2025). At its core, the system includes a central reasoning agent that breaks down complex tasks into smaller, manageable steps and coordinates with specialised tools to carry them out. By using natural language as the primary interface, this approach allows different users, such as ship owners, captains, engineers, and technicians, to interact with the digital twin in a simple and intuitive manner. Each user can query and explore ship systems based on their specific needs, without requiring technical expertise in AI or system modelling.

We developed a suite of tools specifically designed for industrial applications, enhancing agentic decision-making by providing a comprehensive view of real-time ship operations along with contextual insights. These tools integrate structured information from a domain-specific knowledge graph, real-time sensor data, and technical specifications, including potential failure data. By leveraging LangGraph,[1] the framework's modular and adaptable design allows it to meet diverse operational

[1] https://langchain-ai.github.io/langgraph/

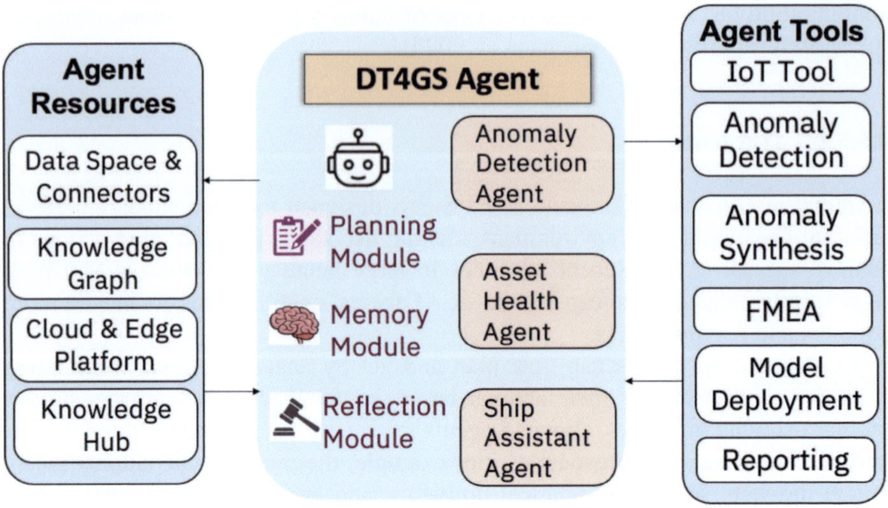

Fig. 3.6 The DT4GS Agent acts as a central orchestrator that coordinates specialised sub-agents and tools to support key operational tasks. It includes a planning module that decomposes tasks and delegates them to appropriate agents (e.g. anomaly detection, asset health, and ship assistant). The memory and reflection modules track context and assess system performance. On the left, agent resources such as a data space, knowledge graph, and edge/cloud platforms provide contextual information. On the right, agent tools perform domain-specific functions such as anomaly detection, FMEA, and reporting. This modular, language-driven framework enables flexible, autonomous interaction with complex ship systems

requirements and significantly improve decision-making efficiency. LangGraph is a multi-agent orchestration framework built on LangChain that we adopted to manage tool-using LLM agents in DT4GS. A planner in LangGraph (or in similar LLM-based systems) is a critical component that is responsible for task decomposition, decision-making, and execution sequencing. In AI agent frameworks, planners allow agents to break down high-level objectives into manageable subtasks, prioritise them, and decide in what order and manner those subtasks should be executed.

We demonstrate the capabilities of this framework on real-world shipping use cases, showcasing its practical applicability and scalability. Figure 3.6 presents the pipeline that integrates agentic reasoning with a domain-specific knowledge graph and contextual data to streamline the selection, refinement, and execution of digital twin tasks. This framework allows the agent to leverage core digital twin functionalities, including:

- **Dataspace**: a centralised repository encompassing sensor and operational data, providing a comprehensive view of various ship operations
- **Knowledge Graph**: a structured representation capturing the relationships between the vessel, its components, and associated operational variables. The graph interconnects critical elements such as the main engine, auxiliary engine,

and cargo system with key variables like temperature, pressure, and performance metrics, forming a network that reflects the vessel's functional state.
- **Model Builder**: a system designed to handle the selection, construction, and refinement of machine learning models tailored to specific task requirements. It ensures that predictive and prescriptive analytics align with the operational goals of the digital twin.
- **Edge Orchestration Engine**: a component utilising the Linux Foundation's Open Horizon[2] to efficiently manage the deployment and placement of containerised applications across a distributed fleet of edge nodes. This enables real-time processing and decision-making close to the data source.

Agents are equipped with a suite of external tools that extend their ability to interact with digital twin resources and enhance their ability to observe, analyse, and act upon system states. These tools include:

- **IoT Tool** facilitates integration with IoT-enabled systems, allowing the agent to retrieve and process real-time sensor data.
- **Knowledge Graph Tool** grants agent access to a domain-specific knowledge graph, enabling complex queries on ship operations, asset relationships, and historical data for enriched decision support.
- **Data Quality Tool** conducts initial data quality assessments by evaluating expected behaviours, identifying inconsistencies, and ensuring data integrity before further processing. Within an agentic workflow, it can generate metrics on data quality, providing users with insights for further refinement and iteration.
- **Anomaly Detection Tool**: an online unsupervised anomaly detection approach that leverages an efficient formulation of the optimal transport (OT) problem in one dimension to detect anomalies in noisy seasonal time series data, as described in Langbridge et al. (2024).
- **Failure Modes and Effects Analysis (FMEA) Tool**: FMEA is a systematic method that is used to identify and assess possible failure modes within a system, process, or product. It focuses on understanding the causes and effects of each failure mode and evaluating their potential impact (Stamatis 2003). FMEA helps prioritise risks based on factors such as severity, occurrence, and detectability, enabling preventive measures to be taken. Typical information included in an FMEA consists of the component or system location, potential failure modes, causes of failure, severity of the failure, and the resulting effects of the failure on the system. In industrial sectors like shipping, Failure Mode and Effects Analysis (FMEA) is often mandated by classification societies, including the American Bureau of Shipping (American Bureau of Shipping 2015).
- **Time Series Forecasting Tool** configures a time series forecasting model to analyse sensor data identified by the agent. This enables predictive assessments of component health based on historical data and relevant contributory features.

[2] https://lfedge.org/projects/open-horizon/

The model can be deployed in zero-shot mode for immediate inference or further fine-tuned using historical data for improved accuracy.

- **Edge Orchestration Tool** automates the generation of JSON policy documents to manage the deployment of applications at the edge on a specific ship. Deployment decisions are based on parameters such as ship name, geographic location, company affiliation, or integration with an IoT tool, ensuring adaptive and policy-driven orchestration of edge computing resources.
- **Work Order Generation Tool** automates failure response planning by evaluating identified anomalies and leveraging FMEA insights to suggest potential root causes and recommend mitigation actions. This tool streamlines maintenance workflows and enhances predictive maintenance strategies.

3.5.3.4 Deployment of the DT4GS Agent

Shipboard asset management and maintenance require a structured workflow that integrates data-driven diagnostics with AI-powered automation. This process involves data retrieval, model inference, results visualisation, and actionable diagnosis. Agentic workflows facilitate an end-to-end solution, reducing manual intervention while enhancing predictive capabilities for shipboard systems. The process begins with a user prompt that the planning agent receives. This agent decomposes the prompt into a series of subtasks and sends them to domain-specialised agents based on the nature of each task.

Each specialised agent, such as those responsible for navigation, data exploration, failure analysis (FMEA), and ship systems, has access to the tailored tools and APIs required for task execution. Once an agent completes its task, its output is returned to the *Re-router*, which assesses the response in the context of the original goal. If additional actions are needed, the *Re-router* invokes further agents or returns the final result.

This architecture supports sequential and iterative agent interactions, enabling the system to handle complex, multi-step workflows with minimal human intervention. By encapsulating functionality into modular agents and enabling flexible coordination, the framework offers a scalable and extensible approach to digital twin interactions in maritime operations.

The entire workflow of extracting pertinent sensor data, deploying anomaly detection models, diagnosing root causes is automated by the AI agent (Timms et al. 2024). It is insightful to view the trajectory or reasoning process adopted by the LLM to resolve the user query. Figure 3.7 presents the trajectory including task decomposition, tool calling, and natural language understanding to resolve the task.

Agent capabilities include accurately identifying relevant IoT variables associated with asset components (guided by a knowledge graph), detecting exogenous variables that may influence system dynamics (leveraging pre-trained knowledge from an LLM), applying off-the-shelf anomaly detection algorithms to these variables, and presenting the results to the user through visualisations or natural language summaries.

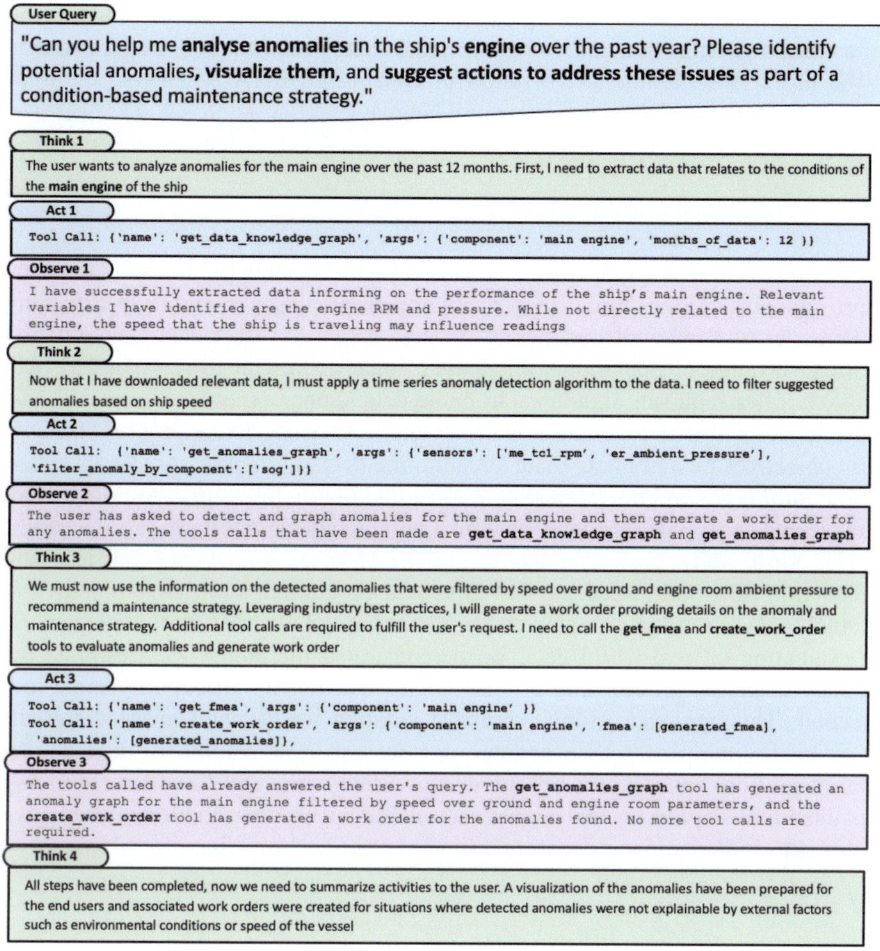

Fig. 3.7 Sample trajectory illustrating the step-by-step process of responding to a user query regarding anomalies in a ship engine. It demonstrates the integration of multiple tools—such as data retrieval, anomaly detection, and failure mode analysis—to detect, visualise, and recommend maintenance strategies. The process culminates in generating a work order based on detected anomalies, filtered by external factors like speed and ambient pressure, ensuring informed decision-making for condition-based maintenance

The agent planner module processes the user query and decomposes it into a series of subtasks. It identifies pertinent tool calls and the associated arguments. For example, it identifies that it needs to extract data related to the main engine, so it correctly calls the tool *get_data_knowledge_graph* with the appropriate arguments for *component* and *months_of_data*.

The agent proceeds through four iterations of think–act–observe (TAO in AI agents' parlance), to accomplish required subtasks such as data retrieval, applying anomaly detection model, getting failure modes and effects for the component, generating a visualisation and work order, and finally returning all the information to the user. This highlights the powerful capabilities of AI agents in terms of automation and user interaction in natural language.

3.5.3.5 Discussion and Conclusion

Decarbonising the shipping industry is a critical societal goal, with efficiency improvements through digital twin technologies playing a key role in achieving it. However, the inherent complexity of ship systems, vessel heterogeneity, and the diverse range of operational challenges make it difficult to scale digital twin solutions, often requiring significant in-house data science expertise. Agentic LLMs, when augmented with external tools, can significantly accelerate progress towards these objectives by automating and streamlining decision-making processes across various ship types and operational scenarios, making digital twin capabilities more scalable and accessible through a generic natural language interface.

This case study presented an agentic LLM-based framework for anomaly detection in shipping, leveraging real-time data integration, knowledge graphs, and external tools to support complex decision-making in industrial asset management. By evaluating on real-world data, we demonstrated the ability to streamline the anomaly detection process and provide contextual insights through dynamic tool selection. This approach offers a scalable solution to the challenges of digitalisation in the maritime sector, addressing issues such as machinery degradation, data integration, and operational efficiency. Future work will focus on enhancing explainability, improving tool orchestration, and integrating additional predictive analytics to further support decarbonisation and automation efforts in shipping.

3.6 Conclusions

To consolidate the chapter's discussion of machine learning methods, Table 3.2 provides a structured overview of key ML techniques currently applied or emerging in aquaculture. The table highlights representative use cases, the primary benefits these models provide in operational or research settings, and estimated Technology Readiness Levels (TRLs) based on reported deployments and commercial maturity.

The techniques included span both traditional and deep learning models, alongside recent advances such as graph neural networks, large language models, and AI agents. In addition, the table includes foundational approaches in time series modelling, unsupervised learning, and Bayesian inference, which are particularly relevant for data-scarce environments, uncertain decision contexts, or real-time predictive tasks.

Table 3.2 Key machine learning techniques applied in aquaculture, including typical use cases, benefits, and estimated TRLs based on current research and deployments

ML technique	Application in aquaculture	Benefits	TRL
CNNs	Fish behaviour and health monitoring via computer vision	Non-invasive, continuous visual monitoring	6–8
GNNs	Modelling interactions among environmental and biological systems	Captures system connectivity and spatial relationships	2–4
SVR	Forecasting dissolved oxygen using satellite or sensor data	Site-specific, accurate environmental predictions	7–8
LLMs	Natural language interfaces, log analysis, document summarisation	Scalable knowledge extraction across diverse sources	2–4
AI Agents	Autonomous task execution and adaptive response agents	Decision automation and reduced manual intervention	2–4
Random Forests / Decision Trees	Classification and anomaly detection in water quality or sensor data	Robust predictions with explainable outputs	7–9
RNNs/LSTMs	Time series forecasting of DO, temperature, or feeding needs	Handles temporal dependencies in aquaculture processes	5–7
Unsupervised Learning (e.g. Clustering)	Anomaly detection, behaviour clustering, farm typologies	Extracts structure without needing labelled data	4–6
Bayesian Models	Uncertainty-aware forecasting in small or sparse datasets	Probabilistic modelling and uncertainty quantification	3–5

Note: Technology Readiness Levels (TRLs) are indicative estimates based on literature, commercial trials, and reported deployments

By summarising these methods side by side, the table serves as a reference point for selecting appropriate ML approaches depending on the complexity of the aquaculture system, data availability, and deployment goals.

Machine learning is emerging as a key enabler in the digital transformation of aquaculture. From predictive analytics and real-time decision support to autonomous monitoring and control, ML technologies can enhance productivity, sustainability, and resilience across the value chain. However, realising this potential depends on improved data infrastructure, model interpretability, and collaboration between AI developers and aquaculture practitioners.

As aquaculture scales to meet global food demands, ML applications provide critical insights, from optimising feeding schedules and monitoring water quality to predicting potential environmental impacts. Emerging technologies such as Graph Neural Networks (GNNs), large language models (LLMs), and AI agents are driving this shift, enhancing traditional aquaculture practices with data-driven intelligence and paving the way for the digital twin's role in real-time, predictive management.

The use of computer vision, for instance, enables continuous and non-invasive monitoring of fish behaviour and health, providing an unprecedented level of observation that supports better decision-making and aligns with sustainability goals. At the same time, GNNs offer promising capabilities for modelling complex, interconnected systems, such as the relationships between water quality parameters, fish behaviour, and external environmental factors, which are often difficult to capture using conventional data analysis techniques.

LLMs and AI agents bring adaptability and scalability to the aquaculture industry. With their broad applicability across diverse datasets and environments, LLMs or foundation models allow for generalisation, enabling aquaculture operations to adopt sophisticated AI solutions with minimal fine-tuning. AI agents, on the other hand, facilitate autonomous decision-making processes, potentially transforming daily operations and reducing dependency on manual interventions.

As aquaculture embraces the digital twin concept, the role of AI will expand further into predictive analytics, real-time decision support, and complex ecosystem modelling, driving both operational efficiency and environmental responsibility. Ultimately, the convergence of machine learning and digital twin technology presents a robust framework for achieving sustainable, intelligent aquaculture practices capable of meeting the needs of a growing population while preserving aquatic ecosystems for future generations.

Acknowledgments The research presented in this chapter was supported by multiple sources of funding and institutional support. The work received funding from the European Union's Horizon Europe Research and Innovation programme under grant agreement no. 101157369.

The authors of Case Study 1 acknowledge project support and computing resources provided by DeepSense, as well as funding from the Mitacs Accelerate programme under grant IT33656.

Christopher Whidden is supported by a Natural Sciences and Engineering Research Council of Canada (NSERC) Discovery Grant (RGPIN-2021-02988).

The authors of Case Study 3 gratefully acknowledge support from the Horizon Europe programme under grant agreement no. 101056799.

References

Achiam J, Adler S, Agarwal S, Ahmad L, Akkaya I, Aleman FL, Almeida D, Altenschmidt J, Altman S, Anadkat S et al (2023) Gpt-4 technical report. arXiv preprint arXiv:230308774

Adegboye MA, Aibinu AM, Kolo JG, Aliyu I, Folorunso TA, Lee SH (2020) Incorporating intelligence in fish feeding system for dispensing feed based on fish feeding intensity. IEEE Access 8:91948–91960

Alvarez Fanjul E, Ciliberti S, Pearlman J, Wilmer-Becker K, Bahurel P, Ardhuin F, Arnaud A, Azizzadenesheli K, Aznar R, Bell M et al (2024) Promoting best practices in ocean forecasting through an operational readiness level. Front Mar Sci 11:1443284

American Bureau of Shipping (2015) Guidance notes on failure mode and effects analysis (FMEA) for classification. Technical report, American Bureau of Shipping, Texas

Armalivia S, Zainuddin Z, Achmad A, Wicaksono MA (2021) Automatic counting shrimp larvae based you only look once (yolo). In: 2021 international conference on artificial intelligence and mechatronics systems (AIMS). IEEE, pp 1–4

Azmat M, Madondo M, Bawa A, Dipietro K, Horesh R, Jacobs M, Srinivasan R, O'Donncha F (2023) Forecasting soil moisture using domain inspired temporal graph convolution neural networks to guide sustainable crop management. In: IJCAI 2023, vol 6, pp 5897–5905. ISSN: 1045-0823. https://doi.org/10.24963/ijcai.2023/654. https://www.ijcai.org/proceedings/2023/654

Ba A, O'Donncha F, Ploennigs J, Azmat M (2023) Efficient extraction of insights at the edges of distributed systems. In: 2023 IEEE international conference on big data (bigdata). IEEE, pp 1610–1619

Batur E, Maktav D (2018) Assessment of surface water quality by using satellite images fusion based on PCA method in the lake gala, turkey. IEEE Trans Geosci Remote Sens 57(5):2983–2989

Bauer P, Stevens B, Hazeleger W (2021) A digital twin of earth for the green transition. Nat Climate Change 11(2):80–83

Béné C, Arthur R, Norbury H, Allison EH, Beveridge M, Bush S, Campling L, Leschen W, Little D, Squires D et al (2016) Contribution of fisheries and aquaculture to food security and poverty reduction: assessing the current evidence. World Dev 79:177–196

de Bezenac E, Pajot A, Gallinari P (2017) Deep learning for physical processes: incorporating prior scientific knowledge. arxiv preprint arXiv:171107970. http://arxiv.org/abs/1711.07970. 1711.07970

Bi K, Xie L, Zhang H, Chen X, Gu X, Tian Q (2023) Accurate medium-range global weather forecasting with 3d neural networks. Nature 619(7970):533–538

Bommasani R, Hudson DA, Adeli E, Altman R, Arora S, von Arx S, Bernstein MS, Bohg J, Bosselut A, Brunskill E et al (2021) On the opportunities and risks of foundation models. arXiv preprint arXiv:210807258

Bonev B, Kurth T, Hundt C, Pathak J, Baust M, Kashinath K, Anandkumar A (2023) Spherical fourier neural operators: learning stable dynamics on the sphere. arXiv preprint arXiv:230603838

Bui S, Geitung L, Oppedal F, Barrett LT (2020) Salmon lice survive the straight shooter: a commercial scale sea cage trial of laser delousing. Prev Vet Med 181:105063

Burke M, Grant J, Filgueira R, Stone T (2021) Oceanographic processes control dissolved oxygen variability at a commercial Atlantic salmon farm: application of a real-time sensor network. Aquaculture 533:736143

Burt K, Hamoutene D, Mabrouk G, Lang C, Puestow T, Drover D, Losier R, Page F (2012) Environmental conditions and occurrence of hypoxia within production cages of Atlantic salmon on the south coast of newfoundland. Aquac Res 43(4):607–620

Cai K, Miao X, Wang W, Pang H, Liu Y, Song J (2020) A modified yolov3 model for fish detection based on mobilenetv1 as backbone. Aquac Eng 91:102117

Cao X, Liu Y, Wang J, Liu C, Duan Q (2020) Prediction of dissolved oxygen in pond culture water based on k-means clustering and gated recurrent unit neural network. Aquac Eng 91:102122

Chatziantoniou A, Spondylidis SC, Stavrakidis-Zachou O, Papandroulakis N, Topouzelis K (2022) Dissolved oxygen estimation in aquaculture sites using remote sensing and machine learning. Remote Sens Appl Soc Environ 28:100865

Cooney R, Wan AH, O'Donncha F, Clifford E (2021) Designing environmentally efficient aquafeeds through the use of multicriteria decision support tools. Current Opin Environ Sci Health 23:100276

DestinE (2024) Destination earth. https://destination-earth.eu/

Ducournau A, Fablet R (2016) Deep learning for ocean remote sensing: an application of convolutional neural networks for super-resolution on satellite-derived SST data. In: 2016 9th IAPR workshop on pattern recognition in remote sensing (PRRS). IEEE, pp 1–6. https://doi.org/10.1109/PRRS.2016.7867019. http://ieeexplore.ieee.org/document/7867019/

EDITO (2024) EDITO: European digital twin ocean. https://www.edito.eu/

Fernandes AF, Turra EM, de Alvarenga ER, Passafaro TL, Lopes FB, Alves GF, Singh V, Rosa GJ (2020) Deep learning image segmentation for extraction of fish body measurements and prediction of body weight and carcass traits in Nile tilapia. Comput Electron Agric 170:105274

Gaude GS, Borkar S (2019) Fish detection and tracking for turbid underwater video. In: 2019 international conference on intelligent computing and control systems (ICCS). IEEE, pp 326–331

Georgopoulou D, Vouidaskis C, Papandroulakis N (2024) Swimming behavior as a potential metric to detect satiation levels of European seabass in marine cages. Front Mar Sci 11:1350385

Giardina I (2008) Collective behavior in animal groups: theoretical models and empirical studies. HFSP J 2(4):205–219

Godfray HCJ, Beddington JR, Crute IR, Haddad L, Lawrence D, Muir JF, Pretty J, Robinson S, Thomas SM, Toulmin C (2010) Food security: the challenge of feeding 9 billion people. Science 327(5967):812–818

Guo H, Huang JJ, Zhu X, Wang B, Tian S, Xu W, Mai Y (2021) A generalized machine learning approach for dissolved oxygen estimation at multiple spatiotemporal scales using remote sensing. Environ Pollut 288:117734

Guy RD, Fogel ML, Berry JA (1993) Photosynthetic fractionation of the stable isotopes of oxygen and carbon. Plant Physiol 101(1):37–47

Hähnel P, Mareček J, Monteil J, O'Donncha F (2020) Using deep learning to extend the range of air pollution monitoring and forecasting. J Comput Phys 408:109278

Han F, Zhu J, Liu B, Zhang B, Xie F (2020) Fish shoals behavior detection based on convolutional neural network and spatiotemporal information. IEEE Access 8:126907–126926

Heimbach P, O'Donncha F, Garcia-Valdecasas JM, Arnaud A, Wan L (2024) Crafting the future: machine learning for ocean forecasting. State Planet Discuss 2024:1–11

Hu WC, Wu HT, Zhang YF, Zhang SH, Lo CH (2020) Shrimp recognition using ShrimpNet based on convolutional neural network. J Ambient Intell Humaniz Comput, 1–8. https://rdcu.be/eUOP2?

International Maritime Organisation (2023) IMO strategy on reduction of GHG emissions from ships. Technical report, International Maritime Organisation

James SC, Zhang Y, O'Donncha F (2018) A machine learning framework to forecast wave conditions. Coast Eng 137:1–10. https://doi.org/10.1016/J.COASTALENG.2018.03.004. https://www.sciencedirect.com/science/article/pii/S0378383917304969

Jawahar G, Sagot B, Seddah D (2019) What does Bert learn about the structure of language? In: ACL 2019-57th annual meeting of the association for computational linguistics

Jeong J, Revie CW (2020) Appropriate sampling strategies to estimate sea lice prevalence on salmon farms with low infestation levels. Aquaculture 518:734858

Jimenez CE, Yang J, Wettig A, Yao S, Pei K, Press O, Narasimhan K (2023) SWE-bench: Can language models resolve real-world GitHub issues? arXiv preprint arXiv:231006770

Karakaya N, Evrendilek F (2011) Monitoring and validating spatio-temporal dynamics of biogeochemical properties in Mersin Bay (Turkey) using landsat ETM+. Environ Monit Assess 181:457–464

Katsoulakos T, Tsiochantari G, O'Donncha F, Kaklamanis E, Maccari A, Mucharski M (2024) Shipping digital twin landscape. In: State-of-the-art digital twin applications for shipping sector decarbonization. IGI Global, Hershey, pp 1–25

Khan AB, Korus J, Sclodnick T, Whidden C (2024) Applying deep learning to predict adverse environmental conditions in fish aquaculture pens. In: OCEANS 2024-Halifax. IEEE, pp 1–8

Kim YH, Son S, Kim HC, Kim B, Park YG, Nam J, Ryu J (2020) Application of satellite remote sensing in monitoring dissolved oxygen variabilities: a case study for coastal waters in Korea. Environ Int 134:105301

Klinger D, Naylor R (2012) Searching for solutions in aquaculture: charting a sustainable course. Ann Rev Environ Resour 37:247–276

Krom M, Kress N, Brenner S, Gordon L (1991) Phosphorus limitation of primary productivity in the Eastern Mediterranean Sea. Limnol Oceanogr 36(3):424–432

Lam R, Sanchez-Gonzalez A, Willson M, Wirnsberger P, Fortunato M, Pritzel A, Ravuri S, Ewalds T, Alet F, Eaton-Rosen Z et al (2022) GraphCast: learning skillful medium-range global weather forecasting. arXiv preprint arXiv:221212794

Langbridge A, O'Donncha F, Ba A, Lorenzi F, Lohse C, Ploennigs J (2023) Causal temporal graph convolutional neural networks (CTGCN). arXiv preprint arXiv:230309634
Langbridge A, O'Donncha F, Rayfield J, Eck B (2024) Optimal transport for efficient, unsupervised anomaly detection on industrial data. In: 2023 IEEE international conference on big data (BigData), IEEE
Lewis K, Mitchell B, Van Dijken G, Arrigo K (2016) Regional chlorophyll a algorithms in the arctic ocean and their effect on satellite-derived primary production estimates. Deep Sea Res Part II Top Stud Oceanogr 130:14–27
Li X, Shang M, Hao J, Yang Z (2016) Accelerating fish detection and recognition by sharing CNNs with objectness learning. In: OCEANS 2016-Shanghai. IEEE, pp 1–5
Li P, Hua P, Gui D, Niu J, Pei P, Zhang J, Krebs P (2020) A comparative analysis of artificial neural networks and wavelet hybrid approaches to long-term toxic heavy metal prediction. Sci Rep 10(1):13439
Linardatos P, Papastefanopoulos V, Kotsiantis S (2020) Explainable AI: a review of machine learning interpretability methods. Entropy 23(1):18
Liu J, Shen C, O'Donncha F, Song Y, Zhi W, Beck HE, Bindas T, Kraabel N, Lawson K (2025) From RNNs to transformers: benchmarking deep learning architectures for hydrologic prediction. EGUsphere 2025:1–21
Lopes F, Silva H, Almeida J, Pinho C, Silva E (2017) Fish farming autonomous calibration system. In: OCEANS 2017-Aberdeen. IEEE, pp 1–6
Love DC, Allison EH, Asche F, Belton B, Cottrell RS, Froehlich HE, Gephart JA, Hicks CC, Little DC, Nussbaumer EM et al (2021) Emerging covid-19 impacts, responses, and lessons for building resilience in the seafood system. Glob Food Secur 28:100494
Maksimovic M (2018) Greening the future: green internet of things (G-IoT) as a key technological enabler of sustainable development. In: Internet of things and big data analytics toward next-generation intelligence. Springer International Publishing, Cham, pp 283–313
Måløy H, Aamodt A, Misimi E (2019) A spatio-temporal recurrent network for salmon feeding action recognition from underwater videos in aquaculture. Comput Electron Agric 167:105087
Mandal A, Ghosh AR (2024) Role of artificial intelligence (AI) in fish growth and health status monitoring: a review on sustainable aquaculture. Aquac Int, 32:2791–2820. https://doi.org/10.1007/s10499-023-01297-z
Maskell K (2023) The rise of machine learning in weather forecasting. https://www.ecmwf.int/en/about/media-centre/science-blog/2023/rise-machine-learning-weather-forecasting
Matear R, Hirst A (2003) Long-term changes in dissolved oxygen concentrations in the ocean caused by protracted global warming. Glob Biogeochem Cycles 17(4):1–20
Mavropoulou AM, Vervatis V, Sofianos S (2020) Dissolved oxygen variability in the Mediterranean sea. J Mar Syst 208:103348
Mohamed HED, Fadl A, Anas O, Wageeh Y, ElMasry N, Nabil A, Atia A (2020) MSR-YOLO: method to enhance fish detection and tracking in fish farms. Procedia Comput Sci 170:539–546
Monkman GG, Hyder K, Kaiser MJ, Vidal FP (2019) Using machine vision to estimate fish length from images using regional convolutional neural networks. Methods Ecol Evol 10(12):2045–2056
Morimoto T, Zin TT, Itami T (2018) A study on abnormal behavior detection of infected shrimp. In: 2018 IEEE 7th global conference on consumer electronics (GCCE). IEEE, pp 291–292
Mukkavilli SK, Civitarese DS, Schmude J, Jakubik J, Jones A, Nguyen N, Phillips C, Roy S, Singh S, Watson C et al (2023) Ai foundation models for weather and climate: applications, design, and implementation. arXiv preprint arXiv:230910808
Naylor RL, Hardy RW, Buschmann AH, Bush SR, Cao L, Klinger DH, Little DC, Lubchenco J, Shumway SE, Troell M (2021) A 20-year retrospective review of global aquaculture. Nature 591(7851):551–563
O'Donncha F, Grant J (2019) Precision aquaculture. IEEE Int Things Mag 2(4):26–30

O'Donncha F, Zhang Y, Chen B, James SC (2018) An integrated framework that combines machine learning and numerical models to improve wave-condition forecasts. J Mar Syst 186:29–36. https://doi.org/10.1016/J.JMARSYS.2018.05.006. https://www.sciencedirect.com/science/article/pii/S0924796317304827

O'Donncha F, Stockwell CL, Planellas SR, Micallef G, Palmes P, Webb C, Filgueira R, Grant J (2021) Data driven insight into fish behaviour and their use for precision aquaculture. Front Anim Sci 2:695054

O'Donncha F, Hu Y, Palmes P, Burke M, Filgueira R, Grant J (2022) A spatio-temporal LSTM model to forecast across multiple temporal and spatial scales. Ecol Inf 69:101687

O'Donncha F, Akhriev A, Fragoso B, Icely J (2024) Forecasting closures on shellfish farms using machine learning. Aquac Int 32(5):5603–5623

O'Donncha F, Sheehan JD, Touma M, Zemouri S, High RH (2024) Towards intelligent ship-edge computing enabling automated configuration of ship models and adaptive self-learning. In: State-of-the-art digital twin applications for shipping sector decarbonization. IGI Global, Hershey, pp 73–93

Oldham T, Nowak B, Hvas M, Oppedal F (2019) Metabolic and functional impacts of hypoxia vary with size in Atlantic salmon. Comp Biochem Physiol Part A Mol Integr Physiol 231:30–38

Oruche R, O'Donncha F (2023) Attention-based domain adaptation forecasting of streamflow in data-sparse regions. arXiv preprint arXiv:230205386

Oruche R, Egede L, Baker T, O'Donncha F (2021) Transfer learning to improve streamflow forecasts in data sparse regions. arXiv preprint arXiv:211203088

Parra L, Lloret G, Lloret J, Rodilla M (2018) Physical sensors for precision aquaculture: a review. IEEE Sens J 18(10):3915–3923

Raju KRSR, Varma GHK (2017) Knowledge based real time monitoring system for aquaculture using IoT. In: 2017 IEEE 7th international advance computing conference (IACC). IEEE, pp 318–321

Rao C, Sun H, Liu Y (2020) Physics informed deep learning for computational elastodynamics without labeled data. arXiv preprint arXiv:200608472

Rauf HT, Lali MIU, Zahoor S, Shah SZH, Rehman AU, Bukhari SAC (2019) Visual features based automated identification of fish species using deep convolutional neural networks. Comput Electron Agric 167:105075

Remen M, Oppedal F, Torgersen T, Imsland AK, Olsen RE (2012) Effects of cyclic environmental hypoxia on physiology and feed intake of post-smolt Atlantic salmon: initial responses and acclimation. Aquaculture 326:148–155

Remen M, Sievers M, Torgersen T, Oppedal F (2016) The oxygen threshold for maximal feed intake of Atlantic salmon post-smolts is highly temperature-dependent. Aquaculture 464:582–592

Saberioon M, Císař P (2018) Automated within tank fish mass estimation using infrared reflection system. Comput Electron Agricul 150:484–492

Saberioon M, Gholizadeh A, Cisar P, Pautsina A, Urban J (2017) Application of machine vision systems in aquaculture with emphasis on fish: state-of-the-art and key issues. Rev Aquac 9(4):369–387

Selvaraju RR, Cogswell M, Das A, Vedantam R, Parikh D, Batra D (2017) Grad-cam: visual explanations from deep networks via gradient-based localization. In: Proceedings of the IEEE international conference on computer vision, pp 618–626

Shevchenko V, Eerola T, Kaarna A (2018) Fish detection from low visibility underwater videos. In: 2018 24th international conference on pattern recognition (ICPR). IEEE, pp 1971–1976

Stamatis DH (2003) Failure mode and effect analysis: FMEA from theory to execution. Quality Press, Hershey

Stuchtey MR, Vincent A, Merkl A, Bucher M, Haugan PM, Lubchenco J, Pangestu ME (2023) Ocean solutions that benefit people, nature and the economy. In: The blue compendium: from knowledge to action for a sustainable ocean economy. Springer, Berlin, pp 783–906

Tamou AB, Benzinou A, Nasreddine K, Ballihi L (2018) Transfer learning with deep convolutional neural network for underwater live fish recognition. In: 2018 IEEE international conference on image processing, applications and systems (IPAS). IEEE, pp 204–209
Timms A, Langbridge A, O'Donncha F (2024) Agentic anomaly detection for shipping. In: NeurIPS 2024 workshop on open-world agents
Timms A, Langbridge A, Antonopoulos A, Mygiakis A, Voulgari E, O'Donncha F (2025) Agentic AI for digital twin. In: Proceedings of the AAAI conference on artificial intelligence, vol 39, pp 29703–29705
Tunstall L, Von Werra L, Wolf T (2022) Natural language processing with transformers. O'Reilly Media, Sebastopol
Tzilivakis K (2022) A tale of two oceans: scientists are building digital twins of the ocean | research and innovation. Eur Commiss https://ec.europa.eu/research-and-innovation/en/horizon-magazine/tale-two-oceans-scientists-are-building-digital-twins-ocean
United Nations Conference on Trade and Development (2023) Review of maritime transport 2023. In: Annual report, United Nations Conference on Trade and Development, New York
Vapnik V (2013) The nature of statistical learning theory. Springer, Berlin
Vaswani A, Shazeer N, Parmar N, Uszkoreit J, Jones L, Gomez AN, Kaiser Ł, Polosukhin I (2017) Attention is all you need. In: Advances in neural information processing systems, 30
Villon S, Mouillot D, Chaumont M, Darling ES, Subsol G, Claverie T, Villéger S (2018) A deep learning method for accurate and fast identification of coral reef fishes in underwater images. Ecol Inf 48:238–244
Voskakis D, Makris A, Papandroulakis N (2021) Deep learning based fish length estimation. an application for the Mediterranean aquaculture. In: OCEANS 2021: San Diego–Porto. IEEE, pp 1–5
Weng L (2023) LLM Powered autonomous agents. https://lilianweng.github.io/posts/2023-06-23-agent/. Section: posts
Wiewel S, Becher M, Thuerey N (2018) Latent-space physics: towards learning the temporal evolution of fluid flow. arxiv preprint arXiv:180210123 http://arxiv.org/abs/1802.10123. 1802.10123
Wolf T, Delangue C, Lapeyre M (2025) Hugging Face to sell open-source robots thanks to Pollen Robotics acquisition. https://huggingface.co/blog/hugging-face-pollen-robotics-acquisition
Xingjian S, Chen Z, Wang H, Yeung DY, Wong WK, Woo WC (2015) Convolutional LSTM network: a machine learning approach for precipitation nowcasting. In: Advances in neural information processing systems, pp 802–810. http://papers.nips.cc/paper/5955-convolutional-lstm-network-a-machine-learning-approach-for-precipitation-nowcasting
Yang Y, Dong J, Sun X, Lima E, Mu Q, Wang X (2017) A CFCC-LSTM model for sea surface temperature prediction. IEEE Geosci Remote Sens Lett 15(2):207–211
Yang L, Liu Y, Yu H, Fang X, Song L, Li D, Chen Y (2021) Computer vision models in intelligent aquaculture with emphasis on fish detection and behavior analysis: a review. Arch Comput Methods Eng 28:2785–2816
Zhao J, Gu Z, Shi M, Lu H, Li J, Shen M, Ye Z, Zhu S (2016) Spatial behavioral characteristics and statistics-based kinetic energy modeling in special behaviors detection of a shoal of fish in a recirculating aquaculture system. Comput Electron Agric 127:271–280
Zhao J, Bao W, Zhang F, Zhu S, Liu Y, Lu H, Shen M, Ye Z (2018) Modified motion influence map and recurrent neural network-based monitoring of the local unusual behaviors for fish school in intensive aquaculture. Aquaculture 493:165–175
Zhao S, Zhang S, Liu J, Wang H, Zhu J, Li D, Zhao R (2021) Application of machine learning in intelligent fish aquaculture: a review. Aquaculture 540:736724
Zhou C, Xu D, Chen L, Zhang S, Sun C, Yang X, Wang Y (2019) Evaluation of fish feeding intensity in aquaculture using a convolutional neural network and machine vision. Aquaculture 507:457–465

Chapter 4
Simulation Modelling and Analysis for Aquaculture

Morten Omholt Alver, Michael Hartnett, Francisco Bravo, and Fearghal O'Donncha

Abstract Simulation and computational modelling have become indispensable tools in aquaculture, enabling more informed decision-making across areas such as site selection, environmental interactions, management practices and daily farm operations. This chapter provides an overview of modelling techniques ranging from conceptual models to empirical and data-driven models, both deterministic and stochastic, and their applications within aquaculture. The chapter delves into the use of ocean circulation models, biogeochemical models, models at the farm and cage level, and data assimilation techniques to enhance model accuracy and predictive power. Key applications such as site selection, environmental impact assessment, water quality management, and carrying capacity estimation are examined in detail.

Through detailed case studies, the chapter illustrates the implementation of hybrid epidemiological models, canopy flow simulations, and sea lice population and transport models. These examples highlight the capacity of modelling to capture complex interactions involving aquaculture systems and inform management decisions. The discussion highlights the growing importance of digital twins and real-time data integration in advancing sustainable and resilient aquaculture practices. By synthesising current methodologies and future directions, the chapter underscores the potential of modelling in addressing the complex challenges facing the aquaculture industry.

M. O. Alver (✉)
Department of Engineering Cybernetics, Norwegian University of Science and Technology (NTNU), Trondheim, Norway
e-mail: morten.alver@ntnu.no

M. Hartnett
University of Galway, Galway, Ireland

F. Bravo
Facultad de Ingeniería y Ciencias, Universidad Adolfo Ibáñez, Santiago, Chile

F. O'Donncha
IBM Research Europe, Dublin, Ireland

F. O'Donncha et al. (eds.), *Digital Twin for Marine Fish Farming*, Aquatic Production Systems, https://doi.org/10.1007/978-3-032-10362-8_4

Keywords Mathematical models · Simulation · Ocean circulation models · Biogeochemical model · Environmental impact assessment · Data assimilation

4.1 Introduction

Mathematical models have grown to play a crucial role in deepening our understanding of both natural and industrial processes, and this is increasingly the case also for aquaculture. Models are commonly employed to simulate and predict various environmental factors, such as water temperature, salinity, currents, and nutrient levels, which are crucial for optimal fish growth and health. Other models are used to simulate farm infrastructures or physical and biological processes inside cages or tanks. Models assist in decision-making for site selection, management practices, and understanding the impacts of aquaculture on the surrounding ecosystem, as well as understanding and optimising daily farm operations. Models are appealing due to the reduced cost and ease of deployment compared to extensive field monitoring campaigns or laboratory-based scale modelling. In addition, a numerical model can be more easily modified to investigate different production conditions and scenarios. Perhaps equally important, models serve as valuable training tools for new professionals and play a critical role in preserving institutional memory and learning by encapsulating knowledge and expertise into their frameworks.

In this chapter, we delve into the role of simulation and computational modelling within the realm of aquaculture, considering both current state of the art and future research opportunities. This encapsulates key facets of fish farming, including site planning, the assessment of environmental impacts, and the enhancement of operational decision-making processes. We also discuss the integration of diverse observational data through data assimilation techniques, which serve to significantly augment the fidelity of the models used. A mathematical model is not an exact mathematical description encompassing all aspects of the real system; rather the aim is to incorporate only those features of the problem that are most relevant.

4.2 Modelling Applications in Aquaculture

Various types of models are commonly used in environmental applications, and they can be categorised in multiple ways. Letcher and Jakeman (2009) identified three broad categories: empirical, conceptual, and physics-based models. However, it is also useful to consider the different methodologies and approaches employed in modelling, which include the following:

1. *Empirical Models*: Based on observational data, these models predict future trends or events based on historical patterns. In aquaculture, they could, for instance, be used to forecast algal blooms or oxygen depletion events, helping in pre-emptive management to protect fish stock, or they could describe the

expected growth rate based on feed rate and composition, fish size, and temperature.

2. *Data-Driven Models*: These are models relying on collected data for a system's inputs and outputs, and possibly internal variables as well. Statistical methods can be applied to understand and interpret complex data sets. In aquaculture, statistical models could analyse environmental variables like temperature, salinity, and nutrient levels to optimise feeding schedules and improve yield efficiency. These are being widely extended nowadays with machine learning models that can better take advantage of the increase in observational data. Commonly used data-driven models include linear regression, logistic regression, generalised linear models, random forests, and models based on artificial neural networks. A comprehensive discussion of these models is provided in Chap. 7.
3. *Conceptual Models*: These are simplified representations of systems, providing a framework for understanding complex processes. In the context of aquaculture, a conceptual model might illustrate the interaction between a fish farm and its surrounding ecosystem, helping to assess potential environmental impacts.
4. *Deterministic Models*: These models predict a definite outcome based on a set of initial conditions and mathematical equations. The equations may be based on the laws of physics, with one example being ocean circulation models (OCM). Such models represent real-world physical processes through mathematical equations and simulations. In oceanography, they're crucial for simulating ocean currents, temperature, and salinity. For aquaculture, such models assist in identifying the best sites for farms by analysing conditions like water flow and temperature, which are vital for the health and growth of marine species, and can also describe dynamics at farm or cage scale such as dissolved oxygen conditions or feed distributions, e.g. Alver et al. (2024). Other deterministic models are based on less easily defined processes, such as growth models for fish in aquaculture, e.g. Stavrakidis-Zachou et al. (2019), which typically model the growth process in a highly simplified way.
5. *Stochastic Models*: Incorporating randomness, these models are useful where there is inherent uncertainty or variability. For aquaculture, stochastic models could be used to assess risks associated with environmental factors like storms or disease outbreaks, aiding in the development of risk management strategies. For example, stochastic models may be used to assess disease propagation or economic outputs for aquaculture (Aunsmo et al. 2010; Bravo et al. 2020). Stochastic models commonly include Markov processes, stochastic differential equations, and Monte Carlo simulations.

Physics-based models are central to the digital twin paradigm. Bauer et al. (2021) provide a vision for earth digital twins that combine a physical model with observations using a data assimilation framework. Data assimilation (covered in the Section "Data assimilation") is the process of using observations to correct the state of a mathematical model. Physics- or rule-based models guide environmental decision-making by providing accurate simulations of Earth system processes. For the aquaculture industry, these models can improve decision making in several ways, both within monitoring and forecasting of marine conditions, predicting the spread of pollutants and assessing the impact of environmental changes, and in monitoring processes at the farm level such as feeding, growth, and oxygen levels.

4.2.1 Ocean Circulation Models

Ocean circulation models (OCMs) form the basis for modelling of oceanographic conditions and transport processes around and between aquaculture farms. OCMs are crucial for understanding water circulation patterns and their implications for nutrient distribution and waste dispersal in aquaculture sites (Ibarra et al. 2014; Falconer et al. 2023). These models can assist in site selection, optimising farm design, and managing the risks of disease and harmful algal blooms, which are critical for the health and productivity of aquaculture operations (O'Donncha and Grant 2019).

OCMs are typically based on the primitive equations consisting of the continuity equation (relying on the approximation that water is incompressible), the Navier-Stokes equation (derived from Newton's second law), and a thermal energy equation. Physical variables computed by OCMs generally include temperature, salinity, sea surface elevation and horizontal and vertical current speeds. Figure 4.1 shows an example snapshot of surface elevation and surface currents from a simulation of the ocean model SINMOD (Slagstad and McClimans 2005) set up with 160 m horizontal resolution for the area outside of Trondheimsfjorden in Middle Norway.

OCMs are generally used to simulate large ocean areas at horizontal resolutions ranging from around 100 m to tens of kilometres. To make this computationally feasible, approximations such as hydrostatic conditions are often used, and turbulence closure is handled by using the Reynolds averaged Navier-Stokes equations with eddy viscosity representing the effect of turbulence below the scales represented by the model (Pond and Pickard 1983).

Examples of OCMs include ROMS (Regional Ocean Modelling System) (Shchepetkin and McWilliams 2005), Nucleus for European Modelling of the Ocean (NEMO) (Madec et al. 2024) and SINMOD (Slagstad and McClimans 2005). These models use regular horizontal grids, while other models such as FVCOM (Chen et al. 2012) use unstructured grids with variable resolution. These models use a variety of vertical discretisations ranging from coordinates with fixed vertical layer boundaries, coordinates where changes in surface elevation are distributed over the entire stack of layers and terrain following coordinates where the same number of vertical layers is used in the entire model domain.

As no two water systems are alike, parameters of mathematical models need to be adjusted to suit the local circumstances, or new parameterisations or mechanisms developed to describe the flow dynamics appropriately (Ji 2017). Hence, the accuracy of the numerical model is dependent on the parameter values used and on the formulation of terms in the fundamental equations of motion. It typically

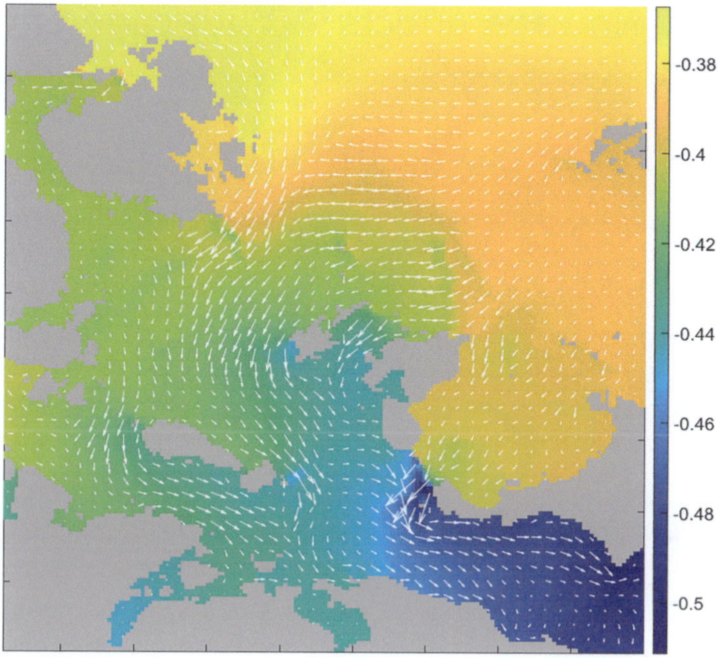

Fig. 4.1 Snapshot of surface elevation (m; heat map) and surface currents (white arrows; for clarity only every fourth point is shown) simulated by the SINMOD ocean model in a setup with 160 m horizontal resolution. The snapshot shows a situation where water is flowing toward and into a fjord (inlet to the lower right), driven by higher tidal elevation on the outside

requires a high degree of user expertise to accurately configure, implement, and interpret model results, as well as adequate observational data for parameterisation and validation.

Models can be used to describe specific interactions between fish cages and their marine environment. For instance, cages affect factors such as hydrodynamics, material transport, and sediment deposition. Figure 4.2 shows an example from a study on farm-flow interactions and the dispersal of material from fish cages (Broch et al. 2020). In the simulations, a tracer was released at the farm location. When the farm's effect on the local flow field is considered by the model, the simulations show higher tracer concentrations in the farm's vicinity, as the locally reduced currents disperse the tracer less efficiently. Such considerations are critical in understanding and mitigating the environmental impacts of marine farming, thereby contributing to the sustainability and efficacy of aquaculture practices.

OCMs may be coupled to biogeochemical models simulating the dispersion-reaction of abiotic and biotic variables such as nutrients, bacteria, phytoplankton, microzooplankton, zooplankton and detritus (Fennel et al. 2022). To simulate primary production, light input and attenuation also need to be modelled. Advection and diffusion processes for biological fields are computed based on model physics,

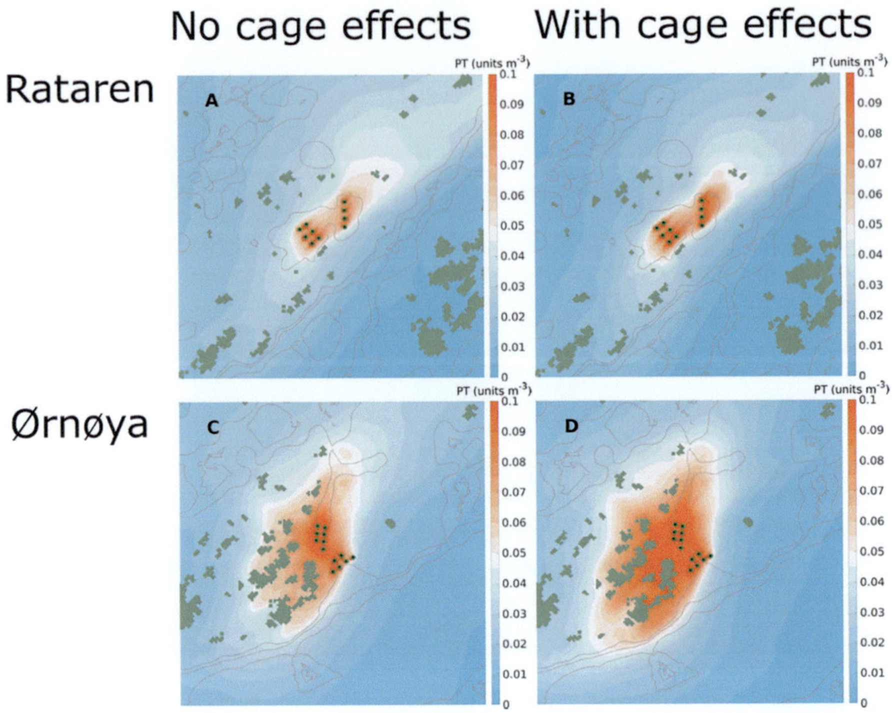

Fig. 4.2 Average simulated dispersal of passive tracer over a 2-month period from two farms in Mid-Norway (Rataren and Ørnøya) in two simulations with the SINMOD model. The "No cage effects" model uses the standard OCM with tracer releases, while the "With cage effects" model uses reduced porosity as well as a drag parameter at fish cage positions to account for the effect of the farms on the flow field. (Figure from Broch et al. (2020), reproduced under the terms of the CC-BY licence)

while biological rates are computed by the equations of the biogeochemical model. Within the large community ocean models, there are many examples of coupled biogeochemical models, such as the model of Fennel et al. (2006) coupled to ROMS (Rocha et al. 2019) or the PISCES model (Aumont and Bopp 2006) coupled to NEMO (Moreau et al. 2016), while the SINMOD model was designed from the start with a biogeochemical module (Wassmann et al. 2006). Marine ecosystems are highly complex, featuring diverse species even at the phytoplankton and zooplankton levels. To address this complexity, models like the CSIRO Environmental Modelling Suite (EMS)[1] adopt simplified yet sophisticated representations. These models incorporate multiple plankton types, spectrally resolved optical processes that simulate light absorption and scattering, variable carbon-to-chlorophyll ratios and remote-sensing reflectance calculations (Baird et al. 2020). Despite these

[1] https://research.csiro.au/cem/software/ems/

advances, many parameters remain challenging to measure experimentally, often spanning broad plausible ranges, which contributes to significant uncertainty in biogeochemical predictions (Fennel et al. 2022).

4.2.2 Site Selection

The process of site selection is integral to aquaculture, impacting production, environmental, economic and social aspects, and potentially affecting human health. Evidence-based site selection frameworks are a fundamental aspect of marine spatial planning (MSP), which is a place-based, multi-sectoral decision-making approach aimed at balancing the location, type, and intensity of ocean user activities to reduce conflicts and environmental impacts (Lester et al. 2018). A prevalent method for site selection is multi-criteria evaluation (MCE), which considers various factors like temperature, depth, and market proximity (Ross et al. 2011; Brigolin et al. 2017). This approach, using a weighted overlay system, involves a spatial assessment of biological, environmental, and socio-economic features relevant to potential sites. The weights in MCE are assigned based on the importance of each parameter, yielding both qualitative and quantitative outputs that are user-friendly for decision-makers (Falconer et al. 2016).

Modelling approaches are exemplified by the Halide et al. (2009) model for cage aquaculture site selection. This tool combines site classification with physical carrying capacity and economic analysis, evaluating parameters like water quality and socio-economic factors, categorising sites as poor, medium, or good. In Norway, the legally mandated Modelling-Ongrowing fish farms-Monitoring (MOM) system integrates environmental impact assessment, monitoring, and quality standards for salmon and trout mariculture. It recommends model-based environmental assessments predicting site impacts (Ervik et al. 1997; Lundebye 2013).

Site selection influences almost all aspects of aquaculture, including production and economic performance, environmental impact, social acceptability, and the location, may even have consequences for human health. It requires spatial assessment of the biological, environmental, and socio-economic characteristics relevant to an aquaculture site; includes consideration of the different levels of importance among parameters; and provides a qualitative and quantitative output which is useful and easy to understand for decision-makers.

4.2.3 Environmental Impact Assessment

Coastal aquaculture farms are often situated in estuaries and depend on the interaction of riverine and oceanic waters for flushing capacity. This capacity is crucial, as well-flushed systems result in lower environmental impacts and better conditions for cultured organisms compared to poorly flushed systems. Understanding the estuary's circulation, including the temporal and spatial variability of water circulation and mass fields like temperature and salinity, is key to choosing optimal aquaculture sites and enhancing the carrying capacity of the basin (Valle-Levinson 2013). Sheldon and Alber (2006) and Lucas (2010) outline various methods to estimate and define flushing characteristics of a bay or estuary. These include concepts like turnover or *flushing time*, *residence time*, *water age*, and *transit time*, each describing different processes affecting water or material renewal (dissolved or suspended) in an estuary.

Physics-based modelling, encompassing hydrodynamic, biogeochemical, and ecophysiological models, plays a crucial role in evaluating impacts of farms (Dabrowski et al. 2013; O'Donncha et al. 2013). These models simulate the movement and dispersal of nutrients, waste, and chemicals from fish farms into the surrounding water bodies, providing insights into potential eutrophication, oxygen depletion, and impacts on local flora and fauna. By simulating different farming scenarios, these models assist in identifying the impacts of operating conditions, thereby helping to minimise environmental impact and develop sustainable aquaculture practices. Some examples of the use of numerical models for environmental impact assessment include the following:

1. *Sediment Impact Modelling*: These models focus on the impact of aquaculture waste, such as uneaten food and faeces, on the sediment quality of oceans or lakes (Cromey et al. 2002; Brigolin et al. 2009; Bravo and Grant 2018). Understanding sediment transport dynamics is essential, as many aquatic contaminants are hydrophobic and bind to sediments. Numerical modelling of erosion, deposition, and sediment movement is key to quantifying their impact on benthic life and overall aquatic ecosystem health, playing a critical role in environmental assessment (James et al. 2010; Ancey 2020).
2. *Chemical Dispersion Analysis*: Numerical models are essential in aquaculture for predicting the spread of chemicals like medicines and antifoulants, crucial in assessing chemical pollution risk and its impact on nontarget species. These models are particularly important for understanding the dispersion of Veterinary Medicinal Products (VMPs) used in intensive aquaculture to combat bacterial and parasitic infestations. There are two main types of Environmental Risk Assessment (ERA) models for marine cage systems: one assessing the dilution and dispersal of bath treatment chemicals (antifungals and some antiparasitics) and another focusing on particle tracking for in-feed medications (antiparasitics, antimicrobials), considering their dispersal through both water and sediment. This is extensively reviewed in Rico et al. (2019).
3. *Disease Spread Modelling*: In fish farming, the spread of diseases and parasites is a significant concern. Models are used to simulate the transmission dynamics of pathogens, both within individual farms and between farms. This modelling assists in creating effective disease control strategies and biosecurity measures. Farmed finfish are susceptible to various diseases, including viral infections like Infectious Salmon Anemia (ISA); see, for instance, Scheel et al. (2007), Pancreas

Disease (PD)—see Bang Jensen et al. (2021) and example of results in Fig. 4.3, Viral Haemorrhagic Septicaemia (VHS), bacterial diseases such as yersinosis and flavobacteriosis, parasitic infestations including sea lice and Amoebic Gill Disease (AGD), and fungal infections like saprolegniosis (Noble et al. 2018).

4. *Ecological Impact Studies*: Numerical models effectively simulate the extensive ecological effects of fish farming, examining impacts on predator-prey dynamics, competition with wild species, and overall biodiversity. These models facilitate exploration of different farming strategies, including site selection and management, on a broad range of ecological aspects. This encompasses evaluating impacts on native species, biodiversity, habitat restoration, ecosystem services, invasive species, pollutants, multiple stressors, and climate change effects, both spatially and temporally (Schuwirth et al. 2019). Incorporating ecological modelling into environmental decision-making is challenging due to the need for collaboration between decision-makers and modellers for effective knowledge transfer and technical resolution. This complexity often leads to ecological models being confined to research rather than applied in practical environments (Schmolke et al. 2010; Schuwirth et al. 2019).
5. *Scenario Analysis and Management Strategies*: In aquaculture, a key role of modelling is enabling stakeholders to assess different scenarios and the impacts of various management strategies. Central to the digital twin concept, as outlined by Bauer et al. (2021), these models yield reliable predictions and facilitate comprehensive scenario testing. They assume multidimensional representation, covering environmental, ecological, production, and economic impacts. Simulating scenarios like feed-type variations or farm location changes, these models are instrumental in devising and executing more eco-friendly fish farming. They play an important role in aspects such as farm planning (Couture et al. 2021) and welfare management (Urke et al. 2021).

Coastal aquaculture necessitates models of high fidelity and precision. Operationally, this requires advanced skill in model parameterisation and a substantial quantity of observational data for configuration, parameter tuning, and model refinement. Typically, these modifications are executed within a data assimilation framework, highlighting the need for expert proficiency and extensive data for maintaining the model's accuracy and applicability in complex coastal environments (Ragnoli et al. 2012; Edwards et al. 2015; Dabrowski et al. 2016).

4.2.4 Carrying Capacity

Aquaculture management aims to use tools that predict and measure an area's capacity to support cultured species, with a central focus on the concept of carrying capacity. This encompasses the upper limits of production, ecological sustainability, and social acceptability without causing significant change to ecosystems and social structures (Byron and Costa-Pierce 2013; Kluger and Filgueira 2021). Environmental impact assessment is therefore one of several aspects relevant to carrying capacity.

Ferreira et al. (2013) distinguish between carrying capacity, defined as "the biomass of cultured organisms that can be supported without altering system state or function, measured by water or sediment quality standards and processes," and assimilative capacity, described as "the ability of biological systems in the water column or sediments to process organic matter, nutrients, therapeutants, or contaminants without alteration of ecosystem state or function.".

Inglis et al. (2000) and McKindsey et al. (2006) identified four types of aquaculture carrying capacities: physical, production, ecological, and social. While they noted that research predominantly focuses on production carrying capacity, defined as the maximum sustainable yield within a given area, recent efforts have emphasised ecological carrying capacity. This is defined as the level of aquaculture production that does not cause significant changes to ecological processes, species, populations, or communities (Sequeira et al. 2008). Weitzman and Filgueira (2020) trace the evolution and application of carrying capacity concepts, identifying the need for integrative, multidimensional approaches that encompass environmental, economic, and social aspects.

Terms like "ecoaquaculture" and "ecological aquaculture" underscore the growing emphasis on sustainability (Weitzman and Filgueira 2020). Originally applied to bivalve aquaculture, these definitions have also been adapted for finfish cage culture under the term "assimilative capacity" (Geček and Legović 2010).

Modelling plays a significant role in evaluating carrying capacity of water bodies in aquaculture contexts. For instance, McKindsey et al. (2006) reviewed recent carrying capacity models for bivalve culture and provided recommendations for research and management. For non-intensive aquaculture of filter feeders. Additionally, Zhao et al. (2022) analysed ecological carrying capacity of bivalve aquaculture within a management framework. Specific applications, such as sediment assimilative capacity, were defined by Bravo and Grant (2018) in the context of Canadian environmental regulations as "the gross deposition rate of organic wastes that maximises total organic carbon degradation rates while preventing sediment sulfide accumulations in surface sediments above regulatory limits".

These studies highlight the need for context-specific definitions to address the diverse environmental, social, and economic aspects or dimensions involved, and the importance to compute the carrying capacity before fish farms are established in a region (Stigebrandt 2011). Many modelling studies do this through budgeting of nutrient depletion and renewal from marine processes and farm operations (Raillard and Ménesguen 1994; Pilditch et al. 2001; Simpson et al. 2007). Byron and Costa-Pierce (2013) provide a comprehensive evaluation of carrying capacity studies in aquaculture.

4.2.5 Water Quality Management

Numerical models are essential for evaluating and controlling water quality in fish farming, particularly in real-time scenarios where it's vital to maintain water quality not only for optimal fish growth and health but also to comply with regulatory standards. The importance of these models is heightened by the fact that water quality at coastal aquaculture sites is influenced by a combination of on-farm activities, open-water dynamics, weather patterns, and surface runoff from land-based activities. There are numerous ways in which models can contribute to effective water quality management:

1. *Predictive Modelling of Water Parameters*: Models can predict key water quality parameters like dissolved oxygen, temperature, pH, ammonia, nitrates, and phosphates. By forecasting these parameters, fish farmers can anticipate and mitigate potential problems before they become critical (Ascione Kenov et al. 2014).
2. *Real-Time Monitoring and Decision Support Systems*: Integration of numerical models with real-time monitoring systems (like sensors and IoT devices) allows for the continuous assessment of water quality. Decision support systems can use this data to provide instant feedback and recommendations to farmers. For example, automated alerts can be set up for when certain parameters deviate from optimal ranges (O'Donncha et al. 2013; Cerqueira et al. 2024).
3. *Impact Assessment of External Factors*: Numerical models can assess the impact of external factors like weather conditions, runoff from nearby land, and seasonal variations on water quality (Tilman et al. 2001; Webber et al. 2021).
4. *Adaptive Management Strategies*: By continually updating with real-time data, models can help in developing adaptive management strategies that respond to changing conditions in the water, ensuring optimal water quality at all times. The ecosystem approach to aquaculture requires adaptive management approaches that are resilient to the dynamics of change and uncertainty inherent in ecosystems (Dietz et al. 2003).

4.2.6 Processes at Farm and Cage Level

Conditions in cage aquaculture farms are in part governed by the surrounding environment, but there are several processes at the farm or cage scale that are key to understanding the behaviour, welfare, and growth of the fish. In the following, we will discuss those processes briefly and look at how they can be addressed through mathematical models.

The infrastructure at a fish farm serves both to keep the fish in captivity and to house the people working at the farm and the equipment needed to operate the farm. The cages and equipment must be designed to ensure the welfare of the fish even under rough environmental conditions. Several challenges related to net cages can be addressed through mathematical modelling. First, net cages cause a reduction in current speed compared to the outside current, which can be an advantage under high currents and a disadvantage under low currents with limited water exchange. Biofouling can make the cage net less permeable, causing further reduction in inside current speeds. Many studies have sought to quantify these effects, e.g. using computational fluid dynamics (CFD) models (e.g. Cornejo et al. 2020). Second, the interaction between the water flow and the cage net also means that net cages get deformed under strong currents, leading to reduced volume inside the cages with potential negative consequences for fish welfare. Models of net cage deformation (Klebert et al. 2015; Su et al. 2021) can give quantitative estimates of the rate of deformation given the cage design and hydrodynamical conditions (see example in Fig. 4.3).

Feeding is one of the most important processes taking place in an aquaculture farm, as the feed is the single greatest cost factor in fish aquaculture, exceeding 50% of costs for salmon (Asche and Oglend 2016), and efficient transfer of feed into the gut of the fish is the primary driver for biomass growth. A wide range of mathematical models address feeding and growth processes. Growth models include empirically based models like Gompertz, von Bertalanffy, and thermal-unit growth coefficient, and mechanistic energetic models such as dynamic energy budget (DEB) models (Chary et al. 2022). DEB models integrated into aquaculture models help predict how changing temperature and ocean acidification scenarios will affect growth rates and harvest yields (Sarà et al. 2012). To predict growth, one needs to quantify the rate of feed intake, which can be modelled using relations such as the Holling functional responses (Holling 1965), but feed intake is a complex process affected by a large set of variables (Sun et al. 2016). Feeding also has a spatial and temporal aspect, as the feed needs to be distributed to aquaculture units, and the fish needs to catch each pellet within the time window it is available. This is emphasised in a spatially resolved pellet distribution model coupled to either a simplified feed ingestion model or to a spatially explicit individual based behaviour model for Atlantic salmon (Alver et al. 2016; Føre et al. 2016).

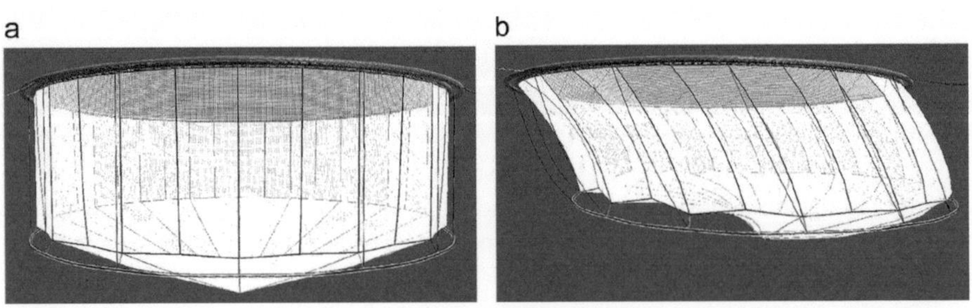

Fig. 4.3 Model simulation of the deformation of an open sea cage: (**a**) initial volume for current speed = 0 m/s and (**b**) current speed = 0.5 m/s (Klebert et al. 2015). (Reproduced with permission)

Fish behaviour is an important factor not just in the feeding process but also when considering farm infrastructure, farming operations, and environmental effects on fish welfare. Individual-based models of behaviour can be formulated based on a set of assumptions about the perception of the fish and their movements based on a set of variables such as nearby fish, feed, light intensity, temperature, and current (Føre et al. 2009, 2013). Both for the purpose of model formulation and tuning, and for validation, behavioural data can be gathered using tools like echosounders and acoustic telemetry, as reviewed by Oppedal et al. (2011).

Environmental conditions in cages are strongly governed by ambient conditions. For variables like temperature and salinity, conditions will be almost equal to the surrounding water. However, dissolved oxygen is different, as there is a strong local effect of the fish's oxygen consumption. It is important to ensure adequate levels of dissolved oxygen, since hypoxia can lead to effects ranging from reduced appetite, stress responses, reduced feed conversion, and reduced growth to acute mortality depending on the oxygen levels encountered (Oppedal et al. 2011; Remen et al. 2013, 2016). The availability of oxygen depends not just on ambient oxygen levels, current speeds, and the permeability of the net cages but also on the distribution and density of the fish biomass, the temperature-dependent oxygen consumption rate of the fish, feeding and size distribution, and on any active addition of oxygen by the farmer. Due to these factors, there is also strong spatial variability of oxygen levels inside cages (Solstorm et al. 2018).

Several studies have sought to establish mathematical models relating these factors to the oxygen levels in sea cages (Alver et al. 2022, 2023; Bergsson et al. 2023); see example in Fig. 4.4. In the study by Alver et al. (2022), the advection-diffusion equation was used for a discretised grid of oxygen concentrations in the cage, coupled to measurements of temperature, current speeds, and directions along with an empirical model estimating the oxygen consumption rate of individual salmon (Grøttum and Sigholt 1998) and a model describing in a simplified way the feeding behaviour of the salmon (Alver et al. 2016).

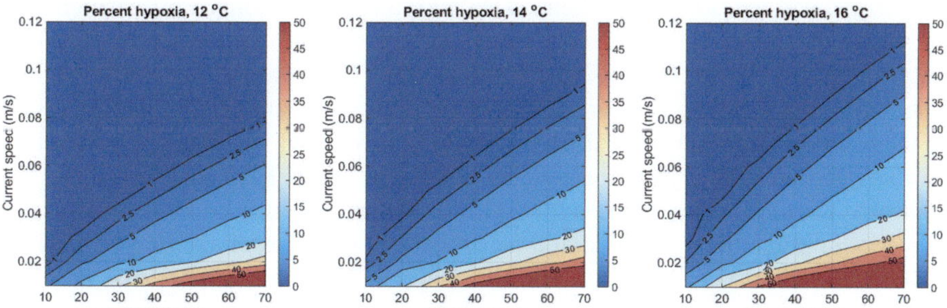

Fig. 4.4 Model-based estimates of dissolved oxygen (DO) in a single sea cage with 100% ambient oxygen saturation. The heat maps show the estimated percentage of the cage volume in hypoxic conditions (here defined as DO level below 6 mg l^{-1}) during feeding for three temperatures as a function of cage scale and current speed. (Reproduced from Alver et al. (2023) under the terms of the CC-BY license)

4.3 Observations and Data Assimilation

Mathematical models have great value as tools for analysing and understanding the dynamics of systems, as well as for making predictive scenarios. However, all models have simplifications and limitations to how well they represent the real systems, and their dynamics will always to a greater or lesser degree deviate from reality. When models are used in an offline setting, for instance, as part of the system design for new aquaculture sites or development of new methods or production concepts, the modeller needs to ensure that the model structure and its parameter values give a sufficiently accurate model that captures the system dynamics that are important in that specific application. This involves the use of observation data for model tuning and for validation.

If the models are used in a real-time setting, e.g. as part of a Digital Twin setup (Rasheed et al. 2020; Føre et al. 2024), a well-designed and validated model is not sufficient. For most systems, even a high-quality model will tend to deviate from the state of the real system over time. To ensure fidelity in model predictions, observational data needs to be transmitted and processed in (near) real time for model correction, enabling the evaluation of field observations as they are captured. This is known as *state estimation* or *data assimilation*. The main task of data assimilation techniques is to provide a mapping from measurements on the real system to corrections that improve the model state. Most systems are undersampled in the sense that they have more state variables than available measurements. For example, a model of the 3D distribution of dissolved oxygen in an aquaculture farm may contain thousands of state variables, but one cannot realistically measure the oxygen level at more than a few points at a time. There are two widely used approaches to data assimilation; the Kalman Filter-based methods and variational assimilation methods such as 4DVAR (Thépaut 1993).

In Kalman Filter-based methods, model corrections to state variables are calculated at the time of one or more observations. The optimal way of relating measured model deviations is found using an estimate of the uncertainty of state variables and the error covariances between them. Also taking the uncertainty of measurements into account, model corrections are calculated to optimise the reduction of model uncertainty. For linear systems, the Kalman Filter (Kalman 1960) can be shown to be an optimal estimator. For nonlinear systems, there are several variants such as the extended Kalman filter (Jazwinski 2007), unscented Kalman filter (Wan and Van Der Merwe 2000), and, for systems with large numbers of state variables, the ensemble Kalman filter (Evensen 2003) illustrated in Fig. 4.5.

Variational methods are based on an optimisation approach, using observations over a time interval to compute the initial condition for the model that best fits both the model forecast and the observed data according to a cost function. The process requires an *adjoint model* derived from the original model equations, which computes the gradient of the cost function with respect to the initial state. Essentially, it tells us how changes in the initial state affect the cost function and thereby can guide the optimisation process.

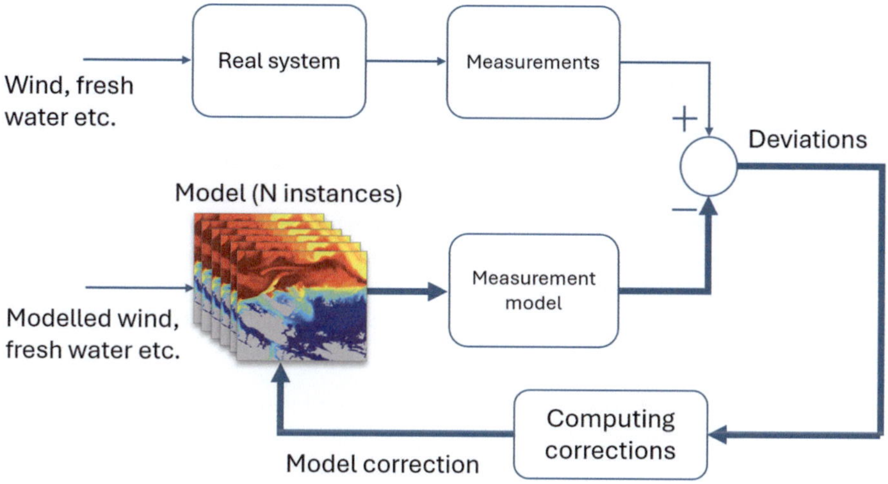

Fig. 4.5 Flow chart of ensemble Kalman filter-based assimilation for an ocean circulation model. The system is simulated in an ensemble of N instances with perturbations representing system and input uncertainty. A measurement model is used to extract model estimates corresponding to the variables measured in the real system, and the ensemble variability is used in the Kalman Filter equations to calculate model corrections based on the recorded deviations

Regardless of which approach is taken, data assimilation significantly increases the computational cost of model simulations, potentially by an order of magnitude or more. For computationally heavy models such as ocean models, this can be a challenge, and one is typically forced into a trade-off of the quality of data assimilation vs. the available time and computational resources. In some cases, simpler *nudging* approaches (Lakshmivarahan and Lewis 2013) may be a good alternative. Nudging is a data assimilation approach that uses the forecast error (the difference between model predictions and measurements) to correct the model by pushing the model variables toward observed values. When applicable, nudging is computationally efficient and straightforward. An important limitation is that it requires sufficient measurements to be available, as there is no simple way to compute corrections to non-measured model variables.

4.4 Case Studies

In this section of the chapter, we delve into three detailed case studies, each focusing on a specific application or application area of mathematical modelling within the aquaculture industry. These case studies aim to provide a comprehensive understanding of how mathematical modelling can be employed to address various challenges and optimise operations in aquaculture. By examining each case study, we highlight the practical implementations, benefits, and potential of mathematical models in improving aquaculture practices.

The three case studies presented are the following:

1. Flow through aquaculture canopies: This case study explores the impacts of suspended aquaculture installations on flows and material transport using amended numerical modelling approach.
2. Risk management systems in the Chilean aquaculture industry: This case study explores epidemiological risk management in salmon farming using a hybrid deterministic and probabilistic approach.
3. Sea lice models: This case study examines the role of mathematical modelling to manage sea lice outbreaks on salmon.

Each case study provides a detailed analysis of the specific problem, the mathematical approach used, and the outcomes achieved. Through these real-world examples, we illustrate the critical role of mathematical modelling in advancing aquaculture and addressing its multifaceted challenges.

4.4.1 Case Study 1: Flow Through Aquaculture Canopies

4.4.1.1 Introduction

This case study builds upon and adapts the work of O'Donncha et al. (2015), which examined hydrodynamic interactions within suspended aquatic canopies using the EFDC model framework. Suspended structures such as longlines, rafts, and cages are widespread in aquaculture, and their influence on local hydrodynamics affects sediment transport, nutrient dispersion, and water quality (Ghisalberti and Nepf 2002). While extensive research has explored flow through submerged or emergent vegetation (Nepf and Vivoni 2000), suspended canopy systems—where flow develops both above and below the structure—remain less well studied.

In the present case study, suspended canopy effects were incorporated into a three-dimensional hydrodynamic model to assess their impact on flow resistance and turbulence generation. The study aimed to parameterise canopy-induced drag and turbulence production and to evaluate the model's predictive capability using laboratory and field-scale applications relevant to aquaculture.

4.4.1.2 Methodology

The hydrodynamic simulations were performed using the US EPA's Environmental Fluid Dynamics Code (EFDC), which solves the three-dimensional, hydrostatic, free-surface Reynolds-averaged Navier–Stokes equations with coupled transport and turbulence closure (Ji 2017). Following the approach detailed in O'Donncha et al. (2015), the model's turbulence representation—originally based on the

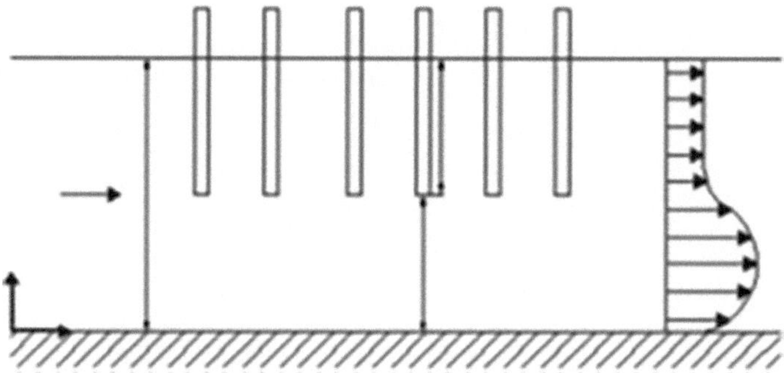

Fig. 4.6 Schematic of suspended canopy and vertical velocity profile

Mellor–Yamada Level 2.5 scheme—was reformulated in terms of the General Length Scale (GLS) framework (Warner et al. 2005). This modification enabled the inclusion of canopy-related turbulence production terms through calibrated weighting coefficients, reflecting the enhanced shear and mixing within and below the canopy layer.

A vertically varying drag coefficient was introduced to represent spatial variations in form drag along canopy elements. Previous experimental studies of flow past submerged vegetation (e.g. Dunn et al. 1996; Ghisalberti and Nepf 2004) have shown that drag is not uniform along the element length but typically peaks within the canopy interior. For suspended canopies, the structure was assumed to behave analogously to an inverted vegetated canopy, with parameters adjusted to reflect boundary conditions and empirical flow data.

The model framework thus incorporated three principal flow regions: (i) a bottom boundary layer adjacent to the bed, (ii) a canopy shear layer where turbulence is generated by shear between canopy and underlying flow, and (iii) an internal canopy layer characterised by local turbulence induced by structural elements. Calibration was undertaken using controlled flume data, ensuring the drag and turbulence coefficients reproduced observed mean velocity and turbulence profiles.

Figure 4.6 presents the general configuration including a schematic representation of flow past a canopy structure.

4.4.1.3 Results

Model performance was evaluated against the laboratory dataset of (Plew 2011), which provides detailed velocity and turbulence measurements for suspended canopy arrays. The revised EFDC model successfully captured key flow structures, including the development of the canopy shear layer and associated turbulence enhancement. For high-density canopy configurations, the model reproduced vertical velocity and shear stress profiles with strong agreement to observations, validating the use of a vertically variable drag profile combined with the modified

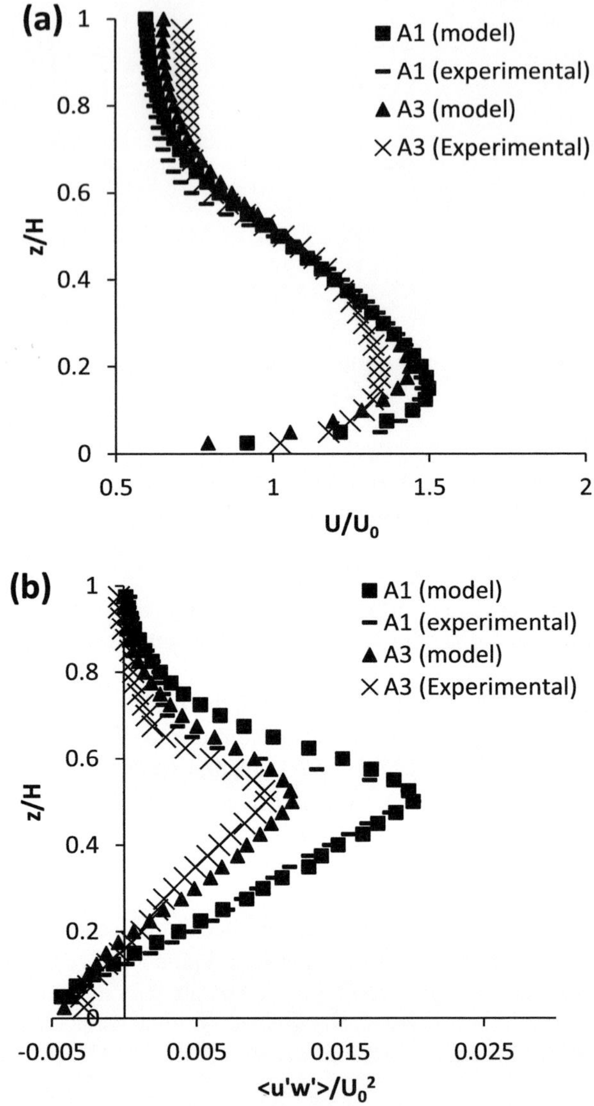

Fig. 4.7 Comparison between model and observed vertical profiles of velocity (top) and turbulent shear stress (bottom) for low- and high-density canopy configurations. (Adapted and redrawn from O'Donncha et al. (2015))

turbulence scheme. Figure 4.7 presents the velocity and turbulent shear stress profiles calculated using this drag profile with the amended turbulence scheme.

Following laboratory-scale validation, the model was applied to Killary Harbour, a fjord on Ireland's west coast that hosts extensive rope mussel aquaculture. The harbour exhibits a partially mixed vertical structure influenced by tidal exchange, freshwater inflow, and seasonal stratification. The simulation domain was constructed from bathymetric and hydrographic datasets, discretised into 20 vertical layers and horizontally at 64 m resolution. River inflows and meteorological forcing were prescribed based on available observations (O'Donohue 2012).

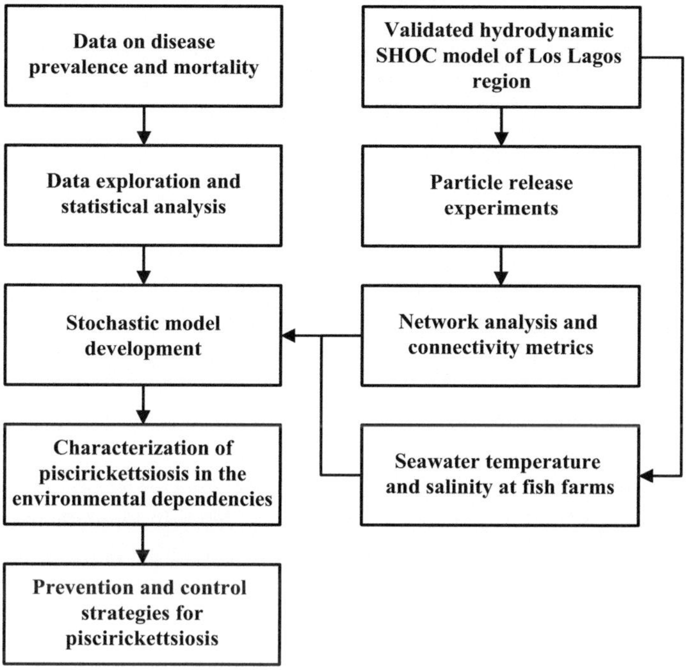

Fig. 4.8 Workflow for a stochastic model development to explore environmental dependencies of piscirickettsiosis in open-cage marine salmon farms. SHOC Sparse Hydrodynamic Ocean Code (Herzfeld 2006). (Source: Bravo et al. (2020))

The model was used to quantify how aquaculture structures influence material transport and flushing. Tracer experiments compared residence times with and without mussel longlines. Results demonstrated that the inclusion of canopy-induced drag significantly increased local residence times within farmed regions, altering transport pathways and mixing efficiency throughout the harbour. Figure 4.8 presents a spatial distribution of the time required to flush 75% of a conservative constituent throughout the fjord without and without mussel lines. Clearly, the presence of the mussel lines affects solute transport and removal.

4.4.1.4 Discussion

Simulations estimated an average residence time of about 15 days under mean river flow conditions, rising to roughly 30 days when river discharge was excluded. Spatial variability was pronounced, with efficient flushing at the outer harbour but significantly longer residence times in mid-harbour regions containing aquaculture installations.

These results highlight the hydrodynamic role of suspended structures in modifying transport and water exchange. Neglecting their drag and turbulence effects can lead to substantial overestimation of system assimilative capacity, which in turn affects assessments of environmental carrying capacity and nutrient dynamics (Dame and Prins 1997). Incorporating canopy effects into hydrodynamic models is therefore essential for reliable evaluation of aquaculture impacts and sustainable management.

4.4.2 *Case Study 2: Hybrid Stochastic-Deterministic Models to Advance Piscirickettsiosis Surveillance and Risk Management Systems in the Chilean Salmon Farming Industry*

4.4.2.1 Introduction

The main challenge of epidemiological risk management in salmon farming is the early detection and suppression of diseases. Hybrid models that combine deterministic (mechanistic) and probabilistic (stochastic) modelling approaches and outputs emerge as a crucial tool in this pursuit, offering an enhanced grasp of disease dynamics. These models go beyond incorporating deterministic trends, such as seawater temperature, salinity, and hydrodynamics patterns around and among farms. They also address the inherent stochastic variability associated with environmental, biological, and farming processes, which may be unknown, uncertain, or poorly understood. Examples include the survival of pathogens in seawater, as well as the influence of fish farm practices at multiple scales, ranging from the cage and farm scale to the bay or other spatial management scales.

Hybrid models are particularly relevant for vector-borne diseases or infections influenced by exogenous variables such as seawater temperature and salinity, as well as endogenous variables associated with diverse farming practices (e.g. stocking date, stocking density, response to disease outbreaks, among others). They can improve the model's ability to recognise patterns in monitoring data and simplify complex and intricate connections.

Integrating deterministic and probabilistic models with real-time data from sanitary and environmental monitoring programs enables prompt responses to outbreaks, preventing their escalation. This approach also supports the development of early warning systems, empowering timely actions to address potential outbreaks effectively.

4.4.2.2 Methodology

As a case study, we examined research conducted by Bravo et al. (2020). The study investigated the influence of oceanographic conditions, including the circulation patterns of seawater among farms (i.e. hydrodynamic connectivity), seawater temperature, and salinity, on the dynamics of piscirickettsiosis (SRS) at salmon

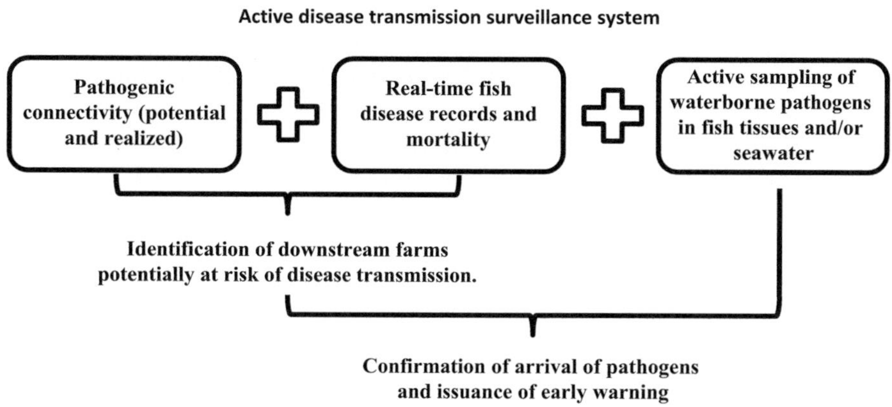

Fig. 4.9 Active disease transmission surveillance system design for Piscirickettsia salmonis at salmon farms in Chile

farms in the Los Lagos region of southern Chile. The research combined aquaculture records of fish mortality due to SRS with oceanographic modelling data within a stochastic, state-based, time-dependent epidemiological model to explore environmental drivers of SRS and provided risk metrics of waterborne transmission of the pathogen among farms (see Fig. 4.9 for a schematic of the approach). The study demonstrated that the number and extent of disease outbreaks in upstream farms, seawater salinity, and temperature were strong determinants of SRS prevalence. Findings indicated that farms downstream from infected farms had close to 100% disease prevalence by the end of the farming cycles, while farms with little or no exposure to upstream infected farms had only around 10% prevalence of SRS.

Building on this case, Fig. 4.9 presents an example of the designing of an active disease transmission surveillance system for piscirickettsiosis in Southern Chile. The system combines knowledge of modelled hydrodynamic connectivity among farms with real-time disease records within a probabilistic modelling framework. This integration aids in identifying downstream farms at risk of waterborne transmission of SRS (realised pathogenic connectivity). The probability of pathogen emergence can guide sampling efforts (Olivares and Marshall 2010) to enhance aquaculture monitoring activities and to improve early warnings.

4.4.2.3 Conclusion

Predicting exposure to aquatic disease transmission at the farm level can guide biosecurity efforts, including early detection of pathogens, prioritisation of vaccination time, depopulation or early harvest (Jones et al. 2015), and treatment. *Piscirickettsia salmonis* is the leading cause for antibiotic treatments in farmed Atlantic salmon in Chile, with treatment efficacy constrained by environmental reinfection rates. The integration of predictive models into real-time monitoring systems can inform pathogen management strategies, including coordination among farms, to improve their efficacy while reducing antibiotic consumption.

Hybrid models can bridge the gap between deterministic and stochastic elements of salmon farming epidemiology. Another valuable aspect of predictive model generation lies in its contribution to fostering industry knowledge and preserving institutional memory. Predictive models act as repositories of knowledge, ensuring that the expertise gained by the industry is retained and remain accessible for new professionals. This knowledge transfer is essential for maintaining a continuum of proficiency within the industry and preventing the loss of valuable insights acquired through years of practical experience and science. Finally, the integration of real-time data streams and continuous updates can ensure that these models remain adaptive to evolving environmental conditions and emerging pathogens.

4.4.3 Case Study 3: Sea Lice Models

4.4.3.1 Introduction

Parasites and pathogens present challenges for all fish production, while for Atlantic salmon aquaculture, sea lice of the species *Lepeophtheirus salmonis* represents one of the greatest challenges (Pike and Wadsworth 1999; Olaussen 2018; Barrett et al. 2020). Controlling infestations is costly in terms of stunted growth and direct costs associated with lice treatments (Abolofia et al. 2017; Overton et al. 2019). Additionally, lice from aquaculture farms can infest migrating wild smolts and potentially cause additional mortality in salmon stocks that are already under pressure (Torrissen et al. 2013). A variety of mathematical modelling tools can be used to aid our understanding of sea lice and their impact on wild salmon. The life cycle of sea lice involves a number of developmental stages (Hamre et al. 2013), with the first two being planktonic stages before the lice settle on host fish to develop further.

Mathematical modelling of sea lice is a prominent feature of Norway's management regime for salmon aquaculture, where the industry's opportunity to grow is regulated by, among other factors, the estimated effect of lice on wild stocks in geographical production areas. Several modules are connected to achieve such estimates according to a generic framework of sea lice model components (Murray et al. 2022). First, the infestations on a single farm or in a single cage are modelled using a stage distributed model that also estimates the production of eggs. Second, ocean circulation models are used to predict dispersion patterns of the planktonic stages, thereby estimating possible infection paths between farms, and a "risk" field of infectious nauplia for migrating smolts. Third, models of the smolt migration are used to estimate the infestation rate and resulting mortality due to the salmon lice dispersed from aquaculture. Here, the discussion will be limited to the first two types of model.

4.4.3.2 Modelling Sea Lice Populations at Aquaculture Farms

A lice infestation in an aquaculture cage consists of lice of various stages attached to individual fish. Adult females produce eggs on egg strings that hatch into planktonic nauplii developing over days to weeks into the second naupliar stage then into copepodids. Copepodids attach to salmon, provided they can find a host, and further develop through chalimus and preadult stages before becoming reproducing adults. The development of the sea lice population and their egg production rate at a farm therefore depends strongly on the population structure, meaning how many lice at each of the development stages are present. In order to make predictions, the population needs to be characterised in terms of its stage distribution.

Mathematical models of sea lice populations need to include an explicit stage distribution, but model stages do not have to correspond directly to biological stages. For instance, the model of (Revie et al. 2005) has four stages—n_1 for all chalimus stages, n_2 for preadult females, n_3 for adult females, and n_4 for gravid females. No model stage is used for nauplii, as the infestation rate is estimated based on the n_4 state with a time delay, in addition to external infestation pressure. As an example, Fig. 4.10, reproduced from (Revie et al. 2005), shows model predictions compared to measurements of the chalimus and mobile stages in a study in Scotland. The values are not from a single farm but are averaged over multiple farms using the same treatment strategy.

These models generally contain a high degree of uncertainty. Even if the internal mechanics of the model correspond well with the actual growth and reproduction rates of the lice at the actual farms—which may or may not be the case—the terms accounting for internal and external infestation pressure are hard to quantify. The first relies on assumptions of how likely each released egg is to infest fish at the same farm, which depends on the flow field around the farm and will be highly variable. The second depends on conditions in the surroundings, particularly eggs released from other farms in the region. One may better estimate these rates if the lice dynamics of the entire region are simulated in an ocean circulation model (see next Section), but even so, a relatively high uncertainty must be assumed. A different approach is to model infestation rates in a network model where sea distances between farms govern the rates of cross infestation, e.g. (Aldrin et al. 2019).

If lice counts are available in near real time at the simulated farm, these can be assimilated into the model in order to improve the estimates and provide a digital twin representation of the lice population at the farm. This can significantly improve the predictive ability of the model, since a good estimate of the current state is important for predicting future states.

4.4.3.3 Modelling Transport of Sea Lice Between Aquaculture Sites

Due to the planktonic stages of sea lice, eggs from one farm can infest fish in other farms, but the rates of infestation depend strongly on the transport patterns of lice in the region around the farm. Planktonic lice can actively swim, but as horizontal

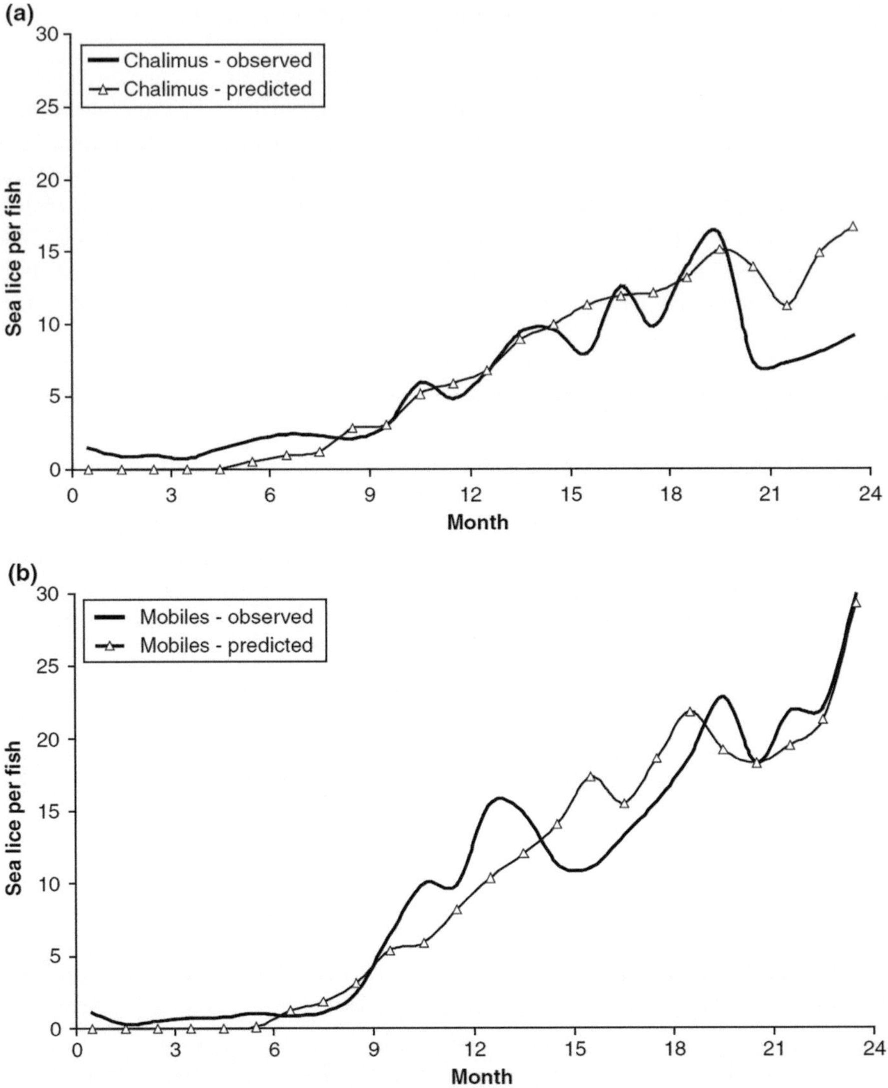

Fig. 4.10 (**a**) Observed and predicted numbers of chalimus stages on Scottish farms where hydrogen peroxide was used as a treatment between 1996 and 2000. (**b**) Observed and predicted numbers of mobile (preadult and adult) stages on the same group of farms. (Reproduced from Revie et al. (2005) with permission)

current speeds in the ocean are generally much larger than vertical speeds, active swimming only enables lice to choose their vertical position. Horizontal transport depends primarily on the current speeds and directions, which can be estimated using ocean circulation models (OCM). Modelling of sea lice dispersal in an OCM can be approached in two ways—the Lagrangian approach where discrete lice particles are released at aquaculture farms and transported as a function of estimated currents, or the Eulerian approach where a concentration field of lice is modelled using the advection-diffusion equation with sources at farm locations. In both cases, one can exploit the fact that sea lice have no effect on the ocean currents, so the dispersal model can be run separately based on output data from the OCM provided the model output has been stored at a sufficiently high time resolution.

For a Lagrangian model, the basic principle is defining the three-dimensional position of each particle as, e.g. $P(t) = [x(t)\ \ y(t)\ \ z(t)]$. If the current field computed by the OCM at time t is available, one can look up the current vector $\begin{bmatrix} v_x & v_y & v_z \end{bmatrix}$ at position $P(t)$. As the OCM may have a relatively coarse grid, and $P(t)$ does not have to be constrained by the grid, the current vector should be found by interpolating the current field to the exact location of the particle. If the lice mode includes active vertical movement, this can be represented by a speed term $w(t)$. Vertical movement is important both because the vertical distribution of lice strongly affects how the lice larvae can interact with fish in an aquaculture farm and because the vertical distribution can significantly affect their horizontal transport in the ocean. The next position of a particle, given a time step of Δt, can be calculated in the simplest way using

$$x\,(t + \Delta t) = x(t) + \Delta t \cdot v_x$$

$$y\,(t + \Delta t) = y(t) + \Delta t \cdot v_y$$

$$z\,(t + \Delta t) = z(t) + \Delta t \cdot (v_z + w(t))$$

Turbulent motion on a subgrid scale cannot be resolved by the OCM. The effect of turbulence is typically represented by adding a random walk term to the transport of the particles.

For a Eulerian model, the simplest approach would be to define a single 3D time-dependent concentration field $c(x, y, z, t)$. However, to represent the development of the lice nauplii toward infective stages and beyond, one can add a stage dimension to the partial differential equation, with biological development represented by advection along the stage dimension. After discretisation, one will have a series of k concentration fields $c_1(x, y, z, t) - c_k(x, y, z, t)$ representing lice stages. The model stages do not need to correspond 1:1 to biological stages, but, for instance, a set of model stages could map to a single biological stage. For each concentration field, the transport process would be modelled using the advection-diffusion equation:

$$\frac{\partial c}{\partial t} + v_x \frac{\partial c}{\partial x} + v_y \frac{\partial c}{\partial y} + (v_z + w) \frac{\partial c}{\partial z} + \kappa_h \left(\frac{\delta^2 c}{\delta x^2} + \frac{\delta^2 c}{\delta y^2} \right) + \kappa_v \frac{\delta^2 c}{\delta z^2} = \alpha - \sigma$$

where v_x, v_y, and v_y are the 3D current fields and w is the field of vertical movement speeds, κ_h and κ_v are horizontal and vertical diffusion coefficients, representing turbulent mixing, and α and σ are source and sink terms, respectively, representing addition and removal of lice. As this model formulation treats the lice collectively rather than as individual particles, the range of active behaviours (vertical movement) is more limited than what can be specified in the particle-based model.

There are several challenges to the accuracy of such models. If it is an objective to model representative concentrations of lice, the number of lice nauplii released from the farms in the region has to be estimated—whether directly based on observations of lice infestation rates or using predictions from site level population models (as discussed in the previous Section) on all farms. If the objective is just to estimate transmission probabilities between the network of farms, it is sufficient to release a standardised amount of lice or particles into the model from each farm to assess survivability rates.

The transport processes are also a source of inaccuracy. First, the currents computed by the OCM are limited by the model's resolution and simplifications and cannot resolve all details of the actual current field. The OCM's initial values, input data, and model parameters also have inaccuracies that contribute to uncertainty in the model outputs (Lermusiaux 2006). An additional challenge is numerical diffusion that adds some inaccuracy when updating the concentration field, but this can be controlled to some degree by using appropriate numerical calculation schemes.

4.4.3.4 Example Studies

In the Norwegian management regime for salmon aquaculture, a model system is used that couples the Norkyst-800 hydrodynamic model (Albretsen et al. 2011) to a Lagrangian lice dispersal model and an on-site population model (Murray et al. 2022) in order to estimate the infestation pressure along the Norwegian coast. The infestation rate of migrating smolts is then estimated using several approaches, including a particle-based model of migrating smolt, and statistical inferences based on observed infestation rates of cultivated smolts in cages along the coast. Three sets of mathematical models are developed and run by different institutions (Institute of Marine Research, Veterinary Institute and SINTEF Ocean) to provide such estimates. Based on those estimates, mortality of wild smolts as a result of infestation rates is estimated using threshold values based on available knowledge of the impact of lice on wild salmon (Taranger et al. 2015). An expert group annually reviews the outputs from the model systems for each of the 14 production areas and formulates management advice regarding the continued growth or reduction of production volumes in the areas.

The study of (Bøhn et al. 2022) is an example of partial validation of lice release and dispersal modelling. They used lice counts from all aquaculture sites along the Norwegian coast as input to a Lagrangian lice model running based on the Norkyst-800 hydrodynamical model system (Albretsen et al. 2011). The modelled lice infection pressure was compared to lice counts on wild sea trout caught in traps at 40 locations along the coast, and the study found that the observed infestations could indeed be explained by the hydrodynamic lice dispersal model.

The Norwegian management regime, discussed above, does not consider mortality of sea trout due to salmon lice, and the study authors suggest that this study could contribute toward the inclusion of this as an additional indicator in the system.

Another study (Adams et al. 2015) coupled the model of (Revie et al. 2005) implemented for nine sites in Loch Fyne, Scotland, with a setup of the Finite Volume Coastal Ocean Model (FVCOM) (Adams et al. 2012), with releases of particles from all the sites modulated based on estimated production of lice larvae. The study investigated the effects of management actions such as bath treatments and in-feed treatments in this connected network. The study found dramatic differences in lice abundances between farms depending on connectivity due to oceanographic conditions and that lice dynamics at a site is dominated by site reproduction and subsequent dispersal once the site is infected. Management actions at sites with the highest rate of incoming connections from other sites were found to have the greatest effect on lice abundances across all sites.

4.4.3.5 Conclusions

This chapter highlights the significant role of simulation and computational modelling in aquaculture, specifically within the context of environmental decision-making and operational optimisation. Through the examination of three key use cases, we see how models are integral to improving decision-making in real-world aquaculture systems.

Looking forward, physics-based models are poised to play an increasingly crucial role in the development of digital twin solutions for aquaculture. By integrating real-world data with simulated models, digital twins offer a dynamic, real-time representation of aquaculture systems. These models, particularly those based on physical principles such as hydrodynamics, biogeochemistry, and ecophysiology, will be essential in optimising farm management. They enable the simulation of various operational scenarios, from feed distribution to environmental conditions, allowing for proactive decision-making and immediate responses to potential risks.

As computational power and data availability continues to improve, the accuracy and applicability of physics-based models will expand, contributing to more efficient and sustainable aquaculture practices. Future advancements in digital twin technology, combined with machine learning and IoT integration, will facilitate the creation of highly adaptive systems capable of simulating and optimising aquaculture operations.

These modelling improvements are not just valuable for operational efficiency—they are critical to the aquaculture industry's ability to respond to larger global challenges, particularly food security and climate change. As the demand for sustainable and scalable food production rises, aquaculture must adapt to both increasing global population pressures and the uncertainties brought about by climate change. Modelling tools help the industry identify optimal locations for farming, manage resource use more efficiently, and reduce environmental impacts, thus improving productivity and sustainability. Additionally, they enable farmers to predict and mitigate risks from climate-induced events such as temperature fluctuations, ocean acidification, and extreme weather events, enhancing the resilience of fish farming systems.

In conclusion, the integration of advanced modelling approaches into aquaculture decision-making is essential for addressing the sector's complex challenges. By enhancing our ability to predict outcomes, optimise resource use, and minimise environmental impacts, modelling serves as a cornerstone for the sustainable development of the aquaculture industry.

References

Abolofia J, Asche F, Wilen JE (2017) The cost of lice: quantifying the impacts of parasitic sea lice on farmed salmon. Mar Resour Econ 32:329–349. https://doi.org/10.1086/691981

Adams T, Black K, MacIntyre C et al (2012) Connectivity modelling and network analysis of sea lice infection in Loch Fyne, west coast of Scotland. Aquac Environ Interact 3:51–63

Adams TP, Proud R, Black KD (2015) Connected networks of sea lice populations: dynamics and implications for control. Aquac Environ Interact 6:273–284

Albretsen J, Sperrevik AK, Staalstrøm A et al (2011) NorKyst-800 Rapport nr. 1: Brukermanual og tekniske beskrivelser. https://hdl.handle.net/11250/113865

Aldrin M, Jansen PA, Stryhn H (2019) A partly stage-structured model for the abundance of salmon lice in salmonid farms. Epidemics 26:9–22. https://doi.org/10.1016/j.epidem.2018.08.001

Alver MO, Skøien KR, Føre M et al (2016) Modelling of surface and 3D pellet distribution in Atlantic salmon (Salmo salar L.) cages. Aquac Eng 72:20–29

Alver MO, Føre M, Alfredsen JA (2022) Predicting oxygen levels in Atlantic salmon (Salmo salar) sea cages. Aquaculture 548:737720

Alver MO, Føre M, Alfredsen JA (2023) Effect of cage size on oxygen levels in Atlantic salmon sea cages: a model study. Aquaculture 562:738831

Alver MO, Føre M, Urke HA, Alfredsen JA (2024) Mathematical modelling of dissolved oxygen levels in a multi-cage salmon farm. Aquaculture 593:741291. https://doi.org/10.1016/j.aquaculture.2024.741291

Ancey C (2020) Bedload transport: a walk between randomness and determinism. Part 1. The state of the art. J Hydraul Res 58:1–17. https://doi.org/10.1080/00221686.2019.1702594

Asche F, Oglend A (2016) The relationship between input-factor and output prices in commodity industries: the case of Norwegian salmon aquaculture. J Commod Mark 1:35–47

Ascione Kenov I, Campuzano F, Franz G et al (2014) Advances in modeling of water quality in estuaries. In: Finkl CW, Makowski C (eds) Remote sensing and modeling: advances in coastal and marine resources. Springer International Publishing, Cham, pp 237–276

Aumont O, Bopp L (2006) Globalizing results from ocean in situ iron fertilization studies. Glob Biogeochem Cycles 20. https://doi.org/10.1029/2005GB002591

Aunsmo A, Valle PS, Sandberg M et al (2010) Stochastic modelling of direct costs of pancreas disease (PD) in Norwegian farmed Atlantic salmon (Salmo salar L.). Prev Vet Med 93:233–241

Baird ME, Wild-Allen KA, Parslow J et al (2020) CSIRO Environmental Modelling Suite (EMS): scientific description of the optical and biogeochemical models (vB3p0). Geosci Model Dev 13:4503–4553. https://doi.org/10.5194/gmd-13-4503-2020

Bang Jensen B, Dean KR, Huseby RB et al (2021) Realtime case study simulations of transmission of Pancreas Disease (PD) in Norwegian salmonid farming for disease control purposes. Epidemics 37:100502. https://doi.org/10.1016/j.epidem.2021.100502

Barrett LT, Oppedal F, Robinson N, Dempster T (2020) Prevention not cure: a review of methods to avoid sea lice infestations in salmon aquaculture. Rev Aquac 12:2527–2543. https://doi.org/10.1111/raq.12456

Bauer P, Stevens B, Hazeleger W (2021) A digital twin of earth for the green transition. Nat Clim Chang 11:80–83. https://doi.org/10.1038/s41558-021-00986-y

Bergsson H, Svendsen MBS, Steffensen JF (2023) Model of oxygen conditions within Aquaculture Sea cages. Biology 12:1408

Bøhn T, Nilsen R, Gjelland KØ et al (2022) Salmon louse infestation levels on sea trout can be predicted from a hydrodynamic lice dispersal model. J Appl Ecol 59:704–714. https://doi.org/10.1111/1365-2664.14085

Bravo F, Grant J (2018) Modelling sediment assimilative capacity and organic carbon degradation efficiency at marine fish farms. Aquac Environ Interact 10:309–328

Bravo F, Sidhu JPS, Bernal P et al (2020) Hydrodynamic connectivity, water temperature, and salinity are major drivers of piscirickettsiosis prevalence and transmission among salmonid farms in Chile. Aquac Environ Interact 12:263–279. https://doi.org/10.3354/aei00368

Brigolin D, Pastres R, Nickell TD et al (2009) Modelling the impact of aquaculture on early diagenetic processes in sea loch sediments. Mar Ecol Prog Ser 388:63–80

Brigolin D, Porporato EMD, Prioli G, Pastres R (2017) Making space for shellfish farming along the Adriatic coast. ICES J Mar Sci 74:1540–1551. https://doi.org/10.1093/icesjms/fsx018

Broch OJ, Klebert P, Michelsen FA, Alver MO (2020) Multiscale modelling of cage effects on the transport of effluents from open aquaculture systems. PLoS One 15:e0228502

Byron CJ, Costa-Pierce BA (2013) Carrying capacity tools for use in the implementation of an ecosystems approach to aquaculture. In: Site selection and carrying capacities for inland and coastal aquaculture. Food and Agriculture Organization of the United Nations, pp 87–101

Cerqueira V, Pimentel J, Korus J et al (2024) Forecasting ocean hypoxia in salmonid fish farms. Front Aquac 3:1365123

Chary K, Brigolin D, Callier MD (2022) Farm-scale models in fish aquaculture–an overview of methods and applications. Rev Aquac 14:2122–2157

Chen C, Beardsley RC, Cowles G et al (2012) An unstructured-grid, finite-volume community ocean model: FVCOM user manual. Sea Grant College Program, Massachusetts Institute of Technology Cambridge

Cornejo P, Guerrero NM, Montes RM et al (2020) Hydrodynamic effect of biofouling in fish cage aquaculture netting. Aquaculture 526:735367. https://doi.org/10.1016/j.aquaculture.2020.735367

Couture JL, Froehlich HE, Buck BH et al (2021) Scenario analysis can guide aquaculture planning to meet sustainable future production goals. ICES J Mar Sci 78:821–831

Cromey CJ, Nickell TD, Black KD (2002) DEPOMOD—modelling the deposition and biological effects of waste solids from marine cage farms. Aquaculture 214:211–239. https://doi.org/10.1016/S0044-8486(02)00368-X

Dabrowski T, Lyons K, Curé M et al (2013) Numerical modelling of spatio-temporal variability of growth of Mytilus edulis (L.) and influence of its cultivation on ecosystem functioning. J Sea Res 76:5–21

Dabrowski T, Lyons K, Cusack C et al (2016) Ocean modelling for aquaculture and fisheries in Irish waters. Ocean Sci 12:101–116

Dame RF, Prins TC (1997) Bivalve carrying capacity in coastal ecosystems. Aquat Ecol 31:409–421. https://doi.org/10.1023/A:1009997011583

Dietz T, Ostrom E, Stern PC (2003) The struggle to govern the commons. Science 302:1907–1912. https://doi.org/10.1126/science.1091015

Dunn C, Lopez F, Garcia MH (1996) Mean flow and turbulence in a laboratory channel with simulated vegatation (HES 51). Hydraulic Engineering Series no. 51. http://hdl.handle.net/2142/12229

Edwards CA, Moore AM, Hoteit I, Cornuelle BD (2015) Regional Ocean data assimilation. Annu Rev Mar Sci 7:21–42. https://doi.org/10.1146/annurev-marine-010814-015821

Ervik A, Hansen PK, Aure J et al (1997) Regulating the local environmental impact of intensive marine fish farming I. The concept of the MOM system (Modelling-Ongrowing fish farms-monitoring). Aquaculture 158:85–94. https://doi.org/10.1016/S0044-8486(97)00186-5

Evensen G (2003) The ensemble Kalman filter: theoretical formulation and practical implementation. Ocean Dyn 53:343–367

Falconer L, Telfer TC, Ross LG (2016) Investigation of a novel approach for aquaculture site selection. J Environ Manag 181:791–804

Falconer L, Ytteborg E, Goris N et al (2023) Context matters when using climate model projections for aquaculture. Front Mar Sci 10:1198451

Fennel K, Wilkin J, Levin J et al (2006) Nitrogen cycling in the Middle Atlantic Bight: results from a three-dimensional model and implications for the North Atlantic nitrogen budget. Glob Biogeochem Cycles 20. https://doi.org/10.1029/2005GB002456

Fennel K, Mattern JP, Doney SC et al (2022) Ocean biogeochemical modelling. Nat Rev Methods Primers 2:76

Ferreira JG, Grant J, Verner-Jeffreys DW, Taylor NG (2013) Carrying capacity for aquaculture, modeling frameworks for determination of. In: Sustainable food production. Springer, New York, pp 417–448

Føre M, Dempster T, Alfredsen JA et al (2009) Modelling of Atlantic salmon (Salmo salar L.) behaviour in sea-cages: a Lagrangian approach. Aquaculture 288:196–204

Føre M, Dempster T, Alfredsen JA, Oppedal F (2013) Modelling of Atlantic salmon (Salmo salar L.) behaviour in sea-cages: using artificial light to control swimming depth. Aquaculture 388:137–146

Føre M, Alver M, Alfredsen JA et al (2016) Modelling growth performance and feeding behaviour of Atlantic salmon (Salmo salar L.) in commercial-size aquaculture net pens: model details and validation through full-scale experiments. Aquaculture 464:268–278. https://doi.org/10.1016/j.aquaculture.2016.06.045

Føre M, Alver MO, Alfredsen JA et al (2024) Digital twins in intensive aquaculture—challenges, opportunities and future prospects. Comput Electron Agric 218:108676

Gečeк S, Legović T (2010) Towards carrying capacity assessment for aquaculture in the Bolinao Bay, Philippines: a numerical study of tidal circulation. Ecol Model 221:1394–1412

Ghisalberti M, Nepf HM (2002) Mixing layers and coherent structures in vegetated aquatic flows. J Geophys Res Oceans 107. https://doi.org/10.1029/2001JC000871

Ghisalberti M, Nepf HM (2004) The limited growth of vegetated shear layers. Water Resour Res 40:2003WR002776. https://doi.org/10.1029/2003WR002776

Grøttum JA, Sigholt T (1998) A model for oxygen consumption of Atlantic salmon (Salmo salar) based on measurements of individual fish in a tunnel respirometer. Aquac Eng 17:241–251

Halide H, Stigebrandt A, Rehbein M, McKinnon AD (2009) Developing a decision support system for sustainable cage aquaculture. Environ Model Softw 24:694–702

Hamre LA, Eichner C, Caipang CMA et al (2013) The salmon louse Lepeophtheirus salmonis (Copepoda: Caligidae) life cycle has only two chalimus stages. PLoS One 8:e73539

Herzfeld M (2006) An alternative coordinate system for solving finite difference ocean models. Ocean Model 14:174–196

Holling CS (1965) The functional response of predators to prey density and its role in mimicry and population regulation. Mem Entomol Soc Can 97:5–60

Ibarra DA, Fennel K, Cullen JJ (2014) Coupling 3-D Eulerian bio-physics (ROMS) with individual-based shellfish ecophysiology (SHELL-E): a hybrid model for carrying capacity and environmental impacts of bivalve aquaculture. Ecol Model 273:63–78

Inglis G, Hayden BJ, Ross AH (2000) An overview of factors affecting the carrying capacity of coastal embayments for mussel culture. National Institute of Water & Atmospheric Research

James SC, Jones CA, Grace MD, Roberts JD (2010) Advances in sediment transport modelling. J Hydraul Res 48:754–763. https://doi.org/10.1080/00221686.2010.515653

Jazwinski AH (2007) Stochastic processes and filtering theory. Courier Corporation

Ji Z-G (2017) Hydrodynamics and water quality: modeling rivers, lakes, and estuaries. John Wiley & Sons

Jones SR, Bruno DW, Madsen L, Peeler EJ (2015) Disease management mitigates risk of pathogen transmission from maricultured salmonids. Aquac Environ Interact 6:119–134

Kalman RE (1960) A new approach to linear filtering and prediction problems. J Basic Eng 82:35–45

Klebert P, Patursson Ø, Endresen PC et al (2015) Three-dimensional deformation of a large circular flexible sea cage in high currents: field experiment and modeling. Ocean Eng 104:511–520. https://doi.org/10.1016/j.oceaneng.2015.04.045

Kluger LC, Filgueira R (2021) Thinking outside the box: embracing social complexity in aquaculture carrying capacity estimations. ICES J Mar Sci 78:435–442

Lakshmivarahan S, Lewis JM (2013) Nudging methods: a critical overview. In: Data assimilation for atmospheric, oceanic and hydrologic applications, vol II. Springer, Berlin, Heidelberg, pp 27–57

Lermusiaux PF (2006) Uncertainty estimation and prediction for interdisciplinary ocean dynamics. J Comput Phys 217:176–199

Lester SE, Stevens JM, Gentry RR et al (2018) Marine spatial planning makes room for offshore aquaculture in crowded coastal waters. Nat Commun 9:945. https://doi.org/10.1038/s41467-018-03249-1

Letcher RA, Jakeman AJ (2009) Types of environmental models. Water Dev 2:131–154

Lucas LV (2010) Implications of estuarine transport for water quality. In: Valle-Levinson A (ed) Contemporary issues in estuarine physics. Cambridge University Press, Cambridge, pp 273–307

Lundebye A-K (2013) Aquaculture site selection and carrying capacity for inland and coastal aquaculture in Northern Europe. In: Site selection and carrying capacities for inland and coastal aquaculture. Food and Agriculture Organization of the United Nations, p 171

Madec G, and the NEMO System Team (2024) NEMO Ocean Engine Reference Manual, Zenodo. https://doi.org/10.5281/zenodo.1464816

McKindsey CW, Thetmeyer H, Landry T, Silvert W (2006) Review of recent carrying capacity models for bivalve culture and recommendations for research and management. Aquaculture 261:451–462

Moreau S, Vancoppenolle M, Bopp L et al (2016) Assessment of the sea-ice carbon pump: insights from a three-dimensional ocean-sea-ice biogeochemical model (NEMO-LIM-PISCES). Elementa 4:000122

Murray AG, Shephard S, Asplin L et al (2022) A standardised generic framework of sea lice model components for application in coupled hydrodynamic-particle models. In: Sea lice biology and control. 5M Books, pp 167–187

Nepf HM, Vivoni ER (2000) Flow structure in depth-limited, vegetated flow. J Geophys Res Oceans 105:28547–28557. https://doi.org/10.1029/2000JC900145

Noble C, Gismervik K, Iversen MH et al (2018) Welfare indicators for farmed Atlantic salmon: tools for assessing fish welfare. In: Aquaculture Books, p 351. http://hdl.handle.net/1893/28435

O'Donncha F, Grant J (2019) Precision aquaculture. IEEE Internet Things Mag 2:26–30

O'Donncha F, Hartnett M, Nash S (2013) Physical and numerical investigation of the hydrodynamic implications of aquaculture farms. Aquac Eng 52:14–26

O'Donncha F, Hartnett M, Plew DR (2015) Parameterizing suspended canopy effects in a three-dimensional hydrodynamic model. J Hydraul Res 53:714–727. https://doi.org/10.1080/00221686.2015.1093036

O'Donohue F (2012) Physical and numerical modelling of impeded tidal flows: effects of aquaculture structures on hydrodynamics and material transport. University of Galway

Olaussen JO (2018) Environmental problems and regulation in the aquaculture industry. Insights from Norway. Mar Policy 98:158–163. https://doi.org/10.1016/j.marpol.2018.08.005

Olivares J, Marshall SH (2010) Determination of minimal concentration of *Piscirickettsia salmonis* in water columns to establish a fallowing period in salmon farms. J Fish Dis 33:261–266. https://doi.org/10.1111/j.1365-2761.2009.01119.x

Oppedal F, Dempster T, Stien LH (2011) Environmental drivers of Atlantic salmon behaviour in sea-cages: a review. Aquaculture 311:1–18

Overton K, Dempster T, Oppedal F et al (2019) Salmon lice treatments and salmon mortality in Norwegian aquaculture: a review. Rev Aquac 11:1398–1417. https://doi.org/10.1111/raq.12299

Pike AW, Wadsworth SL (1999) Sealice on salmonids: their biology and control. In: Baker JR, Muller R, Rollinson D (eds) Advances in parasitology. Academic Press, pp 233–337

Pilditch CA, Grant J, Bryan KR (2001) Seston supply to sea scallops (*Placopecten magellanicus*) in suspended culture. Can J Fish Aquat Sci 58:241–253. https://doi.org/10.1139/f00-242

Plew DR (2011) Depth-averaged drag coefficient for modeling flow through suspended canopies. J Hydraul Eng 137:234–247. https://doi.org/10.1061/(ASCE)HY.1943-7900.0000300

Pond S, Pickard GL (1983) Introductory dynamical oceanography. Gulf Professional Publishing

Ragnoli E, Zhuk S, Donncha FO et al (2012) An optimal interpolation scheme for assimilation of HF radar current data into a numerical ocean model. In: Oceans. IEEE, pp 1–5

Raillard O, Ménesguen A (1994) An ecosystem box model for estimating the carrying capacity of a macrotidal shellfish system. Mar Ecol Prog Ser 115:117–130

Rasheed A, San O, Kvamsdal T (2020) Digital twin: values, challenges and enablers from a modeling perspective. IEEE Access 8:21980–22012

Remen M, Oppedal F, Imsland AK et al (2013) Hypoxia tolerance thresholds for post-smolt Atlantic salmon: dependency of temperature and hypoxia acclimation. Aquaculture 416:41–47

Remen M, Sievers M, Torgersen T, Oppedal F (2016) The oxygen threshold for maximal feed intake of Atlantic salmon post-smolts is highly temperature-dependent. Aquaculture 464:582–592

Revie CW, Robbins C, Gettinby G et al (2005) A mathematical model of the growth of sea lice, Lepeophtheirus salmonis, populations on farmed Atlantic salmon, Salmo salar L., in Scotland and its use in the assessment of treatment strategies. J Fish Dis 28:603–613. https://doi.org/10.1111/j.1365-2761.2005.00665.x

Rico A, Vighi M, Van Den Brink PJ et al (2019) Use of models for the environmental risk assessment of veterinary medicines in European aquaculture: current situation and future perspectives. Rev Aquac 11:969–988. https://doi.org/10.1111/raq.12274

Rocha C, Edwards CA, Roughan M et al (2019) A high-resolution biogeochemical model (ROMS 3.4+ bio_Fennel) of the East Australian Current system. Geosci Model Dev 12:441–456

Ross LG, Falconer LL, Campos Mendoza A, Martinez Palacios CA (2011) Spatial modelling for freshwater cage location in the Presa Adolfo Mateos Lopez (El Infiernillo), Michoacán, México: spatial modelling for freshwater cage location. Aquac Res 42:797–807. https://doi.org/10.1111/j.1365-2109.2010.02689.x

Sarà G, Reid GK, Rinaldi A et al (2012) Growth and reproductive simulation of candidate shellfish species at fish cages in the Southern Mediterranean: Dynamic Energy Budget (DEB) modelling for integrated multi-trophic aquaculture. Aquaculture 324:259–266

Scheel I, Aldrin M, Frigessi A, Jansen PA (2007) A stochastic model for infectious salmon anemia (ISA) in Atlantic salmon farming. J R Soc Interface 4:699–706

Schmolke A, Thorbek P, DeAngelis DL, Grimm V (2010) Ecological models supporting environmental decision making: a strategy for the future. Trends Ecol Evol 25:479–486

Schuwirth N, Borgwardt F, Domisch S et al (2019) How to make ecological models useful for environmental management. Ecol Model 411:108784

Sequeira A, Ferreira JG, Hawkins AJS et al (2008) Trade-offs between shellfish aquaculture and benthic biodiversity: a modelling approach for sustainable management. Aquaculture 274:313–328

Shchepetkin AF, McWilliams JC (2005) The regional oceanic modeling system (ROMS): a split-explicit, free-surface, topography-following-coordinate oceanic model. Ocean Model 9:347–404. https://doi.org/10.1016/j.ocemod.2004.08.002

Sheldon JE, Alber M (2006) The calculation of estuarine turnover times using freshwater fraction and tidal prism models: a critical evaluation. Estuar Coasts 29:133–146. https://doi.org/10.1007/BF02784705

Simpson JH, Berx B, Gascoigne J, Saurel C (2007) The interaction of tidal advection, diffusion and mussel filtration in a tidal channel. J Mar Syst 68:556–568

Slagstad D, McClimans TA (2005) Modeling the ecosystem dynamics of the Barents Sea including the marginal ice zone: I. Physical and chemical oceanography. J Mar Syst 58:1–18
Solstorm D, Oldham T, Solstorm F et al (2018) Dissolved oxygen variability in a commercial sea-cage exposes farmed Atlantic salmon to growth limiting conditions. Aquaculture 486:122–129. https://doi.org/10.1016/j.aquaculture.2017.12.008
Stavrakidis-Zachou O, Papandroulakis N, Lika K (2019) A DEB model for European sea bass (Dicentrarchus labrax): parameterisation and application in aquaculture. Ecosyst Based Manag Biosphere New Phase DEB Res 143:262–271. https://doi.org/10.1016/j.seares.2018.05.008
Stigebrandt A (2011) Carrying capacity: general principles of model construction. Aquac Res 42:41–50. https://doi.org/10.1111/j.1365-2109.2010.02674.x
Su B, Kelasidi E, Frank K et al (2021) An integrated approach for monitoring structural deformation of aquaculture net cages. Ocean Eng 219:108424. https://doi.org/10.1016/j.oceaneng.2020.108424
Sun M, Hassan SG, Li D (2016) Models for estimating feed intake in aquaculture: a review. Comput Electron Agric 127:425–438
Taranger GL, Karlsen Ø, Bannister RJ et al (2015) Risk assessment of the environmental impact of Norwegian Atlantic salmon farming. ICES J Mar Sci 72:997–1021. https://doi.org/10.1093/icesjms/fsu132
Thépaut J-N (1993) Assimilation of satellite data by 4D-VAR at ECMWF. In: Proceedings of ECMWF/EUMETSAT seminar on developments in the use of satellite data in numerical weather prediction, 6–10 Sept 1993, Reading
Tilman D, Fargione J, Wolff B et al (2001) Forecasting agriculturally driven global environmental change. Science 292:281–284. https://doi.org/10.1126/science.1057544
Torrissen O, Jones S, Asche F et al (2013) Salmon lice–impact on wild salmonids and salmon aquaculture. J Fish Dis 36:171–194
Urke HA, Daae K, Viljugrein H et al (2021) Improvement of aquaculture management practice by integration of hydrodynamic modelling. Ocean Coast Manag 213:105849. https://doi.org/10.1016/j.ocecoaman.2021.105849
Valle-Levinson A (2013) Some basic hydrodynamic concepts to be considered for coastal aquaculture. In: Site selection and carrying capacities for inland and coastal aquaculture. Food and Agriculture Organization of the United Nations, p 147
Wan EA, Van Der Merwe R (2000) The unscented Kalman filter for nonlinear estimation. In: Proceedings of the IEEE 2000 adaptive systems for signal processing, communications, and control symposium (Cat. No. 00EX373). IEEE, pp 153–158
Warner JC, Sherwood CR, Arango HG, Signell RP (2005) Performance of four turbulence closure models implemented using a generic length scale method. Ocean Model 8:81–113
Wassmann P, Slagstad D, Riser CW, Reigstad M (2006) Modelling the ecosystem dynamics of the Barents Sea including the marginal ice zone: II. Carbon flux and interannual variability. J Mar Syst 59:1–24
Webber JL, Tyler CR, Carless D et al (2021) Impacts of land use on water quality and the viability of bivalve shellfish mariculture in the UK: a case study and review for SW England. Environ Sci Pol 126:122–131. https://doi.org/10.1016/j.envsci.2021.09.027
Weitzman J, Filgueira R (2020) The evolution and application of carrying capacity in aquaculture: towards a research agenda. Rev Aquac 12:1297–1322. https://doi.org/10.1111/raq.12383
Zhao Q, Huang H, Zhu Y et al (2022) Analysing ecological carrying capacity of bivalve aquaculture within the Yellow River Estuary ecoregion through mass-balance modelling. Aquac Environ Interact 14:147–161

Chapter 5
Hybrid Analysis and Modeling for Aquaculture

Adil Rasheed and Martin Føre

Abstract A digital twin (DT) of a system has to cover all relevant aspects describing the system dynamics, and in most cases, the existing knowledge and experience within these aspects will vary. This implies that while it is often possible to describe some phenomenons using knowledge-based models (KBM), other elements will have to be described using data-driven models (DDM). A full digital twin realization will therefore also require the use of methods for integrating KBM and DDM in a common seamless system that exploits the advantages of both approaches. The approach called hybrid analysis and modeling (HAM) can be used to identify methods to achieve this. This chapter describes the concept of HAM and the different categories of HAM methods, all in light of its utility toward building digital twin solutions for aquaculture. Examples of KBM, DDM, and potential HAM methods related to aquaculture production will be provided throughout the chapter to illustrate this connection. In conclusion, it is apparent that the development of digital twins for aquaculture will require the use of HAM methods to bridge knowledge gaps and eventually provide a deeper insight into system dynamics. This will also contribute to bridging the spatial and temporal scales relevant for aquaculture and systems scale analytics, i.e., from individual fish through full farms to regional scales and from daily/weekly operations through full farming cycles to series of multiple farming cycles.

Keywords Hybrid modeling · Physics-informed machine learning · Data-driven modeling · Knowledge-based models · Machine learning

A. Rasheed (✉) · M. Føre
Department of Engineering Cybernetics, NTNU, Trondheim, Norway
e-mail: adil.rasheed@ntnu.no

F. O'Donncha et al. (eds.), *Digital Twin for Marine Fish Farming*,
Aquatic Production Systems, https://doi.org/10.1007/978-3-032-10362-8_5

5.1 Introduction

Hybrid analysis and modeling (HAM) is a practice where knowledge-based models (KBMs) are combined with data-driven models (DDMs) to digitally represent a particular phenomenon or system of interest. The idea behind this is that KBMs and DDMs have different properties and features that in combination can capture the details and main trends in the dynamics of the system better than KBMs or DDMs on their own (San et al. 2021). This method is typically used when the existing knowledge about the system of interest and its dynamics is not sufficient to catch the essential dynamics and is considered a core component in digital twin (DT) technology both in general (Rasheed et al. 2020) and more recently for aquaculture in particular (Føre et al. 2024).

The present chapter elaborates on how the HAM approach can be applied to aquaculture, particularly when developing digital twin technology. This is done by first defining the terms KBMs and DDMs, referring back to other chapters in the book that treat these concepts in more detail, and then discussing the main advantages and drawbacks of these two concepts. The synthesis of KBMs and DDMs through HAM is then outlined in general, detailing different methods for realizing HAM and the advantages that can be harvested through these methods. While these concepts are largely generic, examples related to aquaculture applications will be provided throughout the chapter, thus linking the considerations around the HAM approach to the industrial segment.

5.2 Knowledge-Based Modeling (KBM)

Knowledge-based modeling (KBM), also known as mechanistic modeling, is a concept where a mathematical portrayal of a system and its dynamics based on existing knowledge is sought. This approach involves first observing the phenomenon of interest carefully and then formulating mathematical equations that approximate these observations and that can be solved to estimate the dynamics of this phenomenon. When observing natural systems or processes, a KBM will thus typically end up featuring equations that represent governing laws of nature which are often defined as conservation and constitutive laws. Since most such laws are based on a combination of theoretical development and experimental validation, they usually consist of systems of differential equations that can only be solved/approximated numerically by computers. Wigner (1960) highlighted KMB as a powerful and effective scientific tool, because it exploits our present understanding of a system to predict its future states (Willcox et al. 2021). Examples of KBM range from physics-based modeling (PBM) that specifically rely on physical laws to models describing ecosystem/biological dynamics.

Since Chap. 5 covers the concept of KBMs in aquaculture in detail, this chapter will only consider a sub-selection of KBMs considered particularly suitable for illustrating the concept of HAM in aquaculture. However, the general concept of HAM and the main approach described here will also be relevant to combine any models in aquaculture. As outlined in the following, KBMs that are relevant for aquaculture may be categorized based on their areas of application and what particular dynamics they seek to portray. They may accordingly be broadly divided into models representing the physics of the system or models that mimic the biological processes in the farm.

5.2.1 KBMs for Modeling the Physics of Fish Farms

KBMs in this area are based on well-established physics-based concepts such as Newton's laws and mass balance. These models often focus on branches of physics such as hydrodynamics, materials engineering and thermodynamics, or topics often considered chemistry. These models are typically used to evaluate the structural responses of fish farm systems and quantify the environmental loads imposed by site-specific hydrodynamic conditions and can also be used to generate a realistic virtual environment that in turn can be used as input for biological models or to define boundary conditions for these.

- Hydrodynamics Models: These models simulate water flow patterns within land-based aquaculture systems, such as tanks (Behroozi and Couturier 2019), ponds (Peterson et al. 2000), or raceways (Huggins et al. 2004), or the fluid-structure interactions occurring in marine cage-based fish farms (Klebert et al. 2013; Moe et al. 2010). Such models help understand how water moves through the farming system and affects parameters like oxygen distribution, waste removal, and temperature gradients within the production volume. Moreover, they can also predict how aquaculture structures and systems are affected by hydrodynamic drag and other effects caused by the incident water flow. Many hydrodynamic models are also set up to predict water temperatures and can thus be used to model variations within the production volume in aquaculture facilities due to factors such as solar radiation and heat exchange with the environment. Since understanding temperature dynamics is crucial for maintaining optimal conditions for aquatic organisms, these abilities may be important in a digital twin portraying the full system dynamics of fish production.
- Water Quality Models: These models are designed to predict parameters such as dissolved oxygen, pH, ammonia, and nitrate concentrations and can have a crucial role in assessing these parameters in aquaculture systems. In an aquaculture setting, such models would typically incorporate factors like water chemistry and water exchange rates and would need to be linked with nutrient, bioenergetic, and feed conversion models to also use biological- and farming-related factors such as stocking density, feed inputs, and feces production. Since maintaining adequate oxygen levels is critical in aquaculture production facilities for ensuring the health and productivity of cultured species, models simulating the transport

of dissolved oxygen are of particular interest in aquaculture. These models can be used to both predict and assess the present oxygen states in the farm (Alver et al. 2022) and ultimately to optimize farm/pen design (Alver et al. 2023) or aeration strategies using oxygenation systems. Other models that simulate the cycling of nutrients (e.g., nitrogen, phosphorus) and harmful substances (e.g., ammonia) within a water volume may help in managing nutrient dynamics, which is particularly important for the optimization of nutrient recycling within aquaculture systems (Pedersen and Wik 2020).

- Light Distribution Models: Light distribution models estimate how light distributes within a volume. In aquaculture systems as in other aquatic systems, these models will need to consider factors such as water turbidity, depth, the presence of natural or artificial light sources, and the spectral characteristics of the incident light from these sources (Alver et al. 2014). Understanding light distribution is essential for designing and managing phototrophic organisms like algae and aquatic plants and is known to have an important impact on the behavior of many species of farmed animals.
- Sediment Transport Models: Sediment transport models predict the movement and deposition of particles within a water volume. In aquaculture systems where sediment accumulation is a concern, such as ponds or raceways, these models can aid in designing sedimentation basins and optimizing sediment removal strategies to maintain water quality (Huggins et al. 2004). Such models are also of interest in net-pen fish farms, e.g., when considering the distribution of feed particles or feces in the water column, or the deposition of waste material at the bottom beneath the farm (Cromey et al. 2002).
- Wave Models: There exist various models for predicting and simulating local variations in wave conditions given environmental features such as fetch length and bottom topography. For aquaculture operations in coastal or open-water environments, these models can be used to predict local water movements due to waves (Lader et al. 2017) and their impact on cage structures or floating platforms (Fredriksson et al. 2003) that can be superimposed upon the water movements and responses predicted by hydrodynamic models. This can be crucial for both site selection, structural design, condition-based maintenance (as opposed to periodic maintenance), and operational planning for offshore aquaculture ventures.

5.2.2 *KBMs for Modeling Biological Processes Within Aquaculture*

Some existing knowledge-based models provide valuable insight into biological processes and their interaction with the physical dynamics within aquaculture systems. These models can be used to predict the biological aspects of production and are thus often based on knowledge about the basic biological mechanisms and how they interact with the surrounding environment.

- Bioenergetics Models: Bioenergetics models quantify the energy budget of an organism, accounting for factors like metabolic rates, feed intake, and growth. In aquaculture, such models can be used to optimize feeding regimes and assess how environmental variables can affect the growth and performance of the produced species. They can thus be important tools for maximizing production while minimizing waste generation, which is often a main aim in animal husbandry. Bioenergetics models depend on inputs from physics models, including current and temperature models. Prominent examples of such models that have previously been used in aquaculture are dynamic energy budget (DEB) models (Kooijman 2010; Stavrakidis-Zachou et al. 2019) and Scope For Growth (SFG) models (Filgueira et al. 2011). A simpler approach to bioenergetics modeling is to use the feed conversion ratio (FCR) directly to compute the weight gain a fish obtains when ingesting a certain amount of feed (Sun et al. 2016). The FCR is a number that is typically obtained through experiments and that may link factors such as feed composition and temperature to weight gain. This approach is more commonly used in industry than other models, much due to its simplicity, but comes at the cost of generally being less accurate than DEB or SFG models.
- Behavioral models: Behavioral models seek to portray the behavior of individual animals or groups of animals in light of their internal states and the external conditions they are exposed to. In aquaculture settings, such models can be useful to predict, e.g., distribution patterns within the production unit (Føre et al. 2009), feeding patterns, antagonistic interactions between individuals, and responses to production strategies (Føre et al. 2013). These are all aspects that are highly relevant for both the design and management of fish farms.
- Biofouling Models: Biofouling models predict the accumulation of organic matter, algae, and other fouling organisms on submerged surfaces. Within aquaculture facilities, biofouling on nets, cages, and other infrastructure is important to consider, since it may have strong implications on the flow through the production volume (and hence oxygen supply and waste removal), structural loads through increased drag, and serve as reservoirs for disease pathogens and parasites. Understanding biofouling dynamics thus helps in developing strategies to ensure a production environment within the production units that is conducive of good fish welfare and growth. While there exist few models predicting the development of biofouling in aquaculture facilities, there are models in other marine sectors that could possibly be adapted (Uzun et al. 2019).
- Disease Transmission Models: By augmenting physics-based models that simulate water flow and water movement with other factors such as contact rates and pathogen dynamics, it is possible to simulate the spread of diseases among aquatic organisms (Alaliyat et al. 2019). Such models are of particular interest for aquaculture as they can be used to predict pathogen prevalence and developments within and between aquaculture facilities. This can in turn be used to assess

disease risk, design biosecurity measures, and develop strategies for disease prevention and control.

5.3 Data-Driven Modeling (DDM)

The popularity of data-driven modeling as a research and innovation tool is increasing rapidly, much because new methods within big data processing and machine learning are becoming increasingly available through open-source and easier to use. This development is further stimulated by that the relative costs of the computational infrastructures and resources required to train and solve such models are decreasing. DDM differ from KBM in that they assume that the data produced by a system are a manifestation of both known and unknown physical, biological, and/or chemical dynamics. From this, it follows that a data-driven model, when trained with sufficient relevant data of good quality, can learn the full system dynamics without manual inputs from humans. The emergence and application of this approach have rendered computers able to solve tasks that were previously considered impossible without human intervention. This has resulted in that computers (particularly when using deep learning methods) are now able to solve many such tasks with human-level performance. Since Chap. 5 elaborates on the use of data-driven modeling and machine learning in aquaculture, this chapter will only briefly discuss different types of data-driven models that can haves roles in HAM. Generally, all data-driven models may be categorized into one of the following six categories:

- Supervised linear models: Models that are formulated using linear equations and that are trained using labeled data in which the target variable is known are called supervised linear models. Common examples of models in this category are linear and logistic regression models. Supervised linear models are usually simple and efficient to solve and can be used for various tasks including prediction, classification, and feature selection. Models of this class have seen much use in aquaculture over the years, with some early examples using simple linear regression to assess growth (Van Dam 1990), and later studies have used more complex varieties of supervised linear models for, e.g., flow estimation (Gutiérrez-Estrada et al. 2012).
- Unsupervised linear models: Linear models that can be used when the training data is unlabeled and/or the target variable is unknown are known as unsupervised linear models. Examples of model types pertaining to this category include principal component analysis (PCA) and linear discriminant analysis (LDA). Some of the main applications of such models include reducing dimensionality, extracting features and clustering of datasets. The unsupervised linear model most commonly used in aquaculture is PCA, whose earlier applications have included assessments of water quality (Zhang et al. 2020) and performance and environmental impacts (Abdou et al. 2018).

- Supervised nonlinear models: Linear formulations are sometimes not sufficiently flexible to represent the dynamics of a system, particularly when the relationship between features/inputs and target variables is complex. In cases where the data is labeled, this calls for the use of supervised nonlinear models, examples of which include decision trees, support vector machines (SVMs), and random forests. Previous applications of models from this category in aquaculture include using SVM to predict water quality (Khotimah 2015) and random forest to predict fish survival (Islam et al. 2021).
- Unsupervised nonlinear models: In cases where nonlinear formulation is required but there is no labeled data, unsupervised nonlinear models may be applied. This class of models is designed to be trained using unlabeled data and includes models such as self-organizing maps (SOMs) and autoencoders that are seeing increased use in science and technology, particularly in tasks such as clustering, anomaly detection, and data compression. A relatively early example from this category in aquaculture was the use of SOM to detect skeletal anomalies in candidate species for commercial production (Russo et al. 2011), while more recent studies have used autoencoders for classification (Banerjee et al. 2022).
- Supervised deep learning: Sometimes the system under scrutiny will be too complex and/or feature too many dimensions to be effectively solved using conventional linear or nonlinear approaches. In such cases, it may be necessary to resort to deep learning methods. If there is sufficient labeled training data, supervised deep learning models are usually preferred as they, when properly trained, are often able to mimic complex dynamics through advanced multidimensional architectures. Due to their high flexibility and ability in achieving (or surpassing) state-of-the-art performance, deep learning methods have become very popular tools within both research and industry. Some of the most prominent examples of this is the extensive use of convolutional neural networks (CNNs) in image analysis instead of conventional methods, the use of recurrent neural networks (RNNs) for natural language processing, and regression and classification tasks through fully connected deep neural networks. The rapid development within supervised deep learning methods in the recent decades has also rendered them popular in aquaculture (Yang et al. 2021), example applications including behavioral monitoring (Huang et al. 2022), disease detection (Huang and Khabusi 2023), and many others.
- Unsupervised deep learning: In cases where complex model architectures are required but there is not sufficient labeled training data, it is often necessary to consider unsupervised deep learning methods that are possible to train using large amounts of unlabeled data. Examples of use for such models include image synthesis and unsupervised feature learning through generative adversarial networks (GANs) and using autoencoders to compress and detect anomalies in large datasets. Although applications of models from this category are less common than supervised learning (Yang et al. 2021), there are examples using, e.g., GAN for segmentation of fish in images (Zhou et al. 2022).

5.4 Hybrid Analysis and Modeling (HAM)

The first step in setting up a HAM approach for analyzing a specific system is to identify which types of models to use to portray different parts of the system dynamics. These choices can be made based on different criteria. Following this, the synthesis can be completed resulting in hybrid approaches that may vary in their degree of reliance on KBMs and DDMs. This section will first outline the main criteria for choosing between KBMs and DDMs and then provide descriptions of different possible outcomes of the synthesis.

5.4.1 Availability of System Knowledge and Datasets

The choice of whether to use a KBM or a DDM to describe aspects of system dynamics in HAM is often given by the availability of such models. When there is sufficient knowledge on system dynamics, most will tend toward setting up KBMs describing an as complete system state as possible, while it might be tempting to choose DDMs in cases where input-output datasets from the real system are in abundance. To illustrate this, consider a general case of a dataset associated with a dynamical system (as described by Robinson et al. (2022)):

$$Lx = f\ (g(x), h(x)) \tag{5.1}$$

where L for simplicity is a linear differential operator and $f\ (\cdot)$, $g(\cdot)$, and $h(\cdot)$ are functions of the state. Now, the following scenarios can arise:

1. Equation (5.1) is fully known meaning that the operator L, and the functions $f\ (\cdot)$ (which defines the system response), and $g(\cdot)$, and $h(\cdot)$ (which defines the inputs determining the response) are precisely known.
2. The operator L is known, but one or two of the functions $f\ (\cdot)$, $g(\cdot)$, and $h(\cdot)$ are unknown.
3. The operator L is known, but the functions $f\ (\cdot)$, $g(\cdot)$, and $h(\cdot)$ are all unknown.
4. The operator L and the functions $f\ (\cdot)$, $g(\cdot)$, and $h(\cdot)$ are all unknown.

In the first scenario, one can employ a purely KBM approach based on the known functions. Scenario 2 does not feature sufficient knowledge for a pure KBM approach and hence needs a hybrid solution. However, since much of the system is known, an approach could be to build the system around the KBMs and use DDMs to represent the dynamics that are unknown and thus not represented by the KBMs. The third scenario is a situation where only the operator is known, while the functions are unknown, implying that the problem can in theory be solved entirely using DDM. However, there exists some knowledge on the system dynamics, and it would be unwise to ignore the known part completely, so a hybrid approach where knowledge is incorporated into the DDM may be suggested. This approach may simplify the learning task as well as improve generalization for the DDMs. In the fourth scenario, there is no knowledge on the operator or the functions, meaning that a KBM approach is impossible and that it is necessary to model the process entirely from data.

5.4.2 Characteristics of KBMs and DDMs

Different KBMs and DDMs have different characteristics, and while some of these are endemic to the specific models, others are of a more generic nature pertaining to the basal features of the modeling paradigms. Some important characteristics are (Rasheed et al. 2020) the following:

- Reliability: how strongly one may rely on the outcomes of the model being realistic through explainability and transparency
- Stability: how likely the model is to become numerically unstable during predictions and how well does it tolerate perturbations outside the normal region of operation
- Dealing with unknown physics: how well the model can cope with unknown physics/phenomenons that impact the modeled process
- Adaptability: ability to evolve or adapt when encountering new situations
- Generalizability: to what extent the model can solve a wide variety of problems without problem-specific fine-tuning
- Susceptibility to bias: how likely the model is to be affected by bias in the knowledge base or data used in its creation
- Interpretability: ability to interpret the outcomes of the model and link these with real world phenomena
- Computational cost (online): the computational cost of the model during operation
- Computational cost (offline): the computational cost of the model during development/establishment
- Accuracy: ability to reproduce or simulate the behavior of the real system being modeled
- Uncertainty prediction: ability to quantify uncertainty in predictions of the modeled systems

While most of these characteristics are difficult to quantify using specific metrics and thus compare numerically, they are useful in comparing KBM with DDM and their strengths and weaknesses in a digital twin context (Table 5.1).

Table 5.1 Comparison of knowledge-based modeling (KBM) vs. data-driven modeling (DDM)

Characteristics	Knowledge-based modeling	Data-driven modeling
Reliability	+ Highly reliable as it is grounded in established physical laws, providing predictable and explainable outcomes	– Less transparent due to the opaque nature of many algorithms, making reliability contingent on data quality and algorithm design
Stability	+ Stability is ensured through rigorous physical constraints, although sensitivity issues may arise with complex models	– Prone to instability without sufficient regularization, particularly when extrapolating beyond available data.
Dealing with unknown physics	– Ineffective in scenarios where the physics is not fully understood, limiting applicability in new or uncharted domains	+ Excels in environments where the physical laws are not clear, using patterns in data to make predictions
Adaptability	– Typically static, requiring updates to models to handle new phenomena or changes in the system	+ Highly adaptive, with capabilities such as machine learning continuously updating models based on new data
Generalizability	+ Strong in scenarios where the physical laws are universal, allowing applications across different contexts with similar underlying physics	– Often limited by the range and diversity of the training data, which can restrict application to similar scenarios only
Susceptibility to bias	+ Lower risk of bias from the model itself, though still subject to biases inherent in the assumptions and approximations of the physical models	– High susceptibility to data bias, which can propagate through the model and affect outcomes significantly
Interpretability	+ Generally high, as outcomes can be directly traced back to physical laws and model parameters	– Typically low, especially in complex models like deep learning, where the decision paths are not easily discernible
Computational cost (online)	– Can be high, particularly for real-time simulations of complex systems, potentially requiring simplifications to meet performance requirements	+ Usually lower, with many models capable of delivering fast predictions after initial training, suitable for real-time applications
Computational cost (offline)	+ Minimal unless the model is being developed or refined, with most of the work focused on initial setup	– Considerable, often involving extensive parameter tuning and training on large datasets, which can be resource-intensive
Accuracy	– Dependent on the fidelity and completeness of the physical model, with limitations stemming from simplifications and assumptions	+ Can achieve high levels of accuracy when well-tuned and trained on comprehensive and high-quality data but can also suffer from overfitting or underfitting
Uncertainty prediction	+ Capable of quantifying uncertainties explicitly through analytical methods and error propagation techniques inherent in physical models	+ Some advanced techniques allow for uncertainty quantification, but often not as robust or transparent as in physics-based approaches

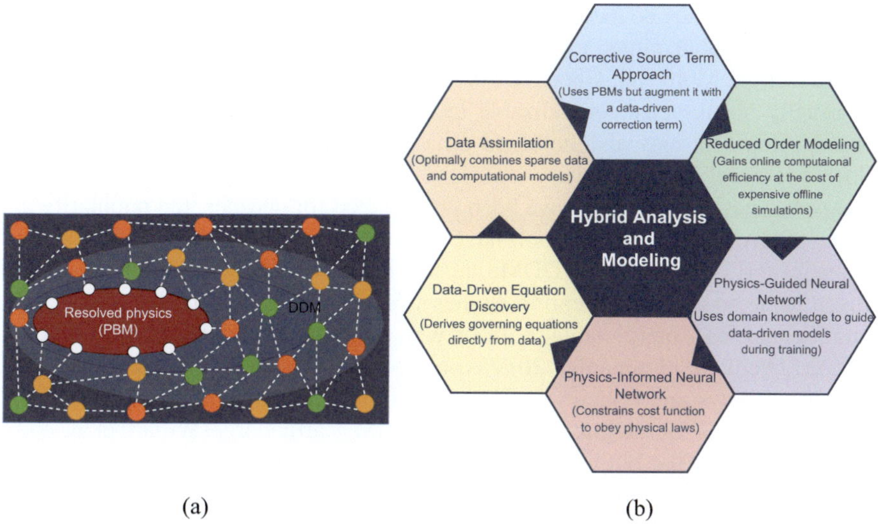

Fig. 5.1 Hybrid analysis and modeling. (**a**) HAM maximizes the utilization of PBM. (**b**) Six broad categories of HAM. (Reproduced from Stadtmann et al. (2023))

5.4.3 Synthesis and Categories of HAM Techniques

Table 5.1 compares the properties of KBM and DDM illustrating how neither of these modeling approaches by themselves are ideal for building a full DT. A new modeling paradigm called hybrid analysis and modeling (HAM, Fig. 5.1a) seeks to fulfill this gap by combining the main perks of KBM (i.e., generalizability, interpretability, robust foundation, and understanding) with those of advanced DDM (i.e., accuracy, computational efficiency, and automatic pattern-identification capabilities). This includes so called "Grey box models" and hybrid semi-parametric models which may be familiar to some.

Although the "perfect" digital twin should score well at all characteristics, this is not always possible. However, to instill physical realism, a DT should strive to at least maintaining the following characteristics: accuracy, computational efficiency, reliability, generalizability, and adaptability. This requires a synthesis of the abilities of KBM and DDM through HAM. KBMs can achieve high accuracy and fidelity but often at a high computational cost. For example, a few seconds of detailed simulations using computational fluid dynamics (CFD) modeling often requires weeks or even months of computation time on high-power computing clusters. Furthermore, while usually being reliable, trustworthy, and generalizable, KBMs are generally not able to adapt or self-evolve but rather fixed to preprogrammed models and mathematical relationships. Conversely, data-driven models are known to be highly adaptable and typically much faster than KBMs, thereby being more likely to suffice for online computation tasks. However, the complexity and black-box nature of DDMs means they do meet the requirements within reliability and trustworthiness for industrial applications, especially in situations that are safety-critical such as fault detection. Moreover, since DDMs are often applied to very specific tasks, they are usually not directly generalizable to new scenarios once trained. In addition, their development may also be based on a trial-and-error approach that cannot be allowed in a real environment.

Despite that their utilities have been investigated for decades, the popularity and impact of HAM approaches as tools have only increased the last few years. The main reason for this is probably that DDMs are now becoming increasingly successful and proven to be so through scientific studies while at the same time becoming more complex and less interpretable, both of which are arguments for the introduction of HAM (Venkatasubramanian 2019; von Stosch et al. 2014). The combined body of recent surveys and reviews on HAM constitutes a comprehensive overview of different methods for integrating DDM with PBM (particularly (Willard et al. 2022) and (San et al. 2021), but also (Willcox et al. 2021; Karpatne et al. 2017; Ihme et al. 2022)). Most of these hybridization techniques (Fig. 5.1b) pertain to one of the following categories in HAM:

Corrective source term approach Corrective source term approach (CoSTA) (Blakseth et al. 2022) methods explicitly address the presence of unknown physical, chemical, or biological dynamics by augmenting the governing equations of a KBM that describe partial system dynamics with a corrective source term that through a DNN is able to account for the remaining unknown dynamics. This approach has the benefit that natural laws can be used to keep a sanity check on the predictions of the DNN, i.e., checking conservation laws for the model outputs. Similar approaches have also been applied to model unresolved turbulent flow physics (Maulik et al. 2019; Pawar et al. 2020a). CoSTA methods can also use data-driven source terms to compensate for inaccuracies due to ignored subgrid scales when development of coarse-scale simulations. This will thus result in more computationally efficient models that are more suited for online operation than high-fidelity models but that suffers less from the accuracy loss often suffered when reducing model resolution. For aquaculture purposes, this approach could be used to enable more efficient predictions of the ambient hydrodynamic conditions around a site that in turn can be used for the planning and execution of operations that are sensitive to the hydrodynamic conditions. The models available for such predictions today are comprehensive oceanographic models that are not suitable for real-time or close to real-time operations. Another relevant example application of CoSTA is its use to achieve accurate wake modeling and fill in the coarse spatiotemporal resolution of measured data when modeling wind farms, which in turn renders the diagnosis of anomalies and future predictions of the system more accurate. This approach could be transferred to aquaculture applications, since the accurate planning of fish farms will require considerations on the hydrodynamic impacts on the farms. In addition to affecting the farm structures, water movements including the formation of wakes and eddies will impact the transport of oxygen, waste materials, and other substances to and from the farm. However, even these approaches assume a specific structure for at least the known part of the equation.

Data assimilation Data assimilation is a methodology that integrates observational data and dynamical models, and that is strongly rooted in numerical weather predictions with a decades long operational history (Lewis and Lakshmivarahan 2006; Lakshmivarahan et al. 2017). Unlike CoSTA, this approach is founded more strongly in DDM due to the need for relatively fast and frequent predictions in typical application areas such as in meteorology. One of the most prominent challenges of using DDM is to find ways of handling distributional shifts or poor generalization in the data, which will often include generalizing the chosen method outside its training conditions. Data assimilation algorithms provide this by using the available sensor measurement streams from the physical system to dynamically correct the model. In addition to resulting in a system that is corrected by real data, this setup can also be used to improve the DDM dynamics by finding parameterizations that give a more accurate model, in turn minimizing the correction term (Pawar et al. 2021a; Pawar and San 2021; Ahmed et al. 2020). Many of the KBM and DDM used to describe aquaculture production processes are subject to large uncertainties, both due to the lack of knowledge and because they often seek to describe complex dynamics within large volumes. The assimilation of data into such models is thus a crucial step in improving their utility in the planning and execution of farming operations. Data assimilation techniques can also be used for the optimization of experimental setups, sensor placement in particular. This can help in both improving the inference algorithms and reducing the costs of data collection (Deng et al. 2021; Uilhoorn 2022). Considering the large volume of a commercial net pen, this method is highly relevant as it is practically impossible to fully measure factors such as oxygen or temperature throughout the entire volume. See (Gettelman et al. 2022; Geer 2021; Brajard et al. 2021; Buizza et al. 2022) for further outlines and discussions on model-data fusion and the integration of DDM and data assimilation approaches.

Data-driven discovery of new dynamics In some cases, the understanding of the physical, chemical, or biological dynamics of a system in question is incomplete. It may then be more challenging to derive a digital twin representation of the system as the equations describing the system dynamics cannot be written down directly from first principles. Although some of the other HAM approaches (e.g., COSTA) seek to compensate this through the use of DDM methods where knowledge is too scarce to derive KBM equations, these do not contribute to improving the understanding of this phenomenon and discovering the new dynamics. This calls for a more targeted approach where the DDM outputs are used to identify the underlying mechanisms. Examples of such practices include sparse regression through l_1 regularization and symbolic regression based on gene expression programming, and there are several concrete examples of their use (Brunton et al. 2016; Champion et al. 2019; Vaddireddy et al. 2020). Other approaches include using high-resolution LIDAR data to directly parameterize mathematical models of wakes and the detection of anomalies in gearboxes using equations derived based on infrared thermography data. This method is also relevant in aquaculture, particularly to derive models of phenomenons such as complex hydrodynamics (representing wakes, eddies, and vortex shedding). Alternatively, this technique could be used as an exploratory tool for discovering new insights into fish biology. A possible case related to the latter of these could be to derive DDM from images or video footage of fish that are experiencing reduced welfare (e.g., due to disease or parasites). If successful, the resulting DDM could provide a way of diagnosing the reduced welfare state at an early stage through hitherto unknown external features or behaviors. However, methods in this category may be limited by other factors such as the need to handcraft additional features (in the case of sparse regression) or overfitting or instability (in the case of symbolic regression).

Physics-guided neural network Physics-guided NN (PGNN) or physics-guided ML (PGML) is a rapidly expanding field featuring methods designed to generalize data-driven models by incorporating neural networks with information from simplified KBMs (Robinson et al. 2022; Pawar et al. 2021b). This increasing interest has also led to the release of the Theseus library (Pineda et al. 2023) that contributes to further increasing the research in this area. In brief terms, a PGNN/ML fuses data and different types of models by embedding the knowledge from one or several simplified KBM into the intermediate layers of a neural network (Pawar et al. 2021c, 2022). The injection of this knowledge into the neural network ensures that the network only seeks to learn the elements not covered by the KBM (and that hence is known a priori) rather than seeking to parameterize a full system model from scratch. This also contributes to rendering the resulting model more generalizable than a purely data-driven approach, since the knowledge from the KBM is based on the laws of nature. An illustrative example of using this approach can be the prospect of predicting the flow patterns around an airfoil, which is a high-dimensional and nonlinear problem. While this is a problem that can be solved purely based on CFD, such a solution would be infeasible for online purposes. In applications requiring real-time or close to real-time operation such as the determination of aerodynamic characteristic of turbines in power plant control systems, PGNN/PGML could be a more suitable solution. The solutions from CFD analyses could then be obtained offline and used to insert system knowledge into the intermediate layers of a NN, thereby targeting the training of the NN more strongly toward the end goal from the start. This example is directly transferrable to aquaculture, as most of the advanced analyses of the flow patterns in commercial units (be they land based or at sea) are done using CFD or other similar high-fidelity methods. As for wind power applications, CFD-type models of aquaculture facilities are not suitable for online use, meaning that other tools would be needed for operational support during critical procedures such as fish handling, delousing, and tank cleaning. Developing solutions based on PGNN/ML for operational support when the flow patterns are crucial would therefore be an important element in enabling decision and operation support in aquaculture.

Physics-informed neural network Another approach for improving DDM performance through the use of physical knowledge is a class of methods called physics-informed neural network (PINN). These methods are designed to bias the DDM toward known physical laws and knowledge by incorporating KBM in the objective function used in the training process. While this may resemble the PGNN/ML approaches mentioned above, moving the knowledge injection from the intermediate layers to the objective function leads to other properties. There are several examples of the use of PINN including the use of a NN to represent the solution of a PDE (Raissi et al. 2019), solving heat transfer problems for parts in a manufacturing process (Zobeiry and Humfeld 2021), enabling control in a state-space setting (Arnold and King 2021) and classifying bearing health by using a NN whose outputs were regularized based on physical thresholds (Shen et al. 2021). Despite their utility, PINN have some challenges, most prominent of which are their need for precise knowledge of the loss term and that the complexity of the cost function may complicate the training process. A state-of-the-art discussion on PINN was recently published, outlining the properties of such models and potential future developments (Cuomo et al. 2022).

Reduced order modeling Reduced order modeling (ROM) is an approach where full-order high-fidelity models (FOM) are simplified into reduced-dimensional spaces. This is typically done using methods such as proper orthogonal decomposition of FOM simulation results (snapshots) (Chinesta et al. 2011; Fonn et al. 2019; Quarteroni and Rozza 2014; Brunton et al. 2020; Taira et al. 2020) and has become a very popular approach for conducting model order reduction (Quarteroni and Rozza 2014). If properly executed, this process can result in a stable ROM featuring several orders of magnitude speedup as long as the dimension truncation does not result in the loss of core information from the original FOM outputs. There are two main limitations of these models when applied to realistic problems, the first of which is that the dimension truncation may destabilize the model. This can be addressed by using eddy-viscosity-based ROMs (Östh et al. 2014; Hijazi et al. 2020; Tsiolakis et al. 2021) or semi-intrusive ROMs (Pawar and San 2021; Ahmed et al. 2019; Pawar et al. 2020b, c; Wang et al. 2020; Ivagnes et al. 2023) that correct the effect of ignored modes using a DDM. The second challenge is that such models tend to be intrusive in the sense that building a ROM requires both the original equations and the data; an issue has led to the development of fully non-intrusive ROMs which are gaining popularity (Xiao et al. 2015; Yu et al. 2019; Rahman et al. 2019; Siddiqui et al. 2019; Ahmed et al. 2021a; Benner et al. 2020). ROM is an attractive approach for aquaculture purposes as many of the models used to describe features such as structural dynamics and water and oxygen flows through the farm tend to be complex and computationally demanding. While the use of PGNN and PINN can be used to get acceptably accurate estimates of this dynamics, it is possible that using ROM to reduce the complexity of the original models will give more accurate assessments of these dynamics. A comprehensive review of the ROM methodology has recently been published (Ahmed et al. 2021b).

5.5 Summary and Conclusion

This chapter has briefly recounted the properties, nature, and use of the concepts of hybrid analysis and modeling (HAM). As discussed above, a successful implementation of a digital twin of a real-world asset or system is likely to need a combination of accuracy/fidelity, computational efficiency, reliability, generalizability, and adaptability that will require exploiting the properties of both KBMs and DDMs. This will call for the application of one or more of the HAM methodologies outlined above. Previous chapters in this book have presented many of the existing components needed to build HAM methods for aquaculture. In addition to highlighting what exists today, this may also guide future developments within KBMs and DDMs as it identifies gaps that need to be filled for the aquaculture sector to follow the same development in HAM as other industries.

The potential impact and utility of using HAM methods in aquaculture is best described through examples of potential implementations within different HAM categories:

- Corrective source term approach: In aquaculture, CoSTA could be used to correct inaccuracies in the use of KBMs for predicting future states. An example here is the modeling of oxygen in modern fish farms that has recently been simulated (Alver et al. 2023). While the main dynamics of this model is founded on the well-established transport equation, interactions with other entities such as adjacent fish farms are harder to define. This might be a case where CoSTA could have been applied using a DDM to correct the boundary conditions for the model and thereby providing a more realistic simulation. While this would require data that describe this interaction, it would not require deep knowledge on how this interaction is manifested.
- Data assimilation: Data assimilation could be used in aquaculture as in other industries for cases where the knowledge on the system is not sufficient for developing KBMs able to representing the full system state with the desired accuracy. An example here could be to use HAM for better integrated simulations of environmental, structural, and biological dynamics on farm scale, as proposed by (Alver et al. 2022).
- Data-driven discovery of new dynamics: Data-driven discovery may also be highly relevant for aquaculture applications. If sufficient data exists for doing this, it is potentially possible to unveil relationships that have hitherto proven difficult to identify. Sometimes such discoveries can even have importance well beyond the specific case for which the exercise is being done. For instance, in a previous study aimed at modeling growth in farmed salmon, the fish in the real asset suffered a disease outbreak during the experimental period, inducing lowered growth rates (Føre et al. 2016). While the KBM used in this study largely captured the fish growth under normal conditions, it did not account for the impacts of disease; hence, there was a deviation between model output and measured growth during the most intense disease period. If both growth data and disease status could be obtained with sufficient level of detail for that study and

other studies describing growth both before, during, and after outbreaks, it could be possible to use data-driven discovery methods to derive ML models able to predict the impacts of disease on fish growth and combine these with the KBMs using other HAM methods.

- Physics guided/informed neural networks: PGNN/ML or PINN are also methods that could be of potential use in aquaculture. In particular, while there is some knowledge on the behavioral dynamics of fish in captivity (Oppedal et al. 2011), the knowledge is not complete, as evidenced by the flaws of KBM seeking to mimic this dynamics (Føre et al. 2016). However, the rapid recent developments in machine learning have also led to an increasing interest in applying ML methods to data obtained with video (Saberioon et al. 2017) or acoustics (Måløy et al. 2019) to automatically detect and categorize different behaviors. This is an area in which physics-guided/informed approaches could be used to refine these methods, so that they become more accurate, and training is steered more specifically toward detecting the desired features early on. Examples of knowledge that could be used in this way could be features extrapolated from theory and qualitative observations like the range of individual movement speed relative to fish size.
- Reduced order modeling: Considering that aquaculture is conducted underwater, hydraulics and hydrodynamics are important concepts that many modeling approaches need to account for and often describe with relatively high fidelity. KBMs portraying such phenomenons tend to be based on computationally intensive methods (e.g., CFD, smoothed hydrodynamic particles) that are not suitable for online or real-time use. This is an area in which ROM could be applied to yield more lightweight models with sufficient accuracy.

In conclusion, it is apparent that various HAM methods will have a use in the development of digital twin technology for aquaculture applications. Some such approaches may be used to cover the gap between existing knowledge on the dynamics in aquaculture production systems, while others may be used to gain new knowledge on system dynamics. HAM could thus represent a crucial component in ensuring that the aquaculture industry can follow other sectors in the development of digital twins for improved control.

References

Abdou K, Ben Rais Lasram F, Romdhane MS, Le Loc'h F, Aubin J (2018) Rearing performances and environmental assessment of sea cage farming in Tunisia using life cycle assessment (LCA) combined with PCA and HCPC. Int J Life Cycle Assess 23:1049–1062

Ahmed SE, Rahman SM, San O, Rasheed A, Navon IM (2019) Memory embedded non-intrusive reduced order modeling of non-ergodic flows. Phys Fluids 31:126602. https://doi.org/10.1063/1.5128374. Publisher: American Institute of Physics

Ahmed SE, Bhar K, San O, Rasheed A (2020) Forward sensitivity approach for estimating eddy viscosity closures in nonlinear model reduction. Phys Rev E 102:043302. https://doi.org/10.1103/PhysRevE.102.043302. Publisher: American Physical Society

Ahmed SE, San O, Rasheed A, Iliescu T (2021a) Nonlinear proper orthogonal decomposition for convection-dominated flows. Phys Fluids 33:121702. https://doi.org/10.1063/5.0074310. Publisher: American Institute of Physics

Ahmed SE et al (2021b) On closures for reduced order models—a spectrum of first-principle to machine-learned avenues. Phys Fluids 33:091301. https://doi.org/10.1063/5.0061577. Publisher: American Institute of Physics

Alaliyat S, Yndestad H, Davidsen PI (2019) An agent-based approach for predicting patterns of pathogen transmission between aquaculture sites in the Norwegian fjords. Aquaculture 505:98–111

Alver MO, Hancke K, Sakshaug E, Slagstad D (2014) A spectrally-resolved light propagation model for aquatic systems: steps toward parameterizing primary production. J Mar Syst 130:134–146

Alver MO, Føre M, Alfredsen JA (2022) Predicting oxygen levels in Atlantic salmon (*Salmo salar*) sea cages. Aquaculture 548:737720. https://doi.org/10.1016/j.aquaculture.2021.737720

Alver MO, Føre M, Alfredsen JA (2023) Effect of cage size on oxygen levels in Atlantic salmon sea cages: a model study. Aquaculture 562:738831. https://doi.org/10.1016/j.aquaculture.2022.738831

Arnold F, King R (2021) State–space modeling for control based on physics-informed neural networks. Eng Appl Artif Intell 101:104195. https://doi.org/10.1016/j.engappai.2021.104195

Banerjee A et al (2022) Carp-DCAE: deep convolutional autoencoder for carp fish classification. Comput Electron Agric 196:106810

Behroozi L, Couturier MF (2019) Prediction of water velocities in circular aquaculture tanks using an axisymmetric CFD model. Aquac Eng 85:114–128

Benner P, Goyal P, Kramer B, Peherstorfer B, Willcox K (2020) Operator inference for non-intrusive model reduction of systems with non-polynomial nonlinear terms. Comput Methods Appl Mech Eng 372:113433. https://doi.org/10.1016/j.cma.2020.113433

Blakseth SS, Rasheed A, Kvamsdal T, San O (2022) Deep neural network enabled corrective source term approach to hybrid analysis and modeling. Neural Netw 146:181–199. https://doi.org/10.1016/j.neunet.2021.11.021

Brajard J, Carrassi A, Bocquet M, Bertino L (2021) Combining data assimilation and machine learning to infer unresolved scale parametrization. Philos Trans R Soc A: Math Phys Eng Sci 379:20200086. https://doi.org/10.1098/rsta.2020.0086

Brunton SL, Proctor JL, Kutz JN (2016) Discovering governing equations from data by sparse identification of nonlinear dynamical systems. Proc Natl Acad Sci 113:3932–3937. https://doi.org/10.1073/pnas.1517384113. Publisher: Proceedings of the National Academy of Sciences

Brunton SL, Noack BR, Koumoutsakos P (2020) Machine learning for fluid mechanics. Annu Rev Fluid Mech 52:477–508. https://doi.org/10.1146/annurev-fluid-010719-060214

Buizza C et al (2022) Data learning: integrating data assimilation and machine learning. J Comput Sci 58:101525. https://doi.org/10.1016/j.jocs.2021.101525

Champion K, Lusch B, Kutz JN, Brunton SL (2019) Data-driven discovery of coordinates and governing equations. Proc Natl Acad Sci 116:22445–22451. https://doi.org/10.1073/pnas.1906995116

Chinesta F, Ladeveze P, Cueto E (2011) A short review on model order reduction based on proper generalized decomposition. Arch Comput Methods Eng 18:395–404. https://doi.org/10.1007/s11831-011-9064-7

Cromey CJ, Nickell TD, Black KD (2002) Depomod—modelling the deposition and biological effects of waste solids from marine cage farms. Aquaculture 214:211–239

Cuomo S et al (2022) Scientific machine learning through physics–informed neural networks: where we are and what's next. J Sci Comput 92:88. https://doi.org/10.1007/s10915-022-01939-z

Deng Z, He C, Liu Y (2021) Deep neural network-based strategy for optimal sensor placement in data assimilation of turbulent flow. Phys Fluids 33:025119. https://doi.org/10.1063/5.0035230

Filgueira R, Rosland R, Grant J (2011) A comparison of scope for growth (SFG) and dynamic energy budget (DEB) models applied to the blue mussel (mytilus edulis). J Sea Res 66:403–410

Fonn E, van Brummelen H, Kvamsdal T, Rasheed A (2019) Fast divergence-conforming reduced basis methods for steady Navier–Stokes flow. Comput Methods Appl Mech Eng 346:486–512. https://doi.org/10.1016/j.cma.2018.11.038

Føre M, Dempster T, Alfredsen JA, Johansen V, Johansson D (2009) Modelling of Atlantic salmon (Salmo salar L.) behaviour in sea-cages: a Lagrangian approach. Aquaculture 288:196–204

Føre M, Dempster T, Alfredsen JA, Oppedal F (2013) Modelling of Atlantic salmon (Salmo salar L.) behaviour in sea-cages: using artificial light to control swimming depth. Aquaculture 388:137–146

Føre M et al (2016) Modelling growth performance and feeding behaviour of Atlantic salmon (*Salmo salar* L.) in commercial-size aquaculture net pens: model details and validation through full-scale experiments. Aquaculture 464:268–278. https://doi.org/10.1016/j.aquaculture.2016.06.045

Føre M et al (2024) Digital Twins in intensive aquaculture — challenges, opportunities and future prospects. Comput Electron Agric 218:108676. https://doi.org/10.1016/j.compag.2024.108676

Fredriksson DW, Swift MR, Irish JD, Tsukrov I, Celikkol B (2003) Fish cage and mooring system dynamics using physical and numerical models with field measurements. Aquac Eng 27:117–146

Geer AJ (2021) Learning earth system models from observations: machine learning or data assimilation? Philos Trans R Soc A Math Phys Eng Sci 379:20200089. https://doi.org/10.1098/rsta.2020.0089

Gettelman A et al (2022) The future of Earth system prediction: advances in model-data fusion. Sci Adv 8:eabn3488. https://doi.org/10.1126/sciadv.abn3488

Gutiérrez-Estrada JC, Pulido-Calvo I, de la Rosa I, Marchini B (2012) Modeling inflow rates for the water exchange management in semi-intensive aquaculture ponds. Aquac Eng 48:19–30

Hijazi S, Stabile G, Mola A, Rozza G (2020) Data-driven POD-Galerkin reduced order model for turbulent flows. J Comput Phys 416:109513. https://doi.org/10.1016/j.jcp.2020.109513

Huang Y-P, Khabusi SP (2023) A cnn-oselm multi-layer fusion network with attention mechanism for fish disease recognition in aquaculture. IEEE Access 11:58729–58744

Huang J et al (2022) Recognizing fish behavior in aquaculture with graph convolutional network. Aquac Eng 98:102246

Huggins DL, Piedrahita RH, Rumsey T (2004) Analysis of sediment transport modeling using computational fluid dynamics (CFD) for aquaculture raceways. Aquac Eng 31:277–293

Ihme M, Chung WT, Mishra AA (2022) Combustion machine learning: principles, progress and prospects. Prog Energy Combust Sci 91:101010. https://doi.org/10.1016/j.pecs.2022.101010

Islam MM, Kashem MA, Uddin J (2021) Fish survival prediction in an aquatic environment using random forest model. IAES International Journal of Artificial Intelligence (IJ-AI) 10:8938

Ivagnes A, Stabile G, Mola A, Iliescu T, Rozza G (2023) Hybrid data-driven closure strategies for reduced order modeling. Appl Math Comput 448:127920. https://doi.org/10.1016/j.amc.2023.127920

Karpatne A et al (2017) Theory-guided data science: a new paradigm for scientific discovery from data. IEEE Trans Knowl Data Eng 29:2318–2331. https://doi.org/10.1109/TKDE.2017.2720168. Conference Name: IEEE Transactions on Knowledge and Data Engineering

Khotimah WN (2015) Aquaculture water quality prediction using smooth SVM. IPTEK J Proc Ser 1:342–345

Klebert P, Lader P, Gansel L, Oppedal F (2013) Hydrodynamic interactions on net panel and aquaculture fish cages: a review. Ocean Eng 58:260–274

Kooijman SALM (2010) Dynamic energy budget theory for metabolic organisation. Cambridge University Press, Cambridge

Lader P, Kristiansen D, Alver M, Bjelland HV, Myrhaug D (2017) Classification of aquaculture locations in Norway with respect to wind wave exposure. In: International conference on

offshore mechanics and arctic engineering, vol 57724, V006T05A005. American Society of Mechanical Engineers
Lakshmivarahan S, Lewis JM, Jabrzemski R (2017) Forecast error correction using dynamic data assimilation. Springer, Cham
Lewis JM, Lakshmivarahan S (2006) Dynamic data assimilation: a least squares approach. Cambridge University Press, New York
Måløy H, Aamodt A, Misimi E (2019) A spatio-temporal recurrent network for salmon feeding action recognition from underwater videos in aquaculture. Comput Electron Agric 167:105087. https://doi.org/10.1016/j.compag.2019.105087
Maulik R, San O, Rasheed A, Vedula P (2019) Subgrid modelling for two-dimensional turbulence using neural networks. J Fluid Mech 858:122–144. https://doi.org/10.1017/jfm.2018.770. Publisher: Cambridge University Press
Moe H, Fredheim A, Hopperstad OS (2010) Structural analysis of aquaculture net cages in current. J Fluids Struct 26:503–516
Oppedal F, Dempster T, Stien LH (2011) Environmental drivers of Atlantic salmon behaviour in sea-cages: a review. Aquaculture 311:1–18. https://doi.org/10.1016/j.aquaculture.2010.11.020
Östh J, Noack BR, Krajnović S, Barros D, Boree J (2014) On the need for a nonlinear subscale turbulence term in POD models as exemplified for a high-Reynolds-number flow over an Ahmed body. J Fluid Mech 747:518–544. https://doi.org/10.1017/jfm.2014.168. Publisher: Cambridge University Press
Pawar S, San O (2021) Data assimilation empowered neural network parametrizations for subgrid processes in geophysical flows. Phys Rev Fluids 6:050501. https://doi.org/10.1103/PhysRevFluids.6.050501. Publisher: American Physical Society
Pawar S, San O, Rasheed A, Vedula P (2020a) A priori analysis on deep learning of subgrid-scale parameterizations for Kraichnan turbulence. Theor Comput Fluid Dyn 34:429–455. https://doi.org/10.1007/s00162-019-00512-z
Pawar S, Ahmed SE, San O, Rasheed A (2020b) Data-driven recovery of hidden physics in reduced order modeling of fluid flows. Phys Fluids 32:036602. https://doi.org/10.1063/5.0002051. Publisher: American Institute of Physics
Pawar S, Ahmed SE, San O, Rasheed A (2020c) An evolve-then-correct reduced order model for hidden fluid dynamics. Mathematics 8:570. https://doi.org/10.3390/math8040570. Number: 4. Publisher: Multidisciplinary Digital Publishing Institute
Pawar S, San O, Rasheed A, Navon IM (2021a) A nonintrusive hybrid neural-physics modeling of incomplete dynamical systems: Lorenz equations. GEM – Int J Geomathematics 12:17. https://doi.org/10.1007/s13137-021-00185-z
Pawar S, San O, Aksoylu B, Rasheed A, Kvamsdal T (2021b) Physics guided machine learning using simplified theories. Phys Fluids 33:011701. https://doi.org/10.1063/5.0038929. Publisher: American Institute of Physics
Pawar S, San O, Nair A, Rasheed A, Kvamsdal T (2021c) Model fusion with physics-guided machine learning: projection-based reduced-order modeling. Phys Fluids 33:067123. https://doi.org/10.1063/5.0053349. Publisher: American Institute of Physics
Pawar S, San O, Vedula P, Rasheed A, Kvamsdal T (2022) Multi-fidelity information fusion with concatenated neural networks. Sci Rep 12:5900. https://doi.org/10.1038/s41598-022-09938-8
Pedersen S, Wik T (2020) A comparison of topologies in recirculating aquaculture systems using simulation and optimization. Aquac Eng 89:102059
Peterson EL, Harris JA, Wadhwa LC (2000) CFD modelling pond dynamic processes. Aquac Eng 23:61–93
Pineda L et al (2023) Theseus: a library for differentiable nonlinear optimization. https://doi.org/10.48550/arXiv.2207.09442. ArXiv:2207.09442 [cs, math]
Quarteroni A, Rozza G (eds) (2014) Reduced order methods for modeling and computational reduction. Springer International Publishing, Cham
Rahman SM, Pawar S, San O, Rasheed A, Iliescu T (2019) Nonintrusive reduced order modeling framework for quasigeostrophic turbulence. Phys Rev E 100:053306. https://doi.org/10.1103/PhysRevE.100.053306. Publisher: American Physical Society

Raissi M, Perdikaris P, Karniadakis G (2019) Physics-informed neural networks: a deep learning framework for solving forward and inverse problems involving nonlinear partial differential equations. J Comput Phys 378:686–707. https://doi.org/10.1016/j.jcp.2018.10.045

Rasheed A, San O, Kvamsdal T (2020) Digital twin: values, challenges and enablers from a modeling perspective. IEEE Access 8:21980–22012. https://doi.org/10.1109/ACCESS.2020.2970143. Conference Name: IEEE Access

Robinson H, Pawar S, Rasheed A, San O (2022) Physics guided neural networks for modelling of non-linear dynamics. Neural Netw 154:333–345. https://doi.org/10.1016/j.neunet.2022.07.023

Russo T, Scardi M, Boglione C, Cataudella S (2011) Application of the self-organizing map to the study of skeletal anomalies in aquaculture: the case of dusky grouper (epinephelus marginatus Lowe, 1834) juveniles reared under different rearing conditions. Aquaculture 315:69–77

Saberioon M, Gholizadeh A, Cisar P, Pautsina A, Urban J (2017) Application of machine vision systems in aquaculture with emphasis on fish: state-of-the-art and key issues. Rev Aquac 9:369–387. https://doi.org/10.1111/raq.12143. _eprint: https://onlinelibrary.wiley.com/doi/pdf/10.1111/raq.12143

San O, Rasheed A, Kvamsdal T (2021) Hybrid analysis and modeling, eclecticism, and multifidelity computing toward digital twin revolution. GAMM-Mitteilungen 44:e202100007. https://doi.org/10.1002/gamm.202100007. _eprint: https://onlinelibrary.wiley.com/doi/pdf/10.1002/gamm.202100007

Shen S et al (2021) A physics-informed deep learning approach for bearing fault detection. Eng Appl Artif Intell 103:104295. https://doi.org/10.1016/j.engappai.2021.104295

Siddiqui MS, Fonn E, Kvamsdal T, Rasheed A (2019) Finite-volume high-fidelity simulation combined with finite-element-based reduced-order modeling of incompressible flow problems. Energies 12:1271. https://doi.org/10.3390/en12071271. Number: 7. Publisher: Multidisciplinary Digital Publishing Institute

Stadtmann F et al (2023) Digital twins in wind energy: emerging technologies and industry-informed future directions. IEEE Access 11:110762–110795. https://doi.org/10.1109/ACCESS.2023.3321320. Conference Name: IEEE Access

Stavrakidis-Zachou O, Papandroulakis N, Lika K (2019) A deb model for European sea bass (Dicentrarchus labrax): parameterisation and application in aquaculture. J Sea Res 143:262–271

Sun M, Hassan SG, Li D (2016) Models for estimating feed intake in aquaculture: a review. Comput Electron Agric 127:425–438

Taira K et al (2020) Modal analysis of fluid flows: applications and outlook. AIAA J 58:998–1022. https://doi.org/10.2514/1.J058462

Tsiolakis V, Kvamsdal T, Rasheed A, Fonn E, van Brummelen H (2021) Reduced order models for finite-volume simulations of turbulent flow around wind-turbine blades. J Phys Conf Ser 2018:012042. https://doi.org/10.1088/1742-6596/2018/1/012042

Uilhoorn F (2022) An algorithm for finding optimal sensor placement and model uncertainty in data assimilation applied to Riemann problems. Appl Math Model 103:649–673. https://doi.org/10.1016/j.apm.2021.11.006

Uzun D, Demirel YK, Coraddu A, Turan O (2019) Time-dependent biofouling growth model for predicting the effects of biofouling on ship resistance and powering. Ocean Eng 191:106432

Vaddireddy H, Rasheed A, Staples AE, San O (2020) Feature engineering and symbolic regression methods for detecting hidden physics from sparse sensor observation data. Phys Fluids 32:015113. https://doi.org/10.1063/1.5136351. Publisher: American Institute of Physics

Van Dam A (1990) Multiple regression analysis of accumulated data from aquaculture experiments: a rice-fish culture example. Aquac Res 21:1–15

Venkatasubramanian V (2019) The promise of artificial intelligence in chemical engineering: is it here, finally? AIChE J 65:466–478. https://doi.org/10.1002/aic.16489. _eprint: https://onlinelibrary.wiley.com/doi/pdf/10.1002/aic.16489

von Stosch M, Oliveira R, Peres J, Feyo de Azevedo S (2014) Hybrid semi-parametric modeling in process systems engineering: past, present and future. Comput Chem Eng 60:86–101. https://doi.org/10.1016/j.compchemeng.2013.08.008

Wang Q, Ripamonti N, Hesthaven JS (2020) Recurrent neural network closure of parametric POD-Galerkin reduced-order models based on the Mori-Zwanzig formalism. J Comput Phys 410:109402. https://doi.org/10.1016/j.jcp.2020.109402

Wigner EP (1960) The unreasonable effectiveness of mathematics in the natural sciences. Richard courant lecture in mathematical sciences delivered at New York University, May 11, 1959. Commun Pure Appl Math 13:1–14. https://doi.org/10.1002/cpa.3160130102

Willard J, Jia X, Xu S, Steinbach M, Kumar V (2022) Integrating scientific knowledge with machine learning for engineering and environmental systems. ACM Comput Surv 55:66:1–66:37. https://doi.org/10.1145/3514228

Willcox KE, Ghattas O, Heimbach P (2021) The imperative of physics-based modeling and inverse theory in computational science. Nat Comput Sci 1:166–168. https://doi.org/10.1038/s43588-021-00040-z

Xiao D et al (2015) Non-intrusive reduced order modelling of the Navier–Stokes equations. Comput Methods Appl Mech Eng 293:522–541. https://doi.org/10.1016/j.cma.2015.05.015

Yang X et al (2021) Deep learning for smart fish farming: applications, opportunities and challenges. Rev Aquac 13:66–90

Yu J, Yan C, Guo M (2019) Non-intrusive reduced-order modeling for fluid problems: a brief review. Proc Inst Mech Eng Part G: J Aerosp Eng 233:5896–5912. https://doi.org/10.1177/0954410019890721. Publisher: IMECHE

Zhang X et al (2020) Evaluation and analysis of water quality of marine aquaculture area. Int J Environ Res Public Health 17:1446

Zhou X et al (2022) Atrous pyramid GAN segmentation network for fish images with high performance. Electronics 11:911

Zobeiry N, Humfeld KD (2021) A physics-informed machine learning approach for solving heat transfer equation in advanced manufacturing and engineering applications. Eng Appl Artif Intell 101:104232. https://doi.org/10.1016/j.engappai.2021.104232

Part III
Precision Fish Farming

Chapter 6
Digital Twin for Aquaculture

Martin Føre, Fearghal O'Donncha, T. Norton, and Jon Grant

Abstract A digital twin of a dynamic system or asset needs to combine monitoring of the states of the process/asset in question with mechanistic/physics based or data-driven models describing system dynamics. The main role of this chapter is to describe how the main aspects from Chaps. 1 to 5 may be synthesized into actionable Digital Twins for aquaculture. We first describe the general process of building a Digital Twin by combining the required components, and how this can be used to construct a digital representation of reality spanning the necessary dimensions to capture the full system dynamics. Following this, we explore how the necessary synthesis can be achieved using the knowledge and case studies presented in preceding chapters. This synthesis is examined in terms of capability levels that Digital Twins can achieve in aquaculture and how these compare with applications in other sectors, such as terrestrial livestock production. Finally, we present example applications of digital twins in aquaculture, founded in the topics presented in Chaps. 1–5 in this book. We conclude with reflections on future developments that will be critical to realize the full potential of digital twins in aquaculture.

Keywords Digital twin synthesis · Aquaculture case studies · Knowledge transfer from livestock production

M. Føre (✉)
Department of Engineering Cybernetics, NTNU, Trondheim, Norway
e-mail: martin.fore@ntnu.no

F. O'Donncha
IBM Research Europe, Dublin, Ireland

T. Norton
KU Leuven, Leuven, Belgium

J. Grant
Department of Oceanography, University of Dalhousie, Halifax, NS, Canada

F. O'Donncha et al. (eds.), *Digital Twin for Marine Fish Farming*,
Aquatic Production Systems, https://doi.org/10.1007/978-3-032-10362-8_6

6.1 Assembly of Digital Twin for Aquaculture

Setting up a Digital Twin of a dynamic system requires the combination of components spanning from instruments for real-time monitoring to predictive data-driven or knowledge-based models (Rasheed et al. 2020; VanDerHorn and Mahadevan 2021). This is also the case for aquaculture applications, as recently outlined by (Føre et al. 2024).

6.1.1 Technical Structure and Synthesis

The preceding chapters have explored key components essential for developing a Digital Twin for aquaculture, with the first two chapters covering systems physically interacting with the farm. Chapter 1 discussed methods for monitoring fish using existing and emerging technologies, while Chap. 2 presented the role of autonomous vehicles in aquaculture. The following three chapters covered the array of digital components necessary for Digital Twin development, with Chap. 3 focusing on machine learning, Chap. 4 on knowledge-based or mechanistic modeling, and Chap. 5 on hybrid approaches that integrate these methods. Assembly of a full digital twin for aquaculture requires combining components from each of these areas to achieve necessary capabilities for practical application. In brief, a digital twin should synthesize real-time data through sensors with existing knowledge through mechanistic models and artificial intelligence using hybrid modeling methods.

The dynamics of purely technical systems such as machineries or manufacturing plants can sometimes be described based on a high-fidelity representation of the system dynamics and the inputs impacting this system. However, this is not the case in aquaculture applications, since the various subsystems comprising a production system are often tightly coupled and interdependent. In aquaculture, Digital Twins must represent not only the dynamics of the system of interest but also the dynamics of adjacent subsystems that provide critical boundary conditions.

For instance, a digital twin used to predict/portray salmon growth dynamics in a net pen will need to define the environmental dynamics at the site and the deformations exhibited by the net structure to obtain the desired realism. This is because variations in temperature and oxygen influence fish appetite and growth (Mørkøre and Rørvik 2001; Solstorm et al. 2018), while the fish's oxygen consumption from feeding and swimming affects oxygen availability in the water (Grøttum and Sigholt 1998). Additionally, net deformation in response to incident water currents and waves and may in turn reduce the flow of water into the pen (Løland 1993).

Given these considerations, a robust Digital Twin for aquaculture should ideally portray environmental, structural, and biological dynamics with high fidelity to support a flexible platform for exploring different system aspects. This can be achieved in phases: first addressing environmental dynamics, then structural responses, and

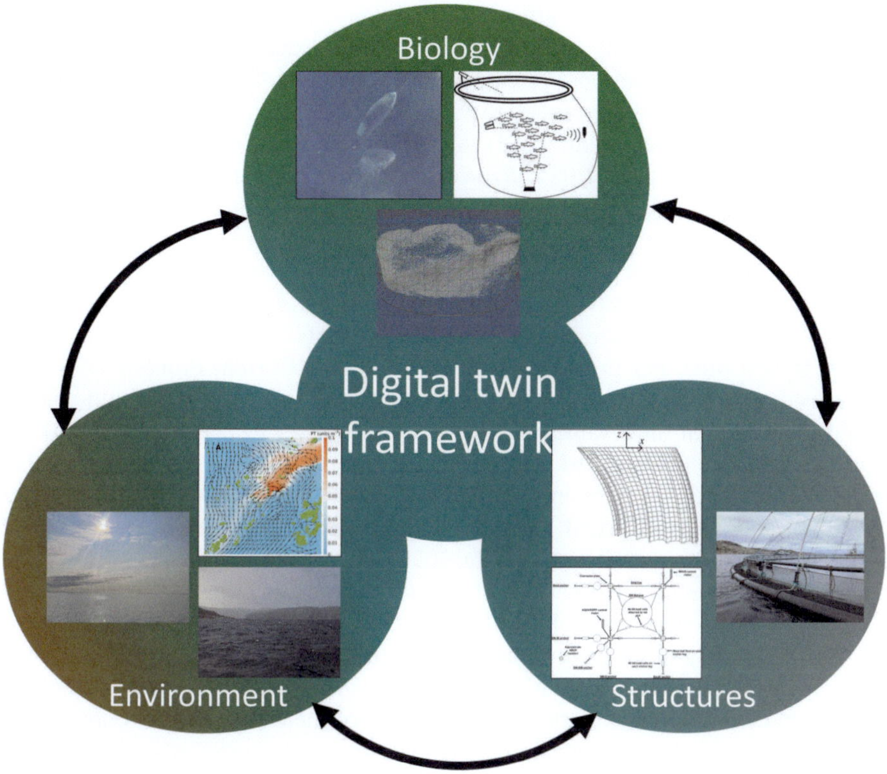

Fig. 6.1 Concept illustration of the interlinked domains a digital twin in aquaculture should portray, with examples of elements within each domain

finally biological interactions, and subsequently integrating these into a cohesive model. The interactions among these components ultimately form the foundation for the Digital Twin (Fig. 6.1).

6.1.2 Environmental Dynamics

In an aquaculture setting, the prevailing environmental dynamics dictate both structural and biological responses in a system. It is thus reasonable that the components comprising the environmental definition in a digital twin provide boundary conditions and inputs for the structural and biological modules. Since marine fish farming sites are exposed to coastal or oceanic environments, the conditions may vary strongly over both short- and long-time horizons, a feature that a digital twin should be equipped to handle.

Some environmental factors affecting fish farms can be considered mechanical in the sense that they typically apply forces that may impact both structures and fish. In marine fish farms, the most prominent of these are waves, currents, and wind. Waves and currents have been found to influence the structures at fish farms (Klebert et al. 2015; Kristiansen et al. 2015), and the behavior (Johansson et al. 2014), welfare/health (Hvas et al. 2021; Szewczyk et al. 2024), and physiology/growth (Solstorm et al. 2015) of farmed fish. Moreover, the water transport into, within, and out from a cage is also a crucial factor behind the influx of unwanted pathogens and parasites within the pen environment (Ferreira et al. 2021), and how intended particles such as feed distribute within the pen volume (Klebert and Su 2020). Several commercially available instruments exist for monitoring these factors in real time. For instance, Acoustic Wave and Current Profilers (AWACs) can effectively monitor both currents and waves at a site and output both spectra over time and real-time values for these factors (e.g., Michelsen et al. 2019). Currents can also be measured using Acoustic Doppler Current Profilers (ADCPs) that provide vertical or horizontal profiles of current in real time (e.g., Klebert et al. 2023), and Vector Current Sensors that give accurate point measurements of 3D water speeds (e.g., Klebert and Su 2020). These are only a selection of the potential data sources for observing the water movement in pens, but despite the plethora of instruments available, measurements will likely not be sufficient to provide a rendering of the environmental conditions precise enough for digital twin applications. A potential solution will be to assimilate data from such sources into existing modeling tools as implied in Chap. 4. For mechanical factors, this could include wave models such as Simulating Waves Nearshore (SWAN) (Booij et al. 1996), large-scale oceanographic models or pen scale current models based on, e.g., computational fluid dynamics (CFD) (Broch et al. 2020). Sophisticated assimilation schemes enable the exploitation of both the real-time properties of sensors and the predictive abilities of models in providing a description that allows reliable real-time assessment and future forecasting of the hydrological conditions at the site at the appropriate spatial scales. Although winds unlike waves and currents do not affect the fish directly, a digital twin should also incorporate wind effects due to the potential impact of winds and weather on the structures in the form of loads, icing, and other elements such as the surface distribution of pellets (e.g., Oehme et al. 2012). A potential combination of weather stations and sensors with forecasts from meteorological models could provide a foundation for this.

Several environmental factors at fish farms, while not directly impacting structural integrity and deformation through mechanical forces, play critical roles in biological dynamics. For simplicity, this discussion will focus on three of these that are often considered important across species, namely, temperature, dissolved oxygen (DO), and natural light intensity. Temperature and oxygen are, together with feed, usually the most important factors behind fish growth in aquaculture operations (Fry 1971). All three factors may also influence or stimulate external events that may influence the fish such as algae/jellyfish blooms (Purcell 2005; Dale et al. 2006), pathogen (Karvonen et al. 2010), and parasite (Brooks 2005) infestations. Additionally, they significantly impact fish behavior, as fish tend to move toward areas where these factors align with their preferred levels.

Several instruments exist to monitor these factors in fish farms. For temperature, point measurements are available either through specialized sensors deployed in the pen or as supplemental data registered by instruments such as submerged cameras. Vertical temperature profiles are also possible to obtain using conductivity temperature depth (CTD) probes or similar instruments, while horizontal variations of surface water temperature in large scale are detectable through satellite imagery (Wentz et al. 2000). Since water currents will influence the spatial temperature distribution over time, such measurements can often be complemented by oceanographic models to portray large-scale changes, and CFD models or similar, to account for local changes.

Similarly, DO can be monitored at specific points through existing tools, though horizontal patterns are more challenging to capture. However, the combined use of point measurements/CTD samples and oceanographic, CFD, or particle distribution models can provide a realistic view of the spatial distribution of DO in sea-based net pen and identify eventual hotspots or hypoxic regions inside the pen (e.g., Alver et al. 2022).

Although light has a less direct impact on the growth of the fish than temperature and dissolved oxygen, it may have an implicit effect influencing fish feeding (Boeuf and Le Bail 1999). Moreover, light will also have impacts on external processes, particularly those governed by photosynthesis such as algal blooms. Light level at points can be measured using simple sensors. These measurements or estimates of surface light can be used to derive vertical variations in the water column using established principles for light attenuation from physics, such as the Beer-Lambert law and thereby obtain a realistic and accurate estimate of the spatial variations in light intensity within the production volume.

6.1.3 Structural Dynamics

Monitoring or estimating the structural dynamics of aquaculture production units is important to avoid structural breakdowns and damages, and by extension fish escapes (Moe Føre and Thorvaldsen 2021). Moreover, structural dynamics also has an important impact on the fish, since movement and deformations will modify the production volume directly (DeCew et al. 2013), which in turn may negatively impact fish welfare (Turnbull et al. 2005). A digital twin for aquaculture therefore needs to assess structural dynamics to present a complete picture of the situation in the production unit, both from an engineering and a biological perspective.

A complete description of structural dynamics needs to cover both how externally induced forces are conveyed to structures and structural components and how they respond to such forces (Løland 1993; Lader et al. 2008; Moe et al. 2010). The main environmental forces affecting marine structures are hydrodynamic and hydraulic forces arising due to the water movement in or around the unit. As outlined in the previous section, the variables causing such forces (e.g., currents and waves) are possible to both measure and assess using various combinations of instruments and model-based tools. However, accurate measurement of the resulting forces is not trivial, since this will require that the component or subsystem in question is isolated from the rest of the system in the sense that none of the induced forces can bypass the sensors.

For a full net pen, this would, for instance, mean that load cells would have to be mounted on all connection points between the pen and the mooring system. While this has been done in both laboratory (e.g., Moe et al. 2010) and full-scale industrial settings (e.g., Fredriksson et al. 2003), this method is probably too complicated and risky to be considered a potential permanent solution for monitoring fish farms in the future. Other principles including the attachment of depth measuring acoustic tags (DeCew et al. 2013) or acoustic locators (Su et al. 2021) to pen components can be used to obtain positional data from these components. However, this will provide only the position or depth of the point at which the device is attached and no further spatial information about the rest of the structure. This situation can be improved by using numerical models designed to predict structural responses toward environmental factors in combination with sensors, as done by Su et al. (2021). It is then possible to derive a prediction of the structural response and dynamics at a site that is corrected by the injected data from the sensors and thereby likely to be more accurate than the numerical models alone.

6.1.4 Biological Dynamics

Since the main aim in aquaculture is to produce biomass, a module describing the biological processes is naturally a critical component in a digital twin designed to support industrial development and research into aquaculture technology. The boundary conditions influencing biological processes are defined by the structural dynamics of the containment unit and the environmental conditions inside this unit, as outlined in the previous sections. Important biological processes are behavior, growth, welfare, and health development for the fish, and external biological factors such as the dynamics of parasites, pathogens, harmful algae blooms, and jellyfish. All these are complex features in themselves, and even more so when studying their interactions. A comprehensive description of these therefore requires an effort where data from the process obtained using sensors and instruments is combined with predictive models able to estimate or simulate the dynamics not captured by the measurements.

From a digital twin perspective, fish behavior, growth, welfare, and health should be considered a single system, since all these aspects are closely integrated in the fish biology and hence should be combined when seeking to realize a digital rendition of the real system (Føre et al. 2024). Several of the interconnections between these elements may be considered internal in a digital twin, since some interactions, e.g., between the gut content and feeding motivation (Olsen and Balchen 1992), are mostly independent of the external environment. Others again, such as the number of pellets ingested by a fish, need to be considered external, since they affect other subsystems being represented by the digital twin (Alver et al. 2016). Based on such considerations, it is possible to derive a description the interface between a fish biology representation in a digital twin and the rest of the system. This also requires identifying the parameters the other systems need to offer to the fish entities for these to have the proper inputs from the environmental and structural representations to describe interactions with these.

While parasite and pathogen dynamics, harmful algae blooms, and jellyfish must be considered separate from the fish-internal elements presented above, they will be a part of the subsystems shaping the production environment facing the fish. Accurately representing these features in a digital twin requires a robust environmental model that characterizes the conditions influencing their development. For instance, the successful prediction of algae blooms will often depend on a combination of a good representation of the ocean current patterns around the farm site, temperature and light in the water column, and satellite observations capable of identifying the presence of algae. In contrast, a good representation of large-scale parasite or pathogen infestation rates will depend more on regional current patterns, the proximity to other farm sites, and the parasite levels at these sites (e.g., Kragesteen et al. 2018). Infestation rates also need to be considered in smaller scale, i.e., at farm/cage level, in which case the farm layout, density of fish in the pens, fish behavior, and host characteristics (e.g., Bui et al. 2020) may be the primary consideration.

6.2 Digital Twin Syntheses: Case Studies and Their Capabilities

Based on the descriptions above, it is possible to illustrate how a digital twin rendition of an aquaculture production system can be composed using components presented in the preceding chapters. While a general digital twin representing the full system dynamics of all biological, physical, and environmental processes featured in an aquaculture operation would be the "holy grail" of digital twins in aquaculture, it is more feasible to initially target specific applications that only need to represent a subset of the system dynamics. This section presents three concrete examples of such targeted digital twin concepts based on the components presented in this book. These illustrate that the path toward a digital twin is incremental and rooted in real-world impact.

Furthermore, as outlined in Føre et al. (2024), the capability levels of a digital twin (Fig. 6.2) may be categorized as one of six different levels:

- Standalone: purely model-based setups that can be used for offline simulation studies

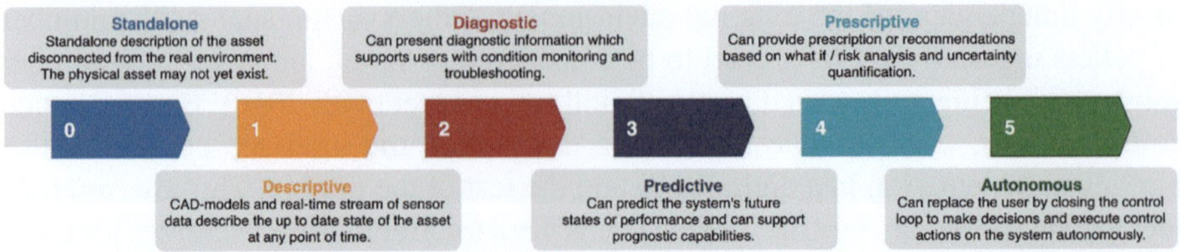

Fig. 6.2 The different capability levels of digital twin technology. Reproduced from Føre et al. (2024) under CC BY 4.0 license

- Descriptive: models combined with real-time/historical sensor data that describe the present state of the system
- Diagnostic: setups that in addition to describing the present state provides causal explanations of observed dynamics for deeper analyses of the system
- Predictive: arrangements with expanded capabilities in predicting future states and providing prognoses of future developments in state
- Prescriptive: setups capable of recommending future actions based on present and expected future states
- Autonomous: hybrid model systems fully connected with the process through sensors and actuators with the ability of proposing and issuing control signals directly to the process

The digital twin examples provided below will also be labeled according to this scale.

6.2.1 Oxygen Distribution in Net Pens

Maintaining sufficient oxygen levels in aquaculture production units is of the utmost importance to ensure acceptable welfare conditions and good growth. As outlined in the case studies presented by Føre et al. (2024), digital twin technology could serve an important purpose in this area. The first step in developing such a tool is to address environmental monitoring, i.e., setting up a digital twin that can provide better insight into the oxygen dynamics within the production unit based on sensor data combined with a hybrid approach using knowledge-based models (KBMs) and data-driven models (DDMs). Although oxygen measurements in open pens and tanks can usually only be obtained as point measurements, such data can be augmented with machine learning methods to obtain a predictive data-driven tool as elaborated in the first case study in Chap. 3. This will provide tools better able to forecast oxygen levels in pens or tanks. Oxygen dynamics in larger scales can be observed using other sensing principles such as satellite-based solutions. One such approach where satellite data is combined with machine learning techniques was presented in the second case study in Chap. 3. Using hybrid analysis and modeling (HAM), these two sensing methods could be fused with KBMs describing the oxygen dynamics in pens while accounting for the consumption by the fish (e.g., Alver et al. 2022). Given that the sensing principles and models used in combination could capture the most important dynamics of oxygen dynamics at an aquaculture site, this would represent a digital twin able to provide observations and forecasts of the oxygen dynamics in the area. As such, this digital twin would be able to achieve at least predictive capabilities in being able to forecast future oxygen conditions in both site and ambient levels. By integrating decision-support logic or optimization frameworks on top of the predictive outputs, the digital twin can be extended to exhibit prescriptive capabilities. For example, it could identify regions of the production volume most at risk of hypoxia and recommend targeted interventions—such as localized oxygenation—to mitigate the issue and enhance fish welfare.

6.2.2 *Impacts of Ammonia from Fish Farms on Ambient Environment*

Fish release ammonia and other nitrogen compounds through urea and feces. In intensive fish farming, the fish are actively fed, stimulating both increased growth and increased urea/feces production, and kept at much higher densities than natural populations would maintain. This means that the aggregated release of waste products from the fish in fish farms can have a notable impact on the local concentration of such substances near the farm, particularly if the ambient environment is naturally low on nutrients and current levels are low. Appropriate sensors could be deployed to detect ammonia levels at points around the farming sites, giving some insight into how the dispersal of the substances around the farm would be expressed. However, to get a better overview of the dynamics of this process and possibly predict such events, models synthesizing existing knowledge would be necessary, for example, by assimilating measurement data into KBMs and DDMs. This approach could be combined with models and measurements for predicting fish growth and possibly the environmental conditions at the site (for instance, similarly to the oxygen setup in the preceding case study). This would result in a digital twin describing the system dynamics of the link between fish growth and ammonia impacts on local marine flora, with predictive capabilities.

6.2.3 *Sea Lice Distribution*

Sea lice are naturally occurring parasites for fish, salmonids in particular, and have arisen to become one of the main challenges for intensive pen-based aquaculture, both due to the direct impact on fish welfare and due to the costs and biological risks of treatment. Being able to predict both the occurrence, scale and development of such outbreaks could be a tremendous aid in countering the industrial challenges with lice. Digital twin technology can be an important factor in achieving a better overview of this phenomenon. Data that describe lice infestations are obtained either counting the number of lice attached to a few individuals manually retrieved from a pen (Thorvaldsen et al. 2019) or by using lice traps that are deployed at strategic locations (e.g., Novales Flamarique et al. 2009). Although useful, both these data forms are scarce in their ability to describe the dynamics over large areas; hence, they need to be supported by models to provide a good picture of the dynamics. Good candidates for this could be a combination of models describing the processes occurring at the individual pen level and the dispersal between farms as outlined in case study 3 in Chap. 4. If complemented by methods from machine learning through HAM principles, these methods would yield a digital twin offering a more reliable, high-resolution, and accurate description of sea lice dynamics at commercial sites than either of these components could by themselves. Such a digital twin would cover the capability levels up to prescriptive, since they would not only provide possibilities of forecasting lice levels at pen and farm level but also implicitly offer advice on when various measures (e.g., lice skirts, delousing procedures) should be applied.

6.3 Parallels from Terrestrial Livestock Production

While this book is mainly focused on the adaptation and development of digital twin technology for applications within aquaculture, it is worthwhile to assess the situation in terrestrial livestock production, which in many aspects may be considered the most comparable industry segment, both in terms of aims and challenges. In practice, however, livestock farming is more amenable toward direct human and animal interaction than what is possible in the marine-based counterpart. Modern fish farms comprise pens with up to 200,000 fish. As farms are typically composed of 10–20 cages, and multiple farms are often co-located in an area, the total number of individual fish is enormous, vastly exceeding animal numbers seen in even the largest industrial livestock farms (e.g., large broiler farms in EU hold up to 100,000 individual animals, van Horne and Leenstra 2010). Furthermore, terrestrial farming does not typically have to manage a three-dimensional rearing environment as with fish farming. This precludes the direct translation of concepts from livestock farming. Moreover, the need to work on the group levels means that in practice, digital twin solutions for the aquacultural area can be seen as more tightly linked to precision livestock farming approaches developed for the pig and poultry sectors than those aimed at the dairy sector. That is, fish are today not managed as individuals as are cows but they are still more complex in management than pigs and poultry because of the additional degrees of freedom they have in their living space.

Where intensive aquaculture is a relatively young industry, terrestrial farming of animals has been a mainstay in human culture and food production for millennia.

Digital twin technology has yet to be fully adapted by the terrestrial livestock industry, mainly due to the lack of maturity in the technology required to realize such solutions. Being producers of commodity products, livestock farmers are typical risk averse when it comes to investing in technology. Nevertheless, the combination of the comparatively longer industrial history, extensive human knowledge on terrestrial animal husbandry, and maturing technologies from adjacent fields, digital technology has started to penetrate the sector. This has resulted in knowledge and technology outputs that are important inputs in the development of digital twin technology for aquaculture. The following will provide a brief overview of the state of the art in digital twin technology in terrestrial animal farming and shed light on some of the main lessons learned in the process of getting there.

6.3.1 Technical Structure and Synthesis

When considering the general constituents of a digital twin describing the dynamics in a terrestrial animal production scenario, the same three main components would be needed: biology, environment, and structures. However, while digital twins for pen-based fish farming require all of these, this need will vary for terrestrial animal husbandry depending on the production facilities. The biological aspects will always have to be present, but the necessity of having accurate descriptions of environmental or structural dynamics will vary from case to case. A main distinction in this area is the difference between outdoors and indoors production practices.

Outdoor Production Animals kept on pastures or in other outdoors production systems are subject to the ambient environment; thus, an accurate environmental description would be necessary to describe various meteorological conditions that may affect production (e.g., wind, air temperature, and humidity, solar loads). While weather forecasts and other public services can provide data describing aspects of the prevailing conditions in the production environment, an accurate digital twin would need data both of a higher spatial and temporal resolution (e.g., farm, field, or even animal level) but also that describe the conditions in real time. This will require the use of sensors and other instruments on site, preferably under farmer control. The environmental description could also describe local ecology to quantify ecosystems services as well as nearby predators considered to pose a risk for the animals (e.g., birds of prey, bears, wolves), and pathogens that could be transferred to the animals from the wild (e.g., bird flu, other diseases).

The structural dynamics representation relevant for this case would in contrast be simpler, as it might only need to describe fences or other structures used to restrain the movements of the animals during production. And this might even be limited to their presence, as such structures are less likely to suffer damage or be adversely affected by the environment during production than sea-based net pens.

Indoor Production Much livestock production is conducted indoors, either through the entire production process (e.g., broiler production) or in certain phases or times of year (e.g., early-life periods or during winter in temperate/subarctic clime zones for dairy and beef stock). This changes the requirements and outline of how a useful digital twin should be designed. Conditions are then largely independent of the ambient environment as the animals are then kept in houses whose interior environment is usually controlled.

The structural integrity and dynamics of the containment unit are of less importance in indoors livestock production than in aquaculture, since it is less likely to suffer damage due to environmental forces. However, the need to accurately control the indoors environment mentioned above means that a digital twin of indoors production would require detailed models and measurements of ventilation and feeding systems used in the facility to maintain appropriate production conditions.

6.3.2 Capability Levels

The models and instruments required for standalone and descriptive capability levels in digital twins for terrestrial farming are generally at the same level of maturity as for aquaculture. Although these levels need to be attained before it is possible to reach the higher capability levels (i.e., diagnostic, predictive, prescriptive, and autonomous), the path toward the higher capabilities may be shorter for terrestrial farming. This is because humanity through millennia has developed a much deeper knowledge on the relationships between cause and effect for terrestrial animals than for fish. In turn, this renders the transition from a description of the present system state into concrete diagnostic (i.e., what does the situation imply?), predictive (i.e., what is a likely future state of the system given the present state and existing knowledge?), or prescriptive (i.e., what can be done now to steer future states in the desired direction?) measures.

Indoors terrestrial farming also has a "head start" on enabling the final and highest capability levels of digital twins, i.e., autonomous capabilities. Since farmers then need to control the conditions in the production units, most of the foundation required to realize an autonomous digital twin would thus be in place, given there are sufficient instruments in the facility to maintain an overview of the conditions in real time.

6.3.3 Example Application in Terrestrial Farming

While there are not many examples of digital twins in the terrestrial livestock sector, some concepts have recently emerged in some production sectors. Early ideas have been explored for pig production, which like aquaculture is applying

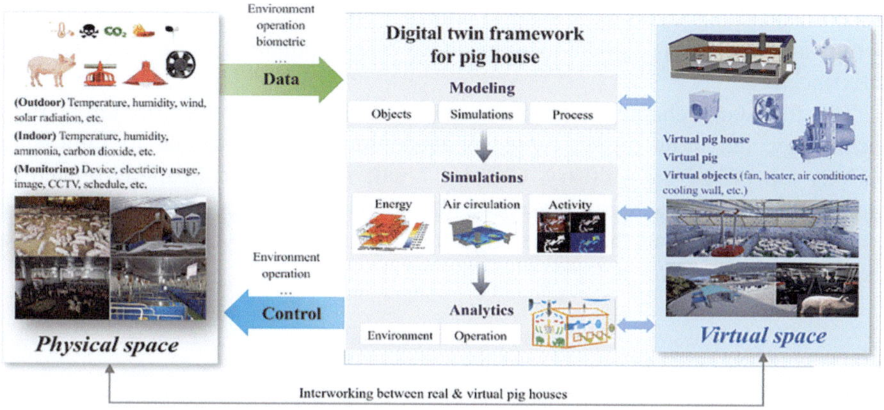

Fig. 6.3 Conceptual digital twin framework for pig production sector. (Reproduced from Jeong et al. 2023)

a batch-based approach to animal management. Jeong et al. (2023) presented a digital twin framework concept for pig houses shown in Fig. 6.3. They considered the energy efficiency of an HVAC (heating ventilation and air conditioning) based control under different climatic scenarios and were from that able to derive an optimal indoor environment control strategy in response to outdoor conditions. They first demonstrated how a digital twin approach can be utilized to estimate the energy demand of heating the pig house with three different HVAC systems using simulations models in combination with sensor data. Specifically, it was found that radiant heating had the highest total energy demand, followed by the heat pump and the fan coil unit. Secondly, using the digital twin, they could explore various weather scenarios to define optimal ventilation and radiant heating control for minimizing energy use. As a result of their work, the researchers suggest that the digital twin framework can contribute to sustainable livestock farming by reducing energy costs and maintaining a comfortable environment for pigs. The biggest limitation of this framework was that it did not incorporate biological factors of production such as animal health and nutrition. Expanding the system to also cope with these features would probably increase the effectiveness and precision of the control system in achieving its goals.

6.4 Conclusions and Future Research

This chapter has outlined how the constituents presented in preceding chapters could be synthesized into a digital twin for aquaculture. Moreover, some cases based on case studies provided earlier in this book have been used to illustrate how such realizations could be manifested and used and which capability levels one could conceivably achieve for these. The considerations have also been seen in light of the corresponding status in terrestrial animal farming, which can be considered a companion industrial segment to aquaculture where complexity, aims, and maturity are concerned.

The following chapters will focus more on the end product of digital twin technology, i.e., how digital twins in aquaculture can contribute to improving various aspects of the industry. Chapter 7 will focus on implications related to climate change; Chap. 8 will present ramifications for the supply chain integration, while Chap. 9 proposes the path onward for intensive pen-based fish culture from a technological perspective.

References

Alver MO, Skøien KR, Føre M, Aas TS, Oehme M, Alfredsen JA (2016) Modelling of surface and 3D pellet distribution in Atlantic salmon (Salmo salar L.) cages. Aquac Eng 72:20–29

Alver MO, Føre M, Alfredsen JA (2022) Predicting oxygen levels in Atlantic salmon (Salmo salar) sea cages. Aquaculture 548:737720

Boeuf G, Le Bail PY (1999) Does light have an influence on fish growth? Aquaculture 177(1-4):129–152

Booij N, Holthuijsen LH, Ris RC (1996) The "SWAN" wave model for shallow water. In: Coastal engineering, pp 668–676

Broch OJ, Klebert P, Michelsen FA, Alver MO (2020) Multiscale modelling of cage effects on the transport of effluents from open aquaculture systems. PLoS One 15(3):e0228502

Brooks KM (2005) The effects of water temperature, salinity, and currents on the survival and distribution of the infective copepodid stage of sea lice (Lepeophtheirus salmonis) originating on Atlantic salmon farms in the Broughton Archipelago of British Columbia, Canada. Rev Fish Sci 13(3):177–204

Bui S, Oppedal F, Nola V, Barrett LT (2020) Where art thou louse? A snapshot of attachment location preferences in salmon lice on Atlantic salmon hosts in sea cages. J Fish Dis 43(6):697–706

Dale B, Edwards M, Reid PC (2006) Climate change and harmful algal blooms. In: Ecology of harmful algae. Springer, Berlin/Heidelberg, pp 367–378

DeCew J, Fredriksson DW, Lader PF, Chambers M, Howell WH, Osienki M, Celikkol B, Frank K, Høy E (2013) Field measurements of cage deformation using acoustic sensors. Aquac Eng 57:114–125

Ferreira JG, Taylor NG, Cubillo A, Lencart-Silva J, Pastres R, Bergh Ø, Guilder J (2021) An integrated model for aquaculture production, pathogen interaction, and environmental effects. Aquaculture 536:736438

Føre M, Alver MO, Alfredsen JA, Rasheed A, Hukkelås T, Bjelland HV, Su B, Ohrem SJ, Kelasidi E, Norton T, Papandroulakis N (2024) Digital Twins in intensive aquaculture – challenges, opportunities and future prospects. Comput Electron Agric 218:108676

Fredriksson DW, Swift MR, Irish JD, Tsukrov I, Celikkol B (2003) Fish cage and mooring system dynamics using physical and numerical models with field measurements. Aquac Eng 27(2):117–146

Fry FEJ (1971) The effect of environmental factors on the physiology of fish. In: Fish physiology, vol 6. Academic, pp 1–98

Grøttum JA, Sigholt T (1998) A model for oxygen consumption of Atlantic salmon (Salmo salar) based on measurements of individual fish in a tunnel respirometer. Aquac Eng 17(4):241–251

Hvas M, Folkedal O, Oppedal F (2021) Fish welfare in offshore salmon aquaculture. Rev Aquac 13(2):836–852

Jeong DY, Jo SK, Lee IB, Shin H, Kim JG (2023) Digital Twin application: making a virtual pig house toward digital livestock farming. IEEE Access 11

Johansson D, Laursen F, Fernö A, Fosseidengen JE, Klebert P, Stien LH, Vågseth T, Oppedal F (2014) The interaction between water currents and salmon swimming behaviour in sea cages. PLoS One 9(5):e97635

Karvonen A, Rintamäki P, Jokela J, Valtonen ET (2010) Increasing water temperature and disease risks in aquatic systems: climate change increases the risk of some, but not all, diseases. Int J Parasitol 40(13):1483–1488

Klebert P, Su B (2020) Turbulence and flow field alterations inside a fish sea cage and its wake. Appl Ocean Res 98:102113

Klebert P, Patursson Ø, Endresen PC, Rundtop P, Birkevold J, Rasmussen HW (2015) Three-dimensional deformation of a large circular flexible sea cage in high currents: field experiment and modeling. Ocean Eng 104:511–520

Klebert P, Su B, Nissen O, Kvæstad B (2023) Response of Atlantic salmon in flexible sea cages to waves and currents: measurements at farm-level and modeling. Appl Ocean Res 138:103599

Kragesteen TJ, Simonsen K, Visser AW, Andersen KH (2018) Identifying salmon lice transmission characteristics between Faroese salmon farms. Aquaculture Environment Interactions 10:49–60

Kristiansen D, Lader P, Jensen Ø, Fredriksson D (2015) Experimental study of an aquaculture net cage in waves and current. China Ocean Engineering 29(3):325–340

Lader P, Dempster T, Fredheim A, Jensen Ø (2008) Current induced net deformations in full-scale sea-cages for Atlantic salmon (Salmo salar). Aquac Eng 38(1):52–65

Løland G (1993) Current forces on, and water flow through and around, floating fish farms. Aquaculture International 1:72–89

Michelsen FA, Klebert P, Broch OJ, Alver MO (2019) Impacts of fish farm structures with biomass on water currents: A case study from Frøya. Journal of Sea Research 154:101806

Moe Føre H, Thorvaldsen T (2021) Causal analysis of escape of Atlantic salmon and rainbow trout from Norwegian fish farms during 2010–2018. Aquaculture 532:736002

Moe H, Fredheim A, Hopperstad OS (2010) Structural analysis of aquaculture net cages in current. Journal of Fluids and Structures 26(3):503–516

Mørkøre T, Rørvik KA (2001) Seasonal variations in growth, feed utilisation and product quality of farmed Atlantic salmon (Salmo salar) transferred to seawater as 0+ smolts or 1+ smolts. Aquaculture 199(1-2):145–157

Novales Flamarique I, Gulbransen C, Galbraith M, Stucchi D (2009) Monitoring and potential control of sea lice using an LED-based light trap. Canadian Journal of Fisheries and Aquatic Sciences 66(8):1371–1382

Oehme M, Aas TS, Sørensen M, Lygren I, Åsgård T (2012) Feed pellet distribution in a sea cage using pneumatic feeding system with rotor spreader. Aquac Eng 51:44–52

Olsen OA, Balchen JG (1992) Structured modeling of fish physiology. Mathematical biosciences 112(1):81–113

Purcell JE (2005) Climate effects on formation of jellyfish and ctenophore blooms: a review. Journal of the Marine Biological Association of the United kingdom 85(3):461–476

Rasheed A, San O, Kvamsdal T (2020) Digital twin: Values, challenges and enablers from a modeling perspective. IEEE Access 8:21980–22012

Solstorm F, Solstorm D, Oppedal F, Fernö A, Fraser TWK, Olsen RE (2015) Fast water currents reduce production performance of post-smolt Atlantic salmon Salmo salar. Aquac Environ Interact 7(2):125–134

Solstorm D, Oldham T, Solstorm F, Klebert P, Stien LH, Vågseth T, Oppedal F (2018) Dissolved oxygen variability in a commercial sea-cage exposes farmed Atlantic salmon to growth limiting conditions. Aquaculture 486:122–129

Su B, Kelasidi E, Frank K, Haugen J, Føre M, Pedersen MO (2021) An integrated approach for monitoring structural deformation of aquaculture net cages. Ocean Eng 219:108424

Szewczyk TM, Morro B, Díaz-Gil C, Gillibrand PA, Hardwick JP, Davidson K, Aleynik D, Planellas SR (2024) Interactive effects of multiple stressors with significant wave height

exposure on farmed Atlantic salmon (Salmo salar) welfare along an inshore-offshore gradient. Aquaculture 579:740184
Thorvaldsen T, Frank K, Sunde LM (2019) Practices to obtain lice counts at Norwegian salmon farms: status and possible implications for representativity. Aquac Environ Interact 11:393–404
Turnbull J, Bell A, Adams C, Bron J, Huntingford F (2005) Stocking density and welfare of cage farmed Atlantic salmon: application of a multivariate analysis. Aquaculture 243(1–4):121–132
van Horne PLM, Leenstra FR (2010) Broiler farm size in relation to sustainability aspects. In: XIIIth European poultry conference, Tours, France, August 23–27
VanDerHorn E, Mahadevan S (2021) Digital Twin: generalization, characterization and implementation. Decis Support Syst 145:113524
Wentz FJ, Gentemann C, Smith D, Chelton D (2000) Satellite measurements of sea surface temperature through clouds. Science 288(5467):847–850

Chapter 7
Climate Change, Marine Aquaculture and the Digital Twin

Gregor Reid, Lynne Falconer, Elisabeth Ytteborg, Brian Helmuth, Hanne Digre, and Torstein Kristensen

Abstract Marine aquaculture is well positioned to help meet global seafood demand, but the reliance of most production on the ambient environment suggests inherent vulnerability to climate change effects. Climate change vulnerability and risk assessments can help determine targeted adaptation needs. Some salmon culture sectors are now adopting variants of these assessments to support sustainability reporting.

Many existing monitoring and modelling methods are already well suited to flag deleterious influences of climate change on marine aquaculture with potential application through digital twins. These may include intensive monitoring approaches like precision aquaculture, and application of ecophysiology and energetic models which aim to predict outcomes that can scale from individual animals to the farm level. Though not without limitations. Digital twin approaches incorporating components of climate change projections are apt to be predictive (a digital twin category) stand-alone applications unconnected to an asset (or 'Digital siblings').

There are important data input considerations for the monitoring of climate change stressors and application of climate data to models, irrespective of digital twin type. Data streams are crucial for most digital twins, and the effort of maintaining monitoring equipment in the marine environment is not trivial. The

G. Reid (✉)
Centre for Marine Applied Research, Dartmouth, NS, Canada
e-mail: greid@perennia.ca

L. Falconer
Institute of Aquaculture, University of Stirling, Stirling, Scotland

E. Ytteborg
Nofima, Tromsø, Norway

B. Helmuth
Coastal Sustainability Institute, Northeastern University, Boston, MA, USA

Proteus Ocean Group, New York, NY, USA

H. Digre · T. Kristensen
Scale AQ Group AS, Trondheim, Norway

F. O'Donncha et al. (eds.), *Digital Twin for Marine Fish Farming*,
Aquatic Production Systems, https://doi.org/10.1007/978-3-032-10362-8_7

present spatial and temporal scale of ocean model climate projections is still too coarse for application at the farm scale. Significant *in situ* data collection is required to correct for projection biases, and this data collection must be long term. There is great opportunity for industry collection of ocean data to contribute to long-term ocean data sets across entire regions.

Keywords Climate change · Aquaculture · Digital twin · Ocean model · Climate projections · Ocean data

7.1 Introduction

The Intergovernmental Panel on Climate Change (IPCC) has repeatedly warned that the world is undergoing increasingly fast and far-reaching changes to climate and weather due to global warming arising from fossil fuels and other greenhouse gas emissions (IPCC 2023). Climate change is no longer a distant threat. Its effects have already been observed in the global food system. Events such as large-scale droughts, floods, storms, and heat waves are becoming more common which can cause significant damage and disruption to global agriculture and seafood supply (FAO 2022b; IPCC 2022a). These impacts are occurring when the global food system is already pressured to produce greater amounts of highly nutritious food, in a sustainable manner (Molotoks et al. 2021). Aquaculture is seen as one viable solution to meet the global demand for seafood. Some studies suggest a near doubling of global fish demand by mid-century (e.g. Naylor et al. 2021). However, the heavy reliance of most global aquaculture on the ambient environment also suggests inherent vulnerability to climate change effects.

The majority of global marine aquaculture production currently occurs in open water, despite recent advances in land-based culture (FAO 2022a, b, 2024). A large proportion of this aquaculture production occurs in freshwater ponds or river deltas, competing with agriculture and urbanization for water needs, and this competition is further exacerbated by climate change (Reid et al. 2019a). There is, however, a large untapped potential to grow marine aquaculture to meet global seafood demands, using a fraction of available culture space (Gentry et al. 2017). While there are many issues to be resolved, marine waters present a promising opportunity for wider-scale aquaculture expansion. The oceans are significantly affected by climate change, with warming water, increasing acidification, freshening, and, changes to stratification, dissolved oxygen, and storms (IPCC 2019b). Consequently, marine aquaculture has not gone unaffected. Climate change impacts reflect heterogeneous patterns of gains and losses on marine aquaculture (Fong et al. 2024) with a possibility of worldwide decline, but with potentially greater decline in bivalve aquaculture compared to finfish production (Froehlich et al. 2018). Despite climate change threats to the global seafood supply, the ability to manage captive stock may give the aquaculture industry an adaptive advantage compared to capture fisheries (Richards et al. 2015; Oyebola and Olatunde 2019), though with an understanding that cultured species do not have the option to relocate away from deleterious conditions as can their free-roaming counterparts.

Proactive planning requires recognizing and learning to respond to climate change effects, to build adaptive capacity (Cinner et al. 2018) and to select the most effective adaptation pathways (Cradock-Henry and Frame 2021). Understanding the effects of climate change on biological responses, resources and economics in aquaculture is therefore a prerequisite (Reid et al. 2019a). As climate change effects on aquaculture intensifies, the need for innovative approaches for aquaculture adaptation through strategic planning, research and management will increase. This suggests a need to develop strategic planning approaches ranging from the simple to the complex. Some aquaculture sectors like large-scale salmon farming and recirculating aquaculture systems (RAS) have begun to apply precision aquaculture approaches which involve sophisticated monitoring and modelling methods (e.g. Føre et al. 2017). Such approaches may already be well positioned to support some planned adaption strategies but are not without challenge.

Monitoring and modelling climate change effects on marine aquaculture needs to be contextualized with risk management. This is because climate change largely exacerbates existing threats, superimposing over natural variability. Many effects result from the changing of ranges, magnitude of extremes, patterns of variability and interactions with other stressors. The magnitude of biological response, whether positive or negative, as well as negligible responses of an aquaculture species or population to changes in climate change stressors (e.g. temperature, oxygen, pH) is reviewed by Reid et al. (2019a). Outcomes will be a function of the following:

(a) Where changes occur relative to optimal ranges and tolerance limits of an organism's life stage and physiological processes
(b) The average magnitude and duration of the stressor over the production cycle
(c) Stressor rate of change
(d) Variation, frequency, duration and magnitude of extremes
(e) Interaction with other stressors
(f) Organism capacity for epigenetic expression or plasticity
(g) Genetic strain and population variation
(h) Diet quality and quantity
(i) General health of the animal

This variety of potential pathways of effects suggests a role for real-time or early warning mechanisms to inform practical management decisions but also the need to run predictive scenarios to support strategic decisions (Mangano et al. 2020).

Climate change effects reach beyond biological responses, with impacts across all aquaculture metrics and scales. Climate change stressors can affect infrastructure integrity, safety, husbandry, management, economics, supply chains and the global seafood economy, suggesting multiple factors must be considered in any assessment (Falconer et al. 2022; Mangano et al. 2022). The complexities of aquaculture and climate change necessitate increased awareness, knowledge and information to support tools for resilience and evidence-based climate action. Depending on how effects are manifested, a range of adaption options could be available.

Aquaculture adaption solutions generally fall into one of several categories. These may include nutritional solutions, relocation, augmented information flow (i.e. real-time alerts, early warning or prediction), genetic and biotechnology applications or engineering solutions (Reid et al. 2019b). While some solutions may be simple and easy to implement, others require substantial resources and capacity building over time. It is generally accepted that to effectively adapt to climate change impacts, there is immense cost–benefit to planned adaptation over reactive coping responses (Shelton 2014).

As with all industries, the aquaculture sector must adapt to changing conditions, while at the same time, there is responsibility to minimize its own contributions to global warming. This additional complexity, along with the vast diversity of species, culture systems, management approaches and regions, suggests one-size-fits-all approaches to impact assessments and adaptation plans which are not viable. Broad generalizations could leave the sector unprepared for future challenges. Consequently, plans need to be site-specific, as climate change effects will have different consequences, exposure, sensitivity and potential adaptive capacity at a given location and system.

Complexities around planned adaption suggest the need for a targeted support approach. Some environmental drivers are relatively straightforward and easy to model if considered individually, and using broad-scale averages, such as mean temperature change on fish growth rate (Iwama and Tautz 1981). However, the convergence of multiple influences, sector needs and underlying variability in drivers makes determining what to model and the cost benefit of such exercises, quite difficult. One way to prioritize needs and consequent efforts is to identify vulnerabilities. Climate change vulnerability assessments are useful prerequisites for planned adaptation.

Vulnerability assessments help to triage priorities around modelling and decision support. Approaches to these assessments may vary from basic to very complex, but they are largely iterations of models. While digital twins have been applied to support components of vulnerability and risk assessments (e.g. Macatulad and Biljecki 2024), they have not been applied to overarching climate change vulnerability assessments, to the best of the authors' knowledge. As such, vulnerability and risk are explored in the following sections, to define needs, areas and opportunities for potential application of digital twins to support climate change adaptation in marine aquaculture.

7.2 Climate Change Vulnerability

Vulnerability is defined by the IPCC (2014) as a susceptibility to harm which can be utilized to identify actionable mitigation to support an industry that may be affected by climate change. Climate change vulnerability assessments have become a common management tool to inform risk and guide planned adaptation over the last decade and a half (e.g. Füssel and Klein 2006; Soares et al. 2012;

Thomas et al. 2019; Dudley et al. 2021). Vulnerability assessments have now been applied to a variety of aquaculture (Brugère 2015; Handisyde et al. 2017; Bell and Bhari 2018; Soto et al. 2018, 2019; Islam et al. 2019; Kim et al. 2019; Stewart-Sinclair et al. 2020; Comte 2021; Winne et al. 2022), although focus and specific approaches vary. While there are several methods to estimate vulnerability, a common approach partitions input data into categories to help ensure greater holistic coverage. This methodology typically considers three indices as components of overall vulnerability: 'Exposure', 'Sensitivity', and 'Adaptive Capacity' (Islam et al. 2019; Krishnan et al. 2019; Soto et al. 2019; Kling et al. 2020; Mafi-Gholami et al. 2020). *Exposure* reflects the magnitude of stressors that are occurring or predicted to manifest at a defined location or scale (Adger 2006; Hunter et al. 2015). For example, it could include the degree of projected ocean warming, ocean acidification and changes in primary production. *Sensitivity* measures a variety of social and economic factors that influence how much a system could be affected by climate-driven changes or events as well as to interacting non-climatic stressors (Sarkodie and Strezov 2019). This could include assessments aspects such as the structural and mooring integrity of sea cage arrays, based on the assumption that well-designed and appropriately resourced culture structures are more likely to withstand extreme storm events. It also can refer to the physiological sensitivity of farmed species to the environmental drivers to which they are exposed (Helmuth et al. 2010). *Adaptive capacity* measures how much a system (e.g. a sector or community) can adjust to climate change in the longer term (Smit and Wandel 2006; IPCC 2014) by reducing *Sensitivity* and *Exposure* (Fig. 7.1). Assessments could account for the economic base of a company or sector, as an indicator of potential access to resources and the capacity to invest in adaptation strategies. Personal understanding or opinion such as the threat perception of climate change may also be important considerations under *Adaptative Capacity*. Motivation to adapt will be limited with a lack of acknowledgement or understanding of a threat. The indices are then combined into an overall relative measure of 'Vulnerability'.

These assessments require a diverse range of data inputs with different spatial and temporal domains and resolutions. Given extensive data diversity, assessing vulnerability must often combine quantitative data such as biophysical and economic value, with more qualitative data such as interviews and expert opinion (e.g. Soto et al. 2018; Bukvic et al. 2020; Dudley et al. 2021). This complexity must ultimately produce manageable outputs to support management recommendations that are easily communicated to decision-makers and interest holders. One way to achieve this is by standardizing data or information ranges from low to high vulnerability. Physical data such as water temperature might be partitioned from high to low vulnerability moving from lethal extremes, near-lethal, stressful, suboptimal and optimal temperature ranges. Answers to a threat perception statement might apply a Likert scale response of strongly agree, agree, neutral, disagree and strongly disagree, with each representing a progressive vulnerability. This approach requires all data sets to have the same number of partitions. Three to five partitions are not uncommon. Each of the three primary indices (*Exposure*, *Sensitivity* and *Adaptive Capacity*) may be an average of several data inputs (e.g. metrics). The overall

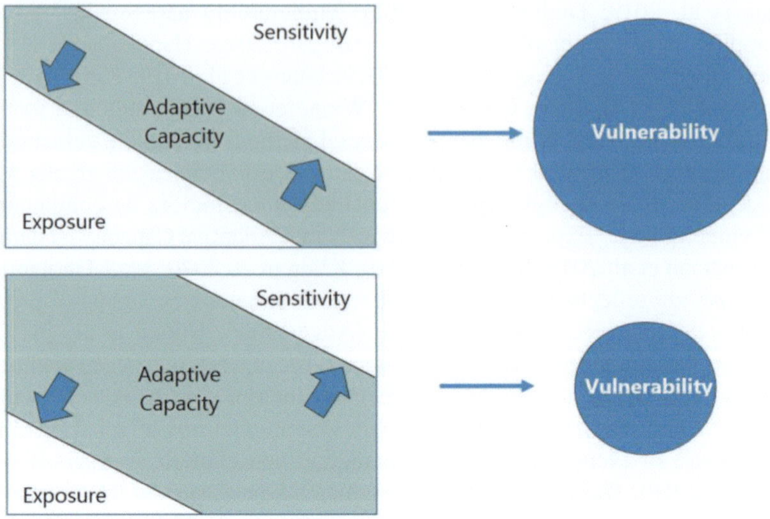

Fig. 7.1 Figurative diagram showing how adaptive capacity can reduce vulnerability by offsetting sensitivity and exposure. (Adapted from Engle (2011) and Thomas et al. (2019))

vulnerability score (sometimes referred to as cumulative vulnerability) is typically calculated by averaging the indices.

To reflect the holistic approach of aquaculture vulnerably assessments, industry input is key to ensure accurate reflection of sector conditions. Vulnerability assessments should be revised regularly as they will become dated as climate conditions and options for adaptive capacity evolve.

7.3 Climate Change Risk

In climate change research, vulnerability is a common component of risk. Depending on the analysis, *Exposure* may be embodied within the vulnerability measure (as described above), or separated out, where *Climate Hazards*, *Exposure* and *Vulnerability* are combined to estimate risk (IPCC 2022b). Regardless of the approach or form, vulnerability or its component indices contribute to risk assessment.

The term *risk* means different things to different audiences. Risk is often used in the vernacular to describe the probability of negative occurrence, such as a destructive weather event (e.g. Saikia and Mahanta 2023). A risk assessment may juxtapose the likelihood of occurrence vs. potential for impact, with risk estimated at the intersection of the two and communicated as a matrix (e.g. Fisheries and Oceans Canada 2019). Vulnerability is typically a relative term reported with dimensionless units. To improve its usefulness for managers, it has been advocated to translate vulnerability into risk by applying absolute values to develop risk thresholds (e.g. low to high risk, Boyce et al. 2022). In the biological and ecological realm, it has been argued that few definitive thresholds exist, which challenges the practical application of threshold-based assessments (Hillebrand et al. 2020). This also extends to aquaculture where there are few absolute thresholds (such as the lethal physiological limits of farmed species), making this approach a challenging prospect. Nevertheless, vulnerability assessments are a viable prerequisite to support risk determination and can have great utility even as a semi-quantitative measure.

7.3.1 *Climate Risk Example from the Salmon Aquaculture Industry Perspective*

Climate change is affecting aquaculture companies and industry media reports, largely acknowledging that producers must be proactive in the face of climate change impacts (e.g. Hanna Gezelius 2022). As an extension of sustainability initiatives, reporting of climate-related risks are starting to become an accountability metric for the aquaculture industry. Within this context, some initiatives have developed to help incentivize the industry to assess and report climate risk (Zitti and Guttormsen 2023). The Task Force on Climate-related Financial Disclosures (TCFD) is one prominent initiative (TCFD 2021). TCFD is a standardized framework, which dictates information that companies should disclose to transparently assess financial risks and opportunities which climate change exposes an organization to.

Within the TCFD framework, climate-related risks are classified into two categories: transition risks and physical risks. Transition risks arise from the shift to a low-carbon economy (climate change mitigation through greenhouse gas reduction) and encompass policy, legal changes, technological shifts, market fluctuations and reputation impacts. Physical risks stem from changes in climate patterns, leading to direct asset damage and indirect impacts, such as storm damage to culture infrastructure causing fish escapes and financial losses. By understanding and addressing physical and transition risks, stakeholders can develop strategies to enhance resilience, adaptability, and sustainability with respect to climate change. This may involve implementing innovative technologies, diversifying production methods, engaging in sustainable practices and collaborating with stakeholders across the value chain, to mitigate risks and seize opportunities in a changing climate.

7.3.2 *Physical-Biological Risk Example in Salmon Farming*

An example is discussed herein based on the ScaleAQ AS integrated annual report for 2023 (Scale Aquaculture AS 2024). Given geographical and regulatory differences in different parts of the world, risk was mapped and evaluated across different salmon-producing countries using publicly available sources. The salmon-producing world was regionalized and data collected on present-day baseline ocean surface temperature data, and short- (2030–2040) and long- (2050–2100) term temperature projections. The IPCC WGI Interactive Atlas (IPCC 2024) was used to obtain baseline and projected ocean surface temperatures, which combines 24 model outcomes to derive projections.

The current grow-out strategy for Atlantic salmon farming uses floating marine cages, which exposes the fish to the prevailing temperature regimens at the farm sites. In Norway, the production period from stocking smolts/post-smolts into cages to harvest at 4–6 kg exceeds more than a year in most cases. This means most fish are exposed to maximum summer sea temperatures at least once during the production cycles. As with most fishes, salmon body temperature is inevitably linked to the ambient temperature in the surrounding water (i.e. they are ectotherms, or 'cold blooded' animals). Temperature is the main regulating factor for metabolism, growth and activity levels, assuming optimal nutritional needs are met. However, salmon have a 'tolerance window' of temperatures they can withstand, as well as an optimal range where growth potential is highest. When temperatures exceed optimum (14–16 °C for growth in Atlantic salmon), blood oxygen levels can drop, straining their circulatory system (gills, blood vessels, heart) which could result in cardiac failure and death in the most severe cases (Pörtner et al. 2017). In less severe conditions, reduced appetite and growth can occur. Risk can be confounded. High water temperatures often coincide with low water oxygen conditions, both from lower physical solubility at higher temperatures and increased microbiological oxygen consumption. Warmer temperatures in combination with poor gill health and suboptimal cardiac function, both conditions commonly observed simultaneously in aquaculture (Jørgensen et al. 2014; Sommerset et al. 2024), may further reduce capacity to endure high temperatures.

Evaluations were based conservatively on working industry knowledge on temperature tolerance levels and critical limits for salmon, to define a simplified three level risk ladder, where:

- Low risk: Maximum summer surface temperatures, likely lower than 16 °C
- Medium risk: Maximum summer surface temperatures, likely between 16 °C and 17 °C
- High risk: Maximum summer surface temperatures, likely higher than 17 °C

This risk ladder was developed from a combination of scientific literature (e.g. Ignatz et al. 2023), applied knowledge based on growth models for salmon and an assumption that all current temperature models lack data and time resolution for predicting extremes. In other words, the models are conservative with respect to days or weeks of extreme heat.

The evaluation used two very different scenarios; one where measures are put in place to limit global warming (SSP1–1.9) and one where GHG emissions continue to increase (SSP3–7.0, IPCC 2021). These scenarios are developed by the Intergovernmental Panel on Climate Change (IPCC), based on different emission pathways and correlated temperature projections. The evolution of future greenhouse gas (GHG) emissions and the underlying driving forces causing emissions is highly uncertain. The IPCC's five scenarios differ in how quickly greenhouse gas emissions can be curbed.

SSP1–1.9 is the only scenario that meets the Paris Agreement goal of limiting global warming to 1.5 °C above pre-industrial temperatures. In this scenario, warming will hit 1.5 °C but then dips back down and stabilizes around 1.4 °C by the end of the century, and the world shifts gradually toward a more sustainable path. Global CO_2 emissions are cut to net zero by 2050 in this scenario, and while extreme weather is more common, the world has avoided the worst impacts of climate change. Societies switch to more sustainable practices, with focus shifting from economic growth to overall well-being. Investments in education and health go up and inequality falls. The SSP3–7.0 scenario has the second highest annual GHG emissions, with CO_2 levels almost doubling from current levels, by 2100. This scenario features an increase in regional rivalry; countries become more competitive with each other, and issues of national and food security are prioritized over stopping climate change.

For Norway, a more detailed assessment was performed. Production site-specific information on temperature (3–5 m depth) is recorded weekly, reported to government agencies and the public (Luse Data 2024). This enables a more in-depth analysis, calculating the temperature mean over the last 5 years (2018–2023) for each location, and choosing the average temperature of the warmest week of the year. Manual data curation for lack of adequate data removed 10 sites from the dataset, leaving 906 locations for analysis. Adding the projected temperature increase of the two SSPs for the short- and long-term timeframes, provided a prognosis for future risk of the 13 productions zones (POs).

7.3.2.1 Results and Discussion

The overall analysis of world regions based on IPCC interactive atlas data indicated several salmon-producing regions are already in high-risk category today (i.e. >17 °C), while several more will enter the high-risk category in the long-term SSP1–1.9 scenarios. In the short term, under the SSP3–7.0 scenario, most regions enter a higher-risk category, with only Iceland and Northern Norway retaining a low-risk category.

This approach enabled risk quantification of direct temperature effects under current and future scenarios for ocean temperatures. It is important to note that neither the global approach nor the site specific Norwegian case have sufficient temperature data resolution for robust modelling of future conditions as of now (Falconer et al. 2024). This rudimentary approach does not include the combined effects of increased temperature and low oxygen conditions nor other biological effects such as algal blooms and mass occurrences of jellyfish. Both the latter are known to occur under rather specific environmental conditions not captured by general climate models, while the former lacks detailed worldwide modelling

(Zhan et al. 2023). While several case studies and modelling efforts address this, the assessment approach requires further development to fully utilize this in a general risk assessment. Further evolution does suggest application for scenario modelling and a role for digital twins. Nonetheless, the current process highlights the importance of accounting for climate risk in future efforts to develop technical solutions (Falconer et al. 2022) to mitigate temperature effects on worldwide and regional production potential for salmon.

With respect to transition risk, to accomplish the net zero emissions by 2050 (i.e. reach the goal of the Paris agreement), new regulations and changes to existing regulations are needed to constrain actions that negatively affect the climate. These regulations will include company reporting, increasing emissions taxes, emission efficiency requirements, circular economy requirements regarding products and credit requirements to fund products that make sustainable contributions. There are currently 137 countries that have committed to carbon neutrality, including salmon-producing countries as Norway, the UK, Australia, Chile and Canada. Carbon taxes are also being introduced in several countries. In addition to increasing the need for sustainability-oriented regulations, an increased focus on sustainability means that more resources will be put into developing lower-emission technology. Higher-resource productivity by reducing the overall consumption of material is needed for reaching the Paris agreement goal. This is likely to affect both availability and cost of the raw materials and occur during an era when increased attention to animal welfare, employee safety, ecosystem preservation and biological diversity will be important.

While a number of climate change digital twin initiatives aim to simulate climate change physical processes (e.g. Hoffmann et al. 2023), there is also advocacy for the role of digital twins (Bonefeld-Dahl 2021) and AI (Chishti et al. 2024) to guide the complexities emissions reduction in multiple sectors. It is reasonable to expect this evolution will make its way to the aquaculture sector, with this potential as a powerful tool to reduce risk.

7.4 Digital Twins as Data-Driven Adaptation Strategies

A 'digital twin' aims to simulate a real system, collect real-time data, interpret, decide and, if advanced enough, act. The application of digital twins across sectors is becoming common, ranging in fields from health care (Sampaio et al. 2024), smart city design (Hananto et al. 2024), manufacturing (Methuselah 2024) and architecture (Smarsly et al. 2024). There has also been application to various ocean technology sectors (e.g. Mali 2024; Zhang et al. 2024), and dynamic simulation of environmental and ecological ocean conditions (Brönner et al. 2023; Masmitja et al. 2024). This is further reflected through creation of digital twin oceans to inform management, and sustainable development (e.g. EU DTO 2024; Iliad consortium 2024).

Fisheries and marine aquaculture management under a changing climate are a major component of ocean sustainably. There has been recent advocacy for data-driven analyses around climate change and aquaculture, through better integration of new technologies to support sustainable aquaculture practices while augmenting global food security (Nor Azra 2025). To meet these needs, there is great potential for digital twins to play a role but also substantial challenges.

Climate change projections under various greenhouse gas emission scenarios are now well established, including the evolution of ocean models that predict temperature and biogeochemical components (IPCC 2019a, 2023). Many global-scale coarse models have been downscaled to higher resolutions as regional models (e.g. Lavoie et al. 2020). While the minimum downscaled resolution is now around several km, this has not quite reached the scale of a typical aquaculture farm (Falconer 2023). Nevertheless, these projections can provide a range of inputs to run scenarios. Consideration of the temporal scale will dictate, in part, the utility of a digital twin. Many temporal climate projections are on the scale of decades, with mid-century and end of century, common endpoints. Though some model outputs are now becoming available in yearly time steps (e.g. Lavoie et al. 2021), future conditions on the scale of years to decades do not necessitate instantaneous decision-making, which is one of the ultimate target roles of a digital twin approach. They also do not capture finer temporal scale variability that is known to drive physiological performance and mortality risk (Helmuth et al. 2010). This begs the question as to what are the potential synergies between long-term projections and digital twin approaches and how can their efficacy potentially be enhanced using finer scale projections (Mangano et al. 2020)?

Digital twin complexity and utility can be thought of as a stepwise continuum of sophistication, moving from stand-alone, descriptive, diagnostic, predictive, prescriptive and, ultimately, an autonomous state (Fig. 7.2). At the current stage of development, application of climate change projections in conjunction with companion approaches like climate change vulnerability or risk assessments is apt to fall in the descriptive, diagnostic and predictive categories. In these circumstances, scenarios run as 'digital sibling' (stand-alone simulation, see Chap. 4) may be more applicable descriptors to support decision support and policy guidance tools.

However, digital twins may arguably have greatest utility where climate change is exacerbating the risk of existing hazards. This could be increased intensity, duration or frequency (IDF) of deleterious environmental extremes, like storms and marine heat waves. In these instances, a digital twin prompting early warning, accounting for IDF metrics to tailor simulated outcomes or recommend action, can be immensely beneficial. Benefits would not only support precision management but also develop adaptive capacity, which is key to reducing sensitivity and exposure to climate change, and ultimately vulnerability.

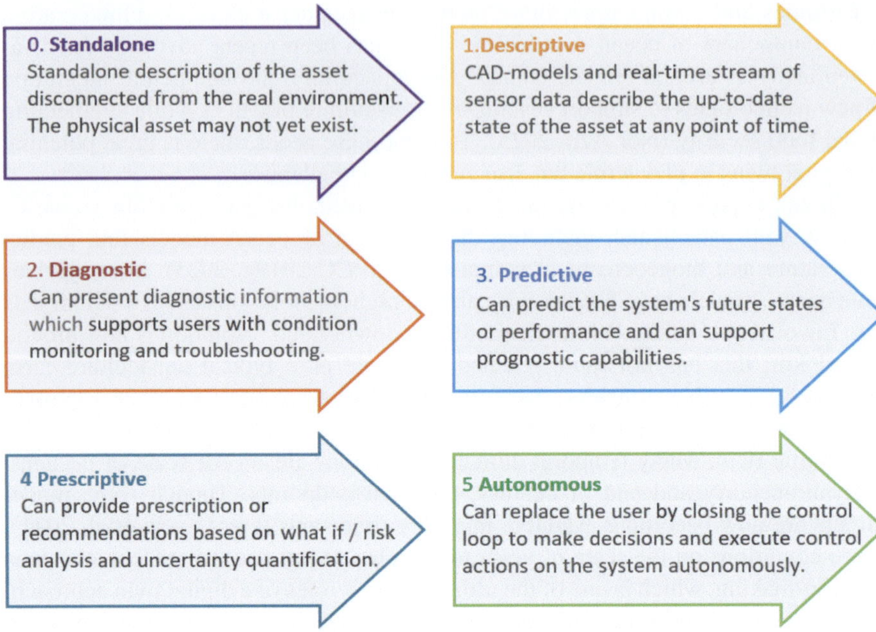

Fig. 7.2 Capability levels of a digital twin. (Modified from Føre et al. 2024)

7.4.1 Descriptive Twins for Real-Time Monitoring

Real-time monitoring is a prerequisite for precision fish farming and digital twins in aquaculture. Not only for basic *Descriptive* digital twins but also for *Diagnostic*, *Predictive* and *Perspective* to fully *Autonomous* variants (Føre et al. 2017; O'Donncha and Grant 2019; Lima et al. 2023). Timely data tracking and interpretation is important in aquaculture, irrespective of climate change. However, the overlay of climate change drivers, augmenting existing stressors through changes in IDF, and also interaction between multiple stressors, real-time monitoring becomes all the more important.

Real-time streaming of sensor data to inform present state of a system (or animal) falls into the *Descriptive* category of digital twin. These systems are now commonplace for management of aquaculture marine cages and essential at modern land-based systems. Common monitoring regimens include feed tracking, water quality (e.g. dissolved oxygen, temperature) and fish tracking (Føre et al. 2017). Water quality monitoring is also a legal requirement for live seafood shipping and holding in some jurisdictions (BC Centre for Disease Control 2013), and this is apt to drive the need for real-time data collection beyond the rearing environment at the farm.

Real-time monitoring in aquaculture has the potential to extend directly to the animal with developments of some microsensor technologies. For example, this can enable heart rate tracking of sentinel individuals in shellfish populations to assess real-time response to environmental or biological stressors (Hellicar et al. 2015; Horricks et al. 2024). Such an approach may be particularly important for species which do not exhibit many visual cues when stressed, thereby limiting the use of visual detection methods.

7.4.2 Diagnostic Twins for Early Warning

Many studies share the goal of trying to forecast what effects future changes will bring to the aquaculture industry (Doubleday et al. 2013; Guyondet et al. 2014; Froehlich et al. 2018). Or at a minimum, provide the mechanistic detail needed to develop early warning systems which can be acted upon through climate adaptation strategies (Reid et al. 2019a, b; Mangano et al. 2020; Maulu et al. 2021).

Every model has a spatial and temporal domain. Ranges can be diverse, depending on the objective and input data availability. Outputs can be very sensitive to scale (Duarte et al. 2005). Early warning in aquaculture suggests time frames within the production cycle, which could range from minutes to months (Reid et al. 2019a). Spatial scales may range from locations within a farm (e.g. Cranford et al. 2008) or at the level of the rearing unit (e.g. Reid and Moccia 2006). At the marine cage scale, there has been recent developments applying machine learning, digital twin approach to predict hypoxia in salmonid cages (Cerqueira et al. 2024). The farm (Hargrave et al. 2022) and regional level (Soto and Norambuena 2004) are also commonly applied scales. In practice, however, many of these models are applied at the individual animal and simply scaled up to various population levels, depending on the model objective (e.g. Duarte et al. 2010).

There are many examples in the scientific literature of eco-physiological modelling and the challenges of scale reconciliation. Montalto et al. (2014) generated model predictions of mussel physiological performance using environmental data at temporal resolutions ranging from 1 to 6 h and discovered that predictions using coarser environmental data significantly underestimated reproductive output. Mangano et al. (2020) used an energetic approach to make forecasts of commercial anchovy stocks in the Mediterranean Sea and reported that while temperature forecasts were likely adequate at the spatial scales required, food levels were not. They argued that rather than being used as a forecast of long-range responses, their method should instead inform an 'early warning system' – essentially a sensitivity analysis that quantifies the suites of conditions most likely to be problematic for growth and survival (Mangano et al. 2020).

Because climatic anomalies at (e.g. seasonal scales) can be, to some degree, forecast in advance (Mills et al. 2017), such systems could be deployed in concert with other methods that instead track animal condition and behaviour (Guarini et al. 2021). In this regard, the application of digital twins and other mechanisms for visualizing current and future high-resolution maps of environmental conditions could be a key decision support tool for siting and selection of species (Montalto et al. 2017; Sarà et al. 2018b). This could include the potential development of adaptation strategies (Galappaththi et al. 2020) such as changing depth (and thus environmental conditions) of operations (Reid et al. 2019a, b).

It should be recognized, however, that not all early warning indicators need support from complex models. Simple approaches such as alert notification, when water qualify safety thresholds are nearing exceedance, can be based on simple trend analysis (e.g. regression). While crude, these can be an effective from of early warning and relatively easy to apply (e.g. Montes et al. 2022).

7.4.3 'Prediction' with Digital Siblings

Digital siblings are stand-alone simulations that do not interact directly with an asset but represent purely hypothetical scenarios using the digital twin to explore the impact of situations or conditions (Føre et al. 2024). Consequently, digital siblings by nature are inherently positioned to explore future scenarios such as effects of changes in climate change stressors on aquaculture systems or species. It should be noted that with overlap of disciplines, there is the potential for terminology confusion. The term 'predictive' in this section is used to describe one of the capacity categories of a digital twin (see Fig. 7.2). While in climate change disciplines, future scenarios are described as 'projections' (IPCC 2023), acknowledging that prediction per se (what will happen) is not possible without knowing which shared socioeconomic pathways will manifest (DeWeerdt 2024).

7.4.3.1 Climate Models

Climate models are used to simulate potential future conditions under a range of greenhouse gas emission scenarios. The IPCC encourages the use of a common set of these scenarios (RCP and SSP) for consistency among applications to enable comparisons across sectors. In most cases, climate models have not been developed specifically for aquaculture purposes and are not intended for use at aquaculture farm level. Ocean climate change trends and projections are largely reported at a global or regional scale. A typical grid cell of a global climate model could be approximately 100 km resolution, though regional downscaling (at time of writing) can achieve a 10 km resolution, still greater than a typical farm (McIntosh et al. 2022). A finer resolution is needed to capture the greater spatial variability close to the coasts (compared to open ocean locations) where most marine aquaculture occurs (Falconer 2023).

A cautionary approach is warranted when applying large-scale averages as model inputs. Large-scale averages may be within tolerance levels for a farmed species, while site-specific conditions may be outside of those limits, or vice versa (Falconer et al. 2020). Comparably, temporally averaged data can mask shorter-term extremes (Montalto et al. 2014). Nevertheless, there are some strategies for applying correction factors. In some circumstances, farm-level data can be used to calibrate climate models to the aquaculture site scale (e.g. sea surface temperature, Falconer et al. 2020; Falconer 2023). Comparison of model outputs with farm observations may also flag differences that suggest the modelled variable may not be suitable for aquaculture decision-making (Falconer 2023). While there are many caveats when comparing models and observations (Skogen et al. 2021), long-term continuous coastal datasets of in situ measurements help to quantify trends and anomalies to support model evaluation (Boero et al. 2015; Dempsey et al. 2023, 2024).

An additional challenge is that current ocean model projections are not yet competent predicting acute events such as marine heat waves, although this capacity may be on the horizon (Jacox et al. 2022), and shorter-term forecasting is developing (Hartog et al. 2023). Marine heatwaves are generally increasing in frequency across the world's oceans, and such acute events can be a risk for all types of fish culture (Reid et al. 2019b). Fortunately, existing energetic models can support scenario analyses to accommodate first-level climate change and mitigation responses.

7.4.3.2 Energetic Models of First-Order Climate Change and Mitigation Responses

Energetic models have been applied to aquaculture production, welfare and carrying capacity for decades. There are a variety of variants depending on the objectives, such as scope for growth (SFG), thermal growth coefficient (TGC), dynamic energy budget (DEB) or nutritional bioenergetics (e.g. Cho and Bureau 1998; Filgueira et al. 2015; Reid et al. 2020). They are practically applied, though the simulation of aquaculture or species systems and in this respect can be considered analogous to *Predictive* digital siblings. The primarily mechanistic nature of these models is well suited to explore potential outcomes of stressor augmentation (e.g. changes to intensity, duration and frequency (IDF) of model inputs) by climate change.

To investigate how ongoing environmental change drives performance of aquatic cultured species, researchers combine biophysical models supported by field measurements at the organism-scale environmental conditions (e.g. Helmuth 2009; Foulk et al. 2024), with temporally and spatially explicit energetics models. For example, Thomas and Bacher (2018) used an individual-based modelling approach to assess the sensitivity of population growth potential in oysters and mussels along a 5000 km stretch of coastline in the Atlantic Ocean using climate projection models. Thomas et al. (2011) similarly explored variability in mussel growth by applying a DEB model supported by satellite collected environmental data. Mangano et al. (2019) used a series of DEB models driven by climate projections to predict maximum size and length at reproductive maturity for oysters, mussels, European anchovies and European seabass in the Mediterranean Sea.

A key advantage of energetics modelling approaches such as DEB (Kooijman et al. 1999) is that they can explicitly incorporate the influence of nutritional state and food supply as a potential counterbalance to environmental stressors such as elevated temperature (van der Meer et al. 2014; Montalto et al. 2017) and even to negative interactions with other species (Matzelle et al. 2015). The ultimate products of these combined modelling approaches are 'high resolution' maps that predict the growth, reproductive output, maximum size, size of reproductive maturity, and risk of mortality of target species (Montalto et al. 2016; Mangano et al. 2020).

Because these models are based on first-principles understanding of the physiological relationship between environmental parameters and organism performance, they can potentially avoid some issues of model stationarity (and especially space for time substitution) inherent in many model projections (Austin 2007; Montalto et al. 2015). Specifically, when physiological relationships with environmental drivers are determined experimentally, they can better capture novel climatic conditions (Matzelle et al. 2015). These approaches are, however, data and time intensive (Teal et al. 2015; Catalán et al. 2019), and their spatial and temporal resolution is generally limited by available environmental data and projections (Montalto et al. 2014). There are, therefore, trade-offs between these data-intensive, process-based modelling approaches and methods that are less complex but also easier to implement over larger scales.

7.5 Data Considerations for Digital Twins

To prepare for the future and minimize risks to aquatic food production, a better understanding of climate change at the farm scale is needed, considering the variability across the sector and site-specific issues and then work to identify appropriate adaptation strategies that reduce vulnerability and enhance resilience. Appropriate data collection, management and interpretation are needed to achieve this.

7.5.1 Importance on On-site Data Collection and Limitations

Real-time monitoring of water quality at marine aquaculture sites is becoming common. Such practices are becoming increasingly a part of farm costs and operational plans, largely to reduce production risks and ensure fish welfare. Though there are many complexities with underwater monitoring, these include challenges around longevity of hardware, connectivity, energy use and battery requirements, memory capacity of loggers, issues with biofouling, sensor drift, ongoing calibration and maintenance (Parra et al. 2018). Nevertheless, with proper data collection and archiving, these datasets could provide a valuable source of long-term data (Falconer et al. 2025).

Long-term datasets are crucial for understanding environmental trends in the marine environment (Boero et al. 2015) and are needed for model development, especially model training in AI approaches. Ongoing environmental data collection at farms could also be used to assemble long-term datasets that support climate change assessment (Falconer et al. 2025). Though, some consideration is needed for parameters that could be influenced by farm activities such as dissolved oxygen, if wider environmental conditions are to be teased out.

One of the challenges establishing a long-term continuous dataset is ensuring that the data is collected using a consistent, reliable method with quantifiable accuracy and precision (Benson et al. 2018; Dempsey et al. 2023). As climate change monitoring must occur over decades, the implications of changing data collection method (e.g. different sensors, different temporal resolutions or spatial densities) must also be considered as differences in accuracy or precision are apt to affect interpretation of results, trend detection and anomalies over time. 'Communities of practice' may be a good resource in this respect, as they recommend measurement protocols and best practises for marine data collection (e.g. GOA-ON 2024; GOOS 2024).

7.5.2 *Remote Sensing*

The spatial and temporal heterogeneity of climate change should be considered when selecting locations to use as 'fixed' monitoring stations for in situ measurements. As it is not practical nor possible to 'instrument the entire ocean' (Geller et al. 2017). For fine spatial coverage, remote sensing technology can provide a viable option for some parameters. Earth observations, such as through satellites, can help researchers monitor difficult to access locations and extend spatial coverage (e.g. SST), otherwise difficult to obtain through in situ sampling alone (Canonico et al. 2019). While the capability of satellite remote sensing has advanced considerably over recent years, there are still challenges in establishing reliable long-term consistent datasets for use in climate change analysis (Yang et al. 2013) and reconciling the timing of image capture with water quality change (Sonnet et al. 2024). Remote sensing data are now being integrated into ocean digital twins, but with an acknowledgement that methods need to be calibrated and validated with in situ measurements (Miedtank et al. 2024).

7.5.3 *Aquaculture Is Well Positioned to Support Climate Change Monitoring*

Long-term monitoring at coastal aquaculture sites can be an opportunistic and powerful tool for monitoring ocean variables under a changing climate (Falconer et al. 2025). Aquaculture farms have technical expertise and good scope for coverage (spatial and temporal), often well beyond the capacity of government and research organizations working in this space. For example, Norway has one of the longest coastlines in the world, and there are more than 1000 Atlantic salmon net-pen sites where conditions are monitored daily, with some of the data publicly available on BarentsWatch (2024). The data in BarentsWatch was not collected specifically for climate change assessment, so caution is warranted, though it can provide insight into some trends such as temperature. However, public accessibility is an important step for the application of FAIR data principles (finable, accessible, interoperable and reusable), which have also been advocated for the development of digital twins in data-intensive research (Schultes et al. 2022).

7.5.4 Experiment Design Based on Future Conditions

Farmed organisms are subject to many different stressors during production, and climate change will compound this. To improve resilience, the sector needs to better understand all the stressors that the species face and how they respond to them. An understanding of past and present challenges can help identify potential risks for the future. The use of data and digital twins to simulate potential future farming conditions enables researchers to design experiments based on future conditions that can generate new information on the biological response of farmed species to climate change (Ytteborg et al. 2023). However, one significant research challenge is accurately and precisely determining how environmental changes will affect the future physical culture environment and therefore what conditions need to be simulated to run valid experiments (Reid et al. 2019b).

7.5.5 The Need to Model Multi-stressor Conditions

Aquaculture will be exposed to multiple stressors under climate change, and hence future aquaculture development should account for this (Sarà et al. 2018b). As is true for all marine organisms, growth, survival, reproductive rates and time to reproductive maturity of farmed species are significantly affected by the interactions of multiple environmental stressors of climate change (Sarà et al. 2012; Giacoletti et al. 2024). Decades of research have explored the physiological tolerances of many cultured species (e.g. Catalán et al. 2019; Brauner and Richards 2020), and in recent years, many studies have experimentally demonstrated the ongoing and often interactive ecophysiological effects of changes in temperature (Martinez et al. 2018), salinity (Bussell et al. 2008; Palmer et al. 2021) and ocean pH (Duarte et al. 2017; Sarà et al. 2018a) on a suite of commercially important species.

Many of these studies share the goal of trying to forecast what effects future changes will bring to the aquaculture industry (Doubleday et al. 2013; Guyondet et al. 2014; Froehlich et al. 2018; Stechele and Lavaud 2024) or at a minimum provide the mechanistic detail needed to develop early warning systems which can be acted upon through climate adaptation strategies (Føre et al. 2017; Reid et al. 2019b; Mangano et al. 2020; Maulu et al. 2021).

Recent work has placed climate change in a multi-stressor framework, suggesting that the synchronicity of multiple stressors in space and time is of great concern (Sarà et al. 2018b). Multi-stressor research is complex, as one must also consider mitigative factors at a metabolic and physiological level. An increasing number of studies have begun to explore the role of nutritional state, where there has been evidence that ample food supply can potentially offset otherwise stressful levels of high temperature and/or low pH. For example, Kroeker et al. (2016) found that the growth of mussels remained elevated under elevated temperatures and lowered pH as long as phytoplankton levels remained sufficiently high, but when high temperature, low pH and low food were coincident, mussel growth declined significantly.

Research on multiple stressors, their ranges and accounting for mitigation factors such as food supply, places a large burden on experimentation. Increasing the number of stressors examined in environmentally relevant climate and mitigative combinations results in a non-linear increase in the number of total culture units (like tanks) needed to support multi-factorial experiments with appropriate statistical replicates (Reid et al. 2019b). This challenge may appear to be a good opportunity for a 'digital twin' to circumvent the need for complex experimental trials. Unfortunately, it seems there are few absolute assumptions for multi-stressor response. Multi-stressor interaction may or may not be fully additive or linear; it may be potentially synergistic, or one dominant driver may have the greatest effect, depending on species and life stage (Reid et al. 2019a). This means that it is unlikely that there are any shortcuts around experimental trials, and these research outcomes will be needed to adequality develop multi-stressor models and respective digital twins.

7.6 Conclusion

Digital twin approaches are well placed to support climate change adaptation strategies in aquaculture within the appropriate context. Climate change largely exacerbates existing aquaculture threats by superimposing influence over natural ranges, variability and interactions of biotic and abiotic environmental parameters. Consequently, many existing monitoring and modelling methods are already well suited to flag deleterious influences of climate change on marine aquaculture with potential application through digital twins. These may include intensive monitoring approaches like precision aquaculture and application of ecophysiology and energetic models which aim to predict outcomes that can scale from individual animals to the farm level. However, there are some limitations for digital twin application to climate change adaptation. Future conditions on the scale of years to decades do not necessitate instantaneous autonomous decision-making, which is one of the ultimate target roles of a digital twin approach. Digital twin approaches incorporating components of climate change projections are apt to be predictive stand-alone applications unconnected to an asset.

There are important data input considerations for the monitoring of climate change stressors and application of climate data to models, irrespective of digital twin type. Data streams are crucial for most digital twins, and the effort of maintaining monitoring equipment in the marine environment is not trivial. The present spatial and temporal scale of ocean model climate projections is still too coarse for application at the farm scale. Significant in situ data collection is required to correct for projection biases, and this type of data collection must occur over the long term. Sufficient long-term ocean data sets are rare, but key to understanding climate change influences. Given the spread of salmon farms throughout some jurisdictions and their ongoing monitoring activities, there is great opportunity for industry collection of ocean data to contribute to long-term ocean data sets. Digital twins are likely to play a key role in modern aquaculture, including the increasingly important task of supporting climate change adaptation solutions.

References

Adger WN (2006) Vulnerability. Glob Environ Change 16(3):268–281. https://doi.org/10.1016/j.gloenvcha.2006.02.006

Austin M (2007) Species distribution models and ecological theory: a critical assessment and some possible new approaches. Ecol Modell 200(1):1–19. https://doi.org/10.1016/j.ecolmodel.2006.07.005

BarentsWatch (2024) BarentsWatch. Accessed October 30, 2024. https://www.barentswatch.no/en/

BC Centre for Disease Control (2013) Fish processing plants guideline for live retail fish holding systems. 21. http://www.bccdc.ca/resource-gallery/Documents/Educational%20Materials/EH/FPS/Food/RetailFishHoldingTankGuidelines_Nov2013trs.pdf

Bell J, Bhari T (2018) A new climate change vulnerability assessment for fisheries and aquaculture. Pacific Community

Benson A, Brooks CM, Canonico G, Duffy E, Muller-Karger F, Sosik HM, Miloslavich P, Klein E (2018) Integrated observations and informatics improve understanding of changing marine ecosystems. Front Mar Sci 5:428

Boero F, Kraberg AC, Krause G, Wiltshire KH (2015) Time is an affliction: why ecology cannot be as predictive as physics and why it needs time series. J Sea Res 101:12–18

Bonefeld-Dahl C (2021) Digital action climate action: 8 ideas to accelerate the twin transition. DigitalEurope

Boyce D, Tittensor D, Garilao C, Henson S, Kaschner K, Kesner-Reyes K, Pigot A, Reyes R, Reygondeau G, Schleit K, Shackell N, Sorongon-Yap P, Worm B (2022) A climate risk index for marine life. Nat Clim Chang 12:854–862

Brauner C, Richards J (2020) Physiological performance in aquaculture: using physiology to help define optimal conditions for growth and environmental tolerance. Academic Press, pp 83–121

Brönner U, Sonnewald M, Visbeck M (2023) Digital twins of the ocean can foster a sustainable blue economy in a protected marine environment. Int Hydrogr Rev 29:26–40

Brugère C (2015) Climate change vulnerability in fisheries and aquaculture: a synthesis of six regional studies. FAO Fisheries Circular. FAO, Rome

Bukvic A, Rohat G, Apotsos A, de Sherbinin A (2020) A systematic review of coastal vulnerability mapping. Sustainability 12:2822

Bussell JA, Gidman EA, Causton DR, Gwynn-Jones D, Malham SK, Jones M, Reynolds B, Seed R (2008) Changes in the immune response and metabolic fingerprint of the mussel, Mytilus edulis (Linnaeus) in response to lowered salinity and physical stress. J Exp Mar Biol Ecol 358:78–85

Canonico G, Buttigieg PL, Montes E, Muller-Karger FE, Stepien C, Wright D, Benson A, Helmuth B, Costello M, Sousa-Pinto I, Saeedi H, Newton J, Appeltans W, Bednaršek N, Bodrossy L, Best BD, Brandt A, Goodwin KD, Iken K, Marques AC, Miloslavich P, Ostrowski M, Turner W, Achterberg EP, Barry T, Defeo O, Bigatti G, Henry L-A, Ramiro-Sánchez B, Durán P, Morato T, Roberts JM, García-Alegre A, Cuadrado MS, Murton B (2019) Global observational needs and resources for marine biodiversity. Front Mar Sci 6:367

Catalán IA, Auch D, Kamermans P, Morales-Nin B, Angelopoulos NV, Reglero P, Sandersfeld T, Peck MA (2019) Critically examining the knowledge base required to mechanistically project climate impacts: a case study of Europe's fish and shellfish. Fish Fish 20:501–517

Cerqueira V, Pimentel J, Korus J, Bravo F, Amorim J, Oliveira M, Swanson A, Grant J, Filgueira R, Torgo L (2024) Forecasting ocean hypoxia in salmonid fish farms. Front Aquac 3:1365123

Chishti MZ, Xia X, Dogan E (2024) Understanding the effects of artificial intelligence on energy transition: the moderating role of Paris Agreement. Energy Econ 131:107388

Cho CY, Bureau DP (1998) Development of bioenergetic models and the Fish-PrFEQ software to estimate production, feeding ration and waste output in aquaculture. Aquat Living Resour 11:199–210

Cinner JE, Adger WN, Allison EH, Barnes ML, Brown K, Cohen PJ, Gelcich S, Hicks CC, Hughes TP, Lau J, Marshall NA, Morrison TH (2018) Building adaptive capacity to climate change in tropical coastal communities. Nat Clim Chang 8:117–123

Comte A (2021) Recent advances in climate change vulnerability/risk assessments in the fisheries and aquaculture sector. FAO Fisheries and Aquaculture Circular:I-67. Food and Agriculture Organization of the United Nations

Cradock-Henry N, Frame B (2021) Balancing scales: enhancing local applications of adaptation pathways. Environ Sci Pol 121:42–48

Cranford P, Li W, Strand O, Strohmeier T (2008) Phytoplankton depletion by mussel aquaculture: high resolution mapping, ecosystem modeling and potential indicators of ecological carrying capacity. International Council for the Exploration of the Sea

Dempsey DP, Reid GK, Lewis-McCrea L, Balch T, Cusack R, Dumas A, Rensel J (2023) Estimating stocking weights for Atlantic salmon to grow to market size at novel aquaculture sites with extreme temperatures. Front Mar Sci. https://doi.org/10.3389/fmars.2023.1094247

Dempsey DP, Torrie NL, Reid GK, Lewis-McCrea LM, McKee A, Woodside R (2024) The makings of a high resolution coastal monitoring program. In: OCEANS 2024 – Halifax. IEEE, pp 1–10

DeWeerdt S (2024) Projections aren't predictions. In: But climate media doesn't always capture the difference. Anthopocene Magazine

Doubleday ZA, Clarke SM, Li X, Pecl GT, Ward TM, Battaglene S, Frusher S, Gibbs PJ, Hobday AJ, Hutchinson N, Jennings SM, Stoklosa R (2013) Assessing the risk of climate change to aquaculture: a case study from south-east Australia. Aquac Environ Interact 3:163–175

Duarte P, Hawkins AJS, Pereira A (2005) How does estimation of environmental carrying capacity for bivalve culture depend upon spatial and temporal scales? In: The comparative roles of suspension-feeders in ecosystems: proceedings of the NATO advanced research workshop on the comparative roles of suspension-feeders in ecosystems Nida, Lithuania 4 9 Oct 2003. Springer Netherlands, Dordrecht, pp 121–135

Duarte P, Fernández-Reiriz MJ, Filgueira R, Labarta U (2010) Modelling mussel growth in ecosystems with low suspended matter loads. J Sea Res 64:273–286

Duarte CM, Wu J, Xiao X, Bruhn A, Krause-Jensen D (2017) Can seaweed farming play a role in climate change mitigation and adaptation? Front Mar Sci 4:100

Dudley PN, Rogers TL, Morales MM, Stoltz AD, Sheridan CJ, Beulke AK, Pomeroy C, Carr MH (2021) A more comprehensive climate vulnerability assessment framework for fisheries social-ecological systems. Front Mar Sci 8:674

Engle NL (2011) Adaptive capacity and its assessment. Glob Environ Chang 21:647–656

EU DTO (2024) European digital twin of the ocean. European Commission

Falconer L (2023) Context matters when using climate model projections for aquaculture. Front Mar Sci 10:1198451

Falconer L, Hjøllo SS, Telfer TC, McAdam BJ, Hermansen Ø, Ytteborg E (2020) The importance of calibrating climate change projections to local conditions at aquaculture sites. Aquaculture 514:734487

Falconer L, Telfer TC, Garrett A, Hermansen Ø, Mikkelsen E, Hjøllo SS, McAdam BJ, Ytteborg E (2022) Insight into real-world complexities is required to enable effective response from the aquaculture sector to climate change. PLOS Clim 1:e0000017

Falconer L, Sparboe L, Dale T, Hjøllo S, Stavrakidis-Zachou O, Bergh Ø, James P, Papandroulakis N, Puvanendran V, Siikavuopio S, Hansen Ø, Ytteborg E (2024) Diversification of marine aquaculture in Norway under climate change. Aquaculture 593:741350

Falconer L, Halstensen S, Rinø SF, Noble C, Dale T, Alvestad R, Ytteborg E (2025) Marine aquaculture sites have huge potential as data providers for climate change assessments. Aquaculture 595:741519

FAO (2022a) Record fisheries and aquaculture production makes critical contribution to global food security. Food and Agriculture Organization of the United Nations

FAO (2022b) The state of the world fisheries and aquaculture 2022: towards blue transformation. FAO, Rome

FAO (2024) Fishery and aquaculture statistics – yearbook 2021. Food and Agriculture Organization of the United Nations. FAO yearbook of fishery and aquaculture statistics

Filgueira R, Comeau LA, Guyondet T, McKindsey CV, Byron CJ (2015) Modelling carrying capacity of bivalve aquaculture: a review of definitions and methods. In: Meyers RA (ed) Encyclopedia of sustainability science and technology. Springer, New York, pp 1–33

Fisheries and Oceans Canada (2019) Framework for aquaculture risk management. https://dfo-mpo.gc.ca/aquaculture/consultations/farm-cgra/farm-cgra-eng.html

Fong CR, DeCesaro J, Clawson G, Frazier M, Halpern BS, Froehlich HE (2024) Winners and losers in U.S. marine aquaculture under climate change. Environ Res Lett 19:114024

Føre M, Frank K, Norton T, Svendsen E, Alfredsen JA, Dempster T, Eguiraun Martinez H, Watson W, Stahl A, Sunde L, Schellewald C, Skøien K, Alver M, Berckmans D (2017) Precision fish farming: a new framework to improve production in aquaculture. Biosyst Eng 173:176–193

Føre M, Alver MO, Alfredsen JA, Rasheed A, Hukkelås T, Bjelland HV, Su B, Ohrem SJ, Kelasidi E, Norton T, Papandroulakis N (2024) Digital twins in intensive aquaculture — challenges, opportunities and future prospects. Comput Electron Agric 218:108676

Foulk A, Gouhier T, Choi F, Torossian J, Matzelle A, Sittenfeld D, Helmuth B (2024) Physiologically informed organismal climatologies reveal unexpected spatiotemporal trends in temperature. Conserv Physiol 12:coae025

Froehlich H, Gentry R, Halpern B (2018) Global change in marine aquaculture production potential under climate change. Nat Ecol Evol 2:1745–1750

Füssel H-M, Klein RJT (2006) Climate change vulnerability assessments: an evolution of conceptual thinking. Clim Chang 75:301–329

Galappaththi EK et al (2020) Climate change adaptation in aquaculture. Rev Aquac 12(4):2160–2176. https://doi.org/10.1111/raq.12427

Geller GN, Halpin PN, Helmuth B, Hestir EL, Skidmore A, Abrams MJ, Aguirre N, Blair M, Botha E, Colloff M, Dawson T, Franklin J, Horning N, James C, Magnusson W, Santos MJ, Schill SR, Williams K (2017) Remote sensing for biodiversity. In: Walters M, Scholes RJ (eds) The GEO handbook on biodiversity observation networks. Springer, pp 187–210

Gentry RR, Froehlich HE, Grimm D, Kareiva P, Parke M, Rust M, Gaines SD, Halpern BS (2017) Mapping the global potential for marine aquaculture. Nat Ecol Evol 1:1317–1324

Gezelius H (2022) 'This is the new normal': Salmon farming industry needs to prepare for further heatwaves. Intrafish

Giacoletti A, Bosch-Belmar M, Mangano M, Tantillo M, Sarà G, Milisenda G (2024) Predicting the effect of fouling organisms and climate change on integrated shellfish aquaculture. Mar Pollut Bull 201:116167

GOA-ON (2024) Global Ocean Acidification Network. Accessed September 15, 2024. https://goa-on.org/home.php

GOOS (2024) The global ocean observing system. The IOC is part of UNESCO. Intergovernmental Oceanographic Commission

Guarini J-M et al (2021) Designing the next generation of condition tracking and early warning systems for shellfish aquaculture. J Mar Sci Eng 9:1084. https://doi.org/10.3390/jmse9101084

Guyondet T, Comeau L, Bacher C, Grant J, Rosland R, Sonier R, Filgueira R (2014) Climate change influences carrying capacity in a coastal embayment dedicated to shellfish aquaculture. Estuar Coasts 38:1593–1618

Hananto A, Tirta A, Herawan S, Idris M, Soudagar ME, Djamari DW, Veza I (2024) Digital twin and 3D digital twin: concepts, applications, and challenges in industry 4.0 for digital twin. Computers 13:100

Handisyde N, Telfer TC, Ross LG (2017) Vulnerability of aquaculture-related livelihoods to changing climate at the global scale. Fish Fish 18:466–488

Hargrave BT, Filgueira R, Grant J, Law B (2022) Combined models of growth, waste production, dispersal and deposition from cage-cultured Atlantic salmon to predict benthic enrichment. Aquac Environ Interact 14:309–328

Hartog J, Spillman C, Smith G, Hobday A (2023) Forecasts of marine heatwaves for marine industries: reducing risk, building resilience and enhancing management responses. Deep-Sea Res II Top Stud Oceanogr 209:105276

Hellicar AD, Rahman A, Smith DV, Smith G, McCulloch J, Andrewartha S, Morash A (2015) An algorithm for the automatic analysis of signals from an oyster heart rate sensor. IEEE Sens J 15:4480–4487

Helmuth B (2009) From cells to coastlines: how can we use physiology to forecast the impacts of climate change? J Exp Biol 212:753–760

Helmuth B, Broitman BR, Yamane L, Gilman SE, Mach K, Mislan KAS, Denny MW (2010) Organismal climatology: analyzing environmental variability at scales relevant to physiological stress. J Exp Biol 213:995–1003

Hillebrand H, Donohue I, Harpole WS, Hodapp D, Kucera M, Lewandowska AM, Merder J, Montoya JM, Freund JA (2020) Thresholds for ecological responses to global change do not emerge from empirical data. Nat Ecol Evol 4:1502–1509

Hoffmann J, Bauer P, Sandu I, Wedi N, Geenen T, Thiemert D (2023) Destination earth – a digital twin in support of climate services. Clim Serv 30:100394

Horricks RA, Clark KF, Watson KL, Lewis-McCrea LM, Reid GK (2024) Effect of acclimation temperature on thermal tolerance between American lobster (Homarus americanus) collected in different lobster fishing areas in Atlantic Canada. Facets 9:1–9

Hunter KL et al (2015) Climate change vulnerability assessment methodology workshop proceedings. In: Canadian Manuscript Report of Fisheries and Aquatic Sciences, vol 3086, 20 pp

Ignatz EH, Sandrelli RM, Vadboncoeur E, Zanuzzo FS, Perry GML, Rise ML, Gamperl AK (2023) The Atlantic salmon's (*Salmo salar*) incremental thermal maximum is a more relevant and sensitive indicator of family-based differences in upper temperature tolerance than its critical thermal maximum. Aquaculture 574:739628

Iliad consortium (2024) Digital twins of the ocean: Access The European Digital Twins of the Ocean Marketplace. Accessed October 30, 2024. https://ocean-twin.eu/

IPCC (2014) Climate change 2014: synthesis report. Intergovernmental Panel on Climate Change (IPCC), Geneva

IPCC (2019a) IPCC special report on the ocean and cryosphere in a changing climate. Cambridge University Press, Cambridge, New York

IPCC (2019b) Summary for policymakers. In: Pörtner H-O, Roberts DC, Masson-Delmotte V, Zhai P, Tignor M, Poloczanska E, Mintenbeck K, Alegría A, Nicolai M, Okem A, Petzold J, Rama B, Weyer NM (eds) IPCC special report on the ocean and cryosphere in a changing climate. IPCC, p 36

IPCC (2021) Climate change 2021: the physical science basis. Contribution of working group I to the sixth assessment report of the intergovernmental panel on climate change. Cambridge University Press, Cambridge, New York

IPCC (2022a) Climate change 2022. In: Impacts, adaptation and vulnerability. Summary for policy makers. Intergovernmental Panel of Climate Change (IPCC), United Nations (UN)

IPCC (2022b) Climate change 2022: impacts, adaptation and vulnerability. Intergovernmental Panel on Climate Change (IPCC), United Nations (U.N.)

IPCC (2023) Synthesis report (SYR) of the IPCC sixth assessment report (AR6): summary for policymakers. IPCC, Geneva

IPCC (2024) IPCC WGI Interactive Atlas. Accessed October 30, 2024. https://interactive-atlas.ipcc.ch/

Islam MM, Barman A, Kundu GK, Kabir MA, Paul B (2019) Vulnerability of inland and coastal aquaculture to climate change: evidence from a developing country. Aquac Fish 4:183–189

Iwama GK, Tautz AF (1981) A simple growth model for salmonids in hatcheries. Can J Fish Aquat Sci 38:649–656

Jacox MG, Alexander MA, Amaya D, Becker E, Bograd SJ, Brodie S, Hazen EL, Pozo Buil M, Tommasi D (2022) Global seasonal forecasts of marine heatwaves. Nature 604:486–490

Jørgensen SM, Castro V, Krasnov A, Torgersen J, Timmerhaus G, Hevrøy EM, Hansen TJ, Susort S, Breck O, Takle H (2014) Cardiac responses to elevated seawater temperature in Atlantic salmon. BMC Physiol 14:2

Kim B-T, Brown CL, Kim D-H (2019) Assessment on the vulnerability of Korean aquaculture to climate change. Mar Policy 99:111–122

Kling MM, Auer SL, Comer PJ, Ackerly DD, Hamilton H (2020) Multiple axes of ecological vulnerability to climate change. Glob Chang Biol 26(5):2798–2813

Kooijman SALM et al (1999) The application of mass and energy conservation laws in physiologically structured population models of heterotrophic organisms. J Theor Biol 197:371–392. https://doi.org/10.1006/jtbi.1998.0881

Krishnan P, Ananthan PS, Purvaja R, Joyson Joe Jeevamani J, Amali Infantina J, Srinivasa Rao C, Ramesh R (2019) Framework for mapping the drivers of coastal vulnerability and spatial decision making for climate-change adaptation: a case study from Maharashtra, India. Ambio 48(2):192–212

Kroeker K, Sanford E, Rose J, Blanchette C, Chan F, Chavez F, Gaylord B, Helmuth B, Hill T, Hofmann G, McManus M, Menge B, Nielsen K, Raimondi P, Russell A, Washburn L (2016) Interacting environmental mosaics drive geographic variation in mussel performance and predation vulnerability. Ecol Lett 19:771–779

Lavoie D, Lambert N, Rousseau S, Dumas J, Chasse J, Long Z, Perrie W, Starr M, Brickman D, Azetsu-Scott K (2020) Projections of future physical and biogeochemical conditions in the Gulf of St. Lawrence, on the Scotian shelf and in the Gulf of Maine. Fisheries and Oceans Canada (DFO), Pelagic and Ecosystem Science Branch, Quebec

Lavoie D, Lambert N, Starr M, Chassé J, Riche O, Le Clainche Y, Azetsu-Scott K, Béjaoui B, Christian JR, Gilbert D (2021) The Gulf of St. Lawrence biogeochemical model: a modelling tool for fisheries and ocean management. Front Mar Sci 8:732269

Lima A, Royer E, Bolzonella M, Pastres R (2023) Digital twins for land-based aquaculture: a case study for rainbow trout (Oncorhynchus mykiss). Open Res Eur 2:16

Luse Data (2024) Luse Data. Accessed October 29, 2024. https://lusedata.no/om-oss

Macatulad E, Biljecki F (2024) Continuing from the Sendai Framework midterm: opportunities for urban digital twins in disaster risk management. Int J Disaster Risk Reduct 102:104310

Mali V (2024) Digital twin for Offshore Windfarm Jacket Lifting Frame. In: International conference on offshore mechanics and arctic engineering, vol 87783. American Society of Mechanical Engineers

Mangano MC, Mieszkowska N, Helmuth B, Domingos T, Sousa T, Baiamonte G, Bazan G, Cuttitta A, Fiorentino F, Giacoletti A, Johnson M, Lucido GD, Marcelli M, Martellucci R, Mirto S, Patti B, Pranovi F, Williams GA, Sarà G (2020) Moving toward a strategy for addressing climate displacement of marine resources: a proof-of-concept. Front Mar Sci 7:408

Mangano MC et al (2019) The role of two non-indigenous serpulid tube worms in shaping artificial hard substrata communities: case study of a fish farm in the central Mediterranean Sea. Aquac Environ Interact 11:41–51. https://www.int-res.com/abstracts/aei/v11/p41-51/

Mangano MC, Berlino M, Corbari L, Milisenda G, Lucchese M, Terzo S, Bosch-Belmar M, Azaza MS, Babarro JMF, Bakiu R, Broitman BR, Buschmann AH, Christofoletti R, Dong Y, Glamuzina B, Luthman O, Makridis P, Nogueira AJA, Palomo MG, Dineshram R, Sanchez-Jerez P, Sevgili H, Troell M, AbouelFadl KY, Azra MN, Britz P, Carrington E, Celić I, Choi F, Qin C, Dionísio MA, Dobroslavić T, Galli P, Giannetto D, Grabowski JH, Helmuth B, Lebata-Ramos MJH, Lim PT, Liu Y, Llorens SM, Mirto S, Pećarević M, Pita C, Ragg N, Ravagnan E, Saidi D, Schultz K, Shaltout M, Tan SH, Thiyagarajan V, Sarà G (2022) The aquaculture supply chain in the time of covid-19 pandemic: vulnerability, resilience, solutions and priorities at the global scale. Environ Sci Pol 127:98–110

Martinez M, Mangano MC, Maricchiolo G, Genovese L, Mazzola A, Sarà G (2018) Measuring the effects of temperature rise on Mediterranean shellfish aquaculture. Ecol Indic 88:71–78

Masmitja I et al (2024). Digital Twin-sustained 4D ecological monitoring of restoration in fishery depleted areas (DIGI4ECO)

Maulu S, Hasimuna OJ, Haambiya L, Monde C, Musuka C, Makorwa T, Munganga B, Phiri K, Nsekanabo J (2021) Climate change effects on aquaculture production: sustainability implications, mitigation, and adaptations. Front Sustain Food Syst 5:16

Mafi-Gholami D et al (2020) Vulnerability of coastal communities to climate change: Thirty-year trend analysis and prospective prediction for the coastal regions of the Persian Gulf and Gulf of Oman. Sci Total Environ 741:140305. https://doi.org/10.1016/j.scitotenv.2020.140305

Matzelle A et al (2015) A bioenergetics framework for integrating the effects of multiple stressors: opening a 'Black Box' in climate change research. Am Malacol Bull 33:1–11. https://doi.org/10.4003/006.033.0107

McIntosh P, Barrett LT, Warren-Myers F, Coates A, Macaulay G, Szetey A, Robinson N, White C, Samsing F, Oppedal F, Folkedal O, Klebert P, Dempster T (2022) Supersizing salmon farms in the coastal zone: a global analysis of changes in farm technology and location from 2005 to 2020. Aquaculture 553:738046

Methuselah J (2024) Digital twin technology for smart manufacturing. J Technol Syst 6:52–65

Miedtank A, Schneider J, Manss C, Zielinski O (2024) Marine digital twins for enhanced ocean understanding. Remote Sens Appl Soc Environ 36:101268

Mills K et al (2017) Forecasting the seasonal timing of Maine's lobster fishery. Front Mar Sci 4. https://doi.org/10.3389/fmars.2017.00337

Molotoks A, Smith P, Dawson TP (2021) Impacts of land use, population, and climate change on global food security. Food Energy Secur 10:e261

Montalto V, Palmeri V, Rinaldi A, Kooijman SALM, Sarà G (2014) Dynamic energy budget parameterisation of Brachidontes pharaonis, a Lessepsian bivalve in the Mediterranean Sea. J Sea Res 94:47–51

Montalto V et al (2015) Life history traits to predict biogeographic species distributions in bivalves. Die Naturwissenschaften 102:61. https://doi.org/10.1007/s00114-015-1313-4

Montalto V et al (2016) A mechanistic approach reveals non linear effects of climate warming on mussels throughout the Mediterranean sea. Clim Change 139(2):293–306. https://doi.org/10.1007/s10584-016-1780-4

Montalto V et al (2017) The effect of the quality of diet on the functional response of Mytilus galloprovincialis (Lamarck, 1819): Implications for integrated multitrophic aquaculture (IMTA) and marine spatial planning. Aquaculture 468:371–377. https://doi.org/10.1016/j.aquaculture.2016.10.030

Montes C, Acharya N, Rumana Hossain P, Amjath Babu TS, Krupnik TJ, Quamrul Hassan SM (2022) Developing a framework for an early warning system of seasonal temperature and rainfall tailored to aquaculture in Bangladesh. Clim Serv 26:100292

Naylor RL, Kishore A, Sumaila UR, Issifu I, Hunter BP, Belton B, Bush SR, Cao L, Gelcich S, Gephart JA, Golden CD, Jonell M, Koehn JZ, Little DC, Thilsted SH, Tigchelaar M, Crona B (2021) Blue food demand across geographic and temporal scales. Nat Commun 12:5413

Nor Azra M (2025) Aquaculture and climate change: a data-driven analysis. Ann Anim Sci. https://doi.org/10.2478/aoas-2024-0085

O'Donncha F, Grant J (2019) Precision aquaculture. IEEE Internet Thing Mag 2:26–30
Oyebola OO, Olatunde OM (2019) Climate change adaptation through aquaculture: ecological considerations and regulatory requirements for tropical Africa. In: Bamutaze Y, Kyamanywa S, Singh B, Nabanoga G, Lal R (eds) Agriculture and ecosystem resilience in sub Saharan Africa. Springer, Cham, pp 435–472
Palmer MR, Shagude YW, Roberts MJ, Popova E, Wihsgott JU, Aswani S, Coupland J, Howe JA, Bett BJ, Osuka KE, Abernethy C, Alexiou S, Painter SC, Kamau JN, Nyandwi N, Sekadende B (2021) Marine robots for coastal ocean research in the Western Indian Ocean. Ocean Coast Manag 212:105805
Parra L, Lloret G, Lloret J, Rodilla M (2018) Physical sensors for precision aquaculture: a review. IEEE Sens J 18(10):3915–3923
Pörtner H-O, Bock C, Mark FC (2017) Oxygen- and capacity-limited thermal tolerance: bridging ecology and physiology. J Exp Biol 220:2685–2696
Reid GK, Moccia RD (2006) Diel fluctuation of soluble phosphorus in the tank water of rainbow trout, *Oncorhynchus mykiss* (Walbaum), and relationships with feed inputs and dissolved oxygen. Aquac Res 15:1606–1610
Reid GK, Gurney-Smith H, Flaherty M, Garber A, Forster I, Brewer-Dalton K, Knowler D, Marcogliese D, Chopin T, Moccia R, Smith C, De Silva S (2019a) Climate change and aquaculture: considering adaptation potential. Aquac Environ Interact 11:603–624
Reid GK, Gurney-Smith H, Marcogliese D, Knowler D, Benfey T, Garber A, Forster I, Chopin T, Brewer-Dalton K, Moccia R, Flaherty M, Smith C, De Silva S (2019b) Climate change and aquaculture: considering biological response and resources. Aquac Environ Interact 11:569–602
Reid GK, Lefebvre S, Filgueira R, Robinson SMC, Broch OJ, Dumas A, Chopin TBR (2020) Performance measures and models for open-water integrated multi-trophic aquaculture. Rev Aquac 12:47–75
Richards RG, Davidson AT, Meynecke J-O, Beattie K, Hernaman V, Lynam T, van Putten IE (2015) Effects and mitigations of ocean acidification on wild and aquaculture scallop and prawn fisheries in Queensland, Australia. Fish Res 161:42–56
Saikia M, Mahanta R (2023) Vulnerability to climate change and its measurement: a survey. Boreal Environ Res 28:111–124
Sampaio R, Aguiar Costa A, Flores-Colen I, Klungseth NJ, Semini M, Nordvik S (2024) An integrated framework for digital twins in hospitals. IOP Conf Ser Earth Environ Sci 1389:012042
Sarà G, Reid GK, Rinaldi A, Palmeri V, Troell M, Kooijman SALM (2012) Growth and reproductive simulation of candidate shellfish species at fish cages in the Southern Mediterranean: Dynamic Energy Budget (DEB) modelling for integrated multi-trophic aquaculture. Aquaculture 324–325:259–266
Sarà G, Gouhier T, Brigolin D, Porporato E, Mangano MC, Mirto S, Mazzola A, Pastres R (2018a) Predicting shifting sustainability tradeoffs in marine finfish aquaculture under climate change. Glob Chang Biol 24:3654–3665
Sarà G, Mangano MC, Johnson M, Mazzola A (2018b) Integrating multiple stressors in aquaculture to build the blue growth in a changing sea. Hydrobiologia 809:5–17
Sarkodie SA, Strezov V (2019) Economic, social and governance adaptation readiness for mitigation of climate change vulnerability: evidence from 192 countries. Sci Total Environ 656:150–164
Scale Aquaculture AS (2024) Scale AQ: Integrated Annual Report 2023. https://scaleaq.no/wp-content/uploads/2024/05/ScaleAQ-Integrated-Annual-Report-2023.pdf
Schultes E, Roos M, da Silva Santos LOB, Guizzardi G, Bouwman J, Hankemeier T, Baak A, Mons B (2022) FAIR digital twins for data-intensive research. Front Big Data 5:883341
Shelton C (2014) Climate change adaptation in fisheries and aquaculture - compilation of initial examples. FAO, Rome
Skogen MD, Ji R, Akimova A, Daewel U, Hansen C, Hjøllo SS, van Leeuwen S, Maar M, Macías D, Mousing E, Almroth-Rosell E, Sailley S, Spence MA, Troost T, Van de Wolfshaar KE

(2021) Disclosing the truth: are models better than observations? Mar Ecol Prog Ser 680:7–13. DYNMOD

Smarsly K, Abadia PP, Dragos K, Ahmad ME, Al-Zuriqat T, Geck CC, Al-Nasser H (2024) A multivocal literature review of digital twins, architectures, and elements in civil engineering. eJNDT 29. https://doi.org/10.58286/29579

Smit B, Wandel J (2006) Adaptation, adaptive capacity and vulnerability. Glob Environ Chang 16:282–292

Soares BM, Gagnon AS, Doherty RM (2012) Conceptual elements of climate change vulnerability assessments: a review. Int J Clim Change Strateg Manag 4:6–35

Sommerset I, Wiik-Nielsen J, Moldal T, Oliveira VHS, Svendsen JC, Haukaas A, Brun E (2024) Norwegian fish health report. Norwegian Veterinary Institute

Sonnet V, Mouw C, Ciochetto A, Carney-Almeida J (2024) Hit or miss? Impact of time series resolution on resolving phytoplankton dynamics at hourly, weekly, and satellite remote sensing frequencies. Limnol Oceanogr Methods 22(4):254–267

Soto D, Norambuena F (2004) Evaluation of salming farming effects on marine systems in the inner seas of southern Chile: a large-scale mensurative experiment. J Appl Ichthyol 20:493–501

Soto D, Ross LG, Handisyde N, Bueno PB, Beveridge MCM, Dabbadie L, Aguilar-Manjarrez J, Cai J, Pongthanapanich T (2018) Climate change and aquaculture: vulnerability and adaptation options. In: Barange M, Bahri T, Beveridge MCM, Cochrane KL, Funge-Smith S, Poulain F (eds) Impacts of climate change on fisheries and aquaculture. Food and Agricultural Organization (FAO), Rome, pp 465–490

Soto D, León-Muñoz J, Dresdner J, Luengo C, Tapia FJ, Garreaud R (2019) Salmon farming vulnerability to climate change in southern Chile: understanding the biophysical, socioeconomic and governance links. Rev Aquac 11:354–374

Stechele B, Lavaud R (2024) Including ocean acidification effects on biocalcification through dynamic energy budget modelling. Ecol Model 496:110816

Stewart-Sinclair PJ, Last KS, Payne BL, Wilding TA (2020) A global assessment of the vulnerability of shellfish aquaculture to climate change and ocean acidification. Ecol Evol 10:3518–3534

TCFD (2021) Task force on climate-related financial disclosures: implementing the recommendations of the task force on climate-related financial disclosures. Task force on climate-related financial disclosures

Teal L et al (2015) Physiology-based modelling approaches to characterize fish habitat suitability: their usefulness and limitations. Estuar Coast Shelf Sci 201. https://doi.org/10.1016/j.ecss.2015.11.014

Thomas Y, Bacher C (2018) Assessing the sensitivity of bivalve populations to global warming using an individual-based modelling approach. Glob Change Biol 24. https://doi.org/10.1111/gcb.14402

Thomas Y et al (2011) Application of a bioenergetic growth model to larvae of the pearl oyster Pinctada margaritifera L. J Sea Res 66(4):331–339. http://dx.doi.org/10.1016/j.seares.2011.04.005

Thomas K, Hardy RD, Lazrus H, Mendez M, Orlove B, Rivera-Collazo I, Roberts JT, Rockman M, Warner BP, Winthrop R (2019) Explaining differential vulnerability to climate change: a social science review. Wiley Interdiscip Rev Clim Change 10:e565

van der Meer J et al (2014) 35 years of DEB research. J Sea Res 94. http://dx.doi.org/10.1016/j.seares.2014.09.004

Winne S, Pritchard A, Kennedy S, O'Gorman S (2022) Review of climate vulnerability assessment of Scotland's marine environment and marine economy. Ramboll Management Consulting

Yang J, Gong P, Fu R, Zhang M, Chen J, Liang S, Xu B, Shi J, Dickinson R (2013) The role of satellite remote sensing in climate change studies. Nat Clim Chang 3:875–883

Ytteborg E, Falconer L, Krasnov A, Johansen L-H, Timmerhaus G, Johansson G, Afanasyev S, Høst V, Hjøllo S, Hansen Ø, Lazado C (2023) Climate change with increasing seawater temperature will challenge the health of farmed Atlantic Cod (Gadus morhua L.). Front Mar Sci 10:1232580

Zhan Y, Ning B, Sun J, Chang Y (2023) Living in a hypoxic world: a review of the impacts of hypoxia on aquaculture. Mar Pollut Bull 194:115207
Zhang H, Li G, Hildre HP (2024) Digital twins for vessel life cycle service. Springer, Singapore
Zitti M, Guttormsen A (2023) Climate risk and financial disclosure in salmon aquaculture. Aquac Econ Manag 27:441–467

Chapter 8
Supply Chain for Aquaculture

Carmen G. Sotelo, Trond A. Henriksen, Nuno Ferreira, M. Angeles Longa-Portabales, Adrian Muwonge, Timothy Robert Wiese, Michael Kozicki, and Fearghal O'Donncha

Abstract The global demand for seafood continues to increase, driven by consumer preferences for nutritious, protein-rich food sources. Aquaculture has emerged as the fastest-growing food production sector, surpassing capture fisheries in aquatic animal production. This chapter explores the critical role of traceability in enhancing the efficiency, sustainability, and transparency of aquaculture supply chains. It discusses the integration of digital traceability systems, which provide accurate and transparent data throughout the supply chain, thereby meeting regulatory requirements and consumer demands for detailed product information. The chapter also examines the challenges and benefits of implementing traceability solutions, including the use of blockchain technology to ensure data integrity and combat illegal, unreported, and unregulated (IUU) fishing. Case studies highlight successful applications of traceability in aquaculture, demonstrating improvements in oper-

C. G. Sotelo (✉)
Instituto de Investigaciones Marinas (CSIC), Food Science and Technology, Vigo, Spain
e-mail: carmen@iim.csic.es

T. A. Henriksen
Atea, FoodTrust and Blockchain, Trondheim, Norway

N. Ferreira
Exporsado—Comércio e Indústria de Produtos do Mar, S.A., Setúbal, Portugal

M. A. Longa-Portabales
Consello Regulador Mexillon de Galicia, I-D Department, Vilagarcia de Arousa, Spain

A. Muwonge · T. R. Wiese
The Digital One Health Laboratory, Division of Epidemiology at The Roslin Institute, University of Edinburgh, Easter Bush, UK

The Blockchain Technology Laboratory, School of Informatics, University of Edinburgh, Edinburgh, UK

M. Kozicki
School of Electrical, Computer and Energy Engineering, Arizona State University, Tempe, AZ, USA

F. O'Donncha
IBM Research Europe, Dublin, Ireland

F. O'Donncha et al. (eds.), *Digital Twin for Marine Fish Farming*,
Aquatic Production Systems, https://doi.org/10.1007/978-3-032-10362-8_8

ational efficiency, product quality, and consumer trust. The findings underscore the importance of traceability as a strategic tool for achieving sustainability and competitiveness in the global seafood market.

Keywords Aquaculture traceability systems · Supply chain transparency · Digital traceability · Blockchain applications

8.1 Introduction

Seafood consumption reports an annual growth rate of 1.2% over the last decade, surpassing global population growth rate of 0.9% (FAO 2024). Driven by consumer demand for nutritious, protein-rich food sources, the global appetite for seafood is expected to increase by 14% by 2030 (Brenan et al. 2023). This rapid growth is driven by factors such as improvements in processing, shipping, and distribution; growth in global incomes; and increasing consumer awareness about the benefits of seafood consumption. This is particularly pronounced in higher-income countries with studies noting a correlation between seafood consumption and income levels (Han et al. 2022). The most recent FAO report (2024) indicates that aquaculture has been the fastest-growing food production sector over the past three decades, with an average annual growth rate of 6.7%, surpassing capture fisheries in aquatic animal production for the first time in 2022.

Aquaculture constitutes 46% of global seafood production, playing a crucial role in ensuring food security. Currently, aquaculture supplies 52% of fish for human consumption. According to FAO projections (2020), this share is expected to rise to 59% by 2030. Global demand for seafood will grow in the next decade due to several factors including "world population and economic growth, increasing urbanisation, dietary diversification and technological developments" (FAO 2020). However, studies have shown that more than 85 percent of the world's fisheries have already been pushed to or beyond their biological limits. Hence global food security demands that aquaculture contribute significantly to meet the growing demand of seafood. This pushes the industry toward enhanced innovation around factors such as operational efficiency, supply chain, and fish feed—particularly those with fishmeal as ingredient (Dewali et al. 2023).

Integration across the seafood supply chain promises a seamless exchange of information between the stakeholders and communication systems active in the domain. Significant differences exist between small scale producers—which are not always able to implement digital information and communication systems—and large companies with considerable resources at their disposal. Seafood supply chains have grown in complexity over the last decades. This is particularly evident in the increasing globalisation of seafood processing. Furthermore, a small number of transnational companies dominate the global production of seafood, some of them involving vertical supply chain integration. These companies rely on sophisticated systems to manage information exchange regarding production or harvest plans, suppliers, transport and distribution logistics, and clients (Österblom et al. 2015).

Consumers increasingly demand detailed labelling information about the provenance, welfare, and sourcing of aquaculture products. Retailers, in turn, require these data from their suppliers to satisfy consumer demands. Consequently, digital traceability systems have become increasingly prominent for providing accurate and transparent data throughout the supply chain (Sterling et al. 2015).

The European Union General Food Law (European Parliament and Council, EC N° 178/2002) defines traceability as "the ability to trace and follow a food, feed, food-producing animal or substance intended to be, or expected to be incorporated into a food or feed, through all stages of the production, processing and distribution". However, this definition doesn't adequately capture all the relevant properties, and it does not reflect the importance of effective data management, storage, and analysis capabilities. Olsen and Borit (2013) proposed an alternative definition as: "the ability to access any or all information relating to what is under consideration, throughout its entire life cycle, by means of recorded information". This provides a more complete and precise definition of traceability. It highlights that traceability-related information should be captured, recorded, and kept in a system and should be able to accompany the food product across its life cycle from production to consumption.

Modern traceability efforts commenced in the early 1960s as a collaboration between a food company in the USA and NASA. At that time, the objective was to provide safe food for the burgeoning manned space program. Initially traceability served as a safety tool, facilitating the detection of safety risks that could be traced one-step back and one-step forward. This ability was crucial to preserve consumers health and save companies costs from product recall costs (Sperder and Stier 2010).

The next significant development in traceability occurred in the 1980s driven by a consumer demand to ensure that the seafood they purchased adhered to animal welfare standards (e.g. Dolphin safe label) or sustainability principles (e.g. MSC certification). This led to the introduction of a chain-of-custody system, which verified that the products could be traced back to their producer. Although these systems still relied on one-step-back and one-step-forward traceability with paper documents, they also incorporated audit systems to ensure the accuracy of the information.

With the advancement and improvement of information technologies in the 2000s, coupled with the globalisation of the seafood trade and the increasing incidence of illegal, unreported, and unregulated fisheries, the digitalisation of traceability information gained momentum. This development made full-chain traceability a reality, allowing consumers to access information about the seafood they purchase (Tamm et al. 2016).

Despite these advancements, the implementation of comprehensive traceability systems remains limited due to the current costs associated with these tools. The costs, which need to be distributed between the industry and final consumers, pose a significant barrier. However, as the costs decrease due to the mass production of digital traceability devices (e.g. smart printed labels, sensors for temperature, and geolocation), the widespread implementation of traceability systems is expected to become the norm rather than the exception.

Seafood traceability is a regulatory requirement in the EU since 2002 (EC 178/2002). Traceability is required for quality certification, safety, and performance management of farms. It also provides consumers assurance about the provenance, quality, and environmental and social aspects of the product. Furthermore, traceability is an essential tool for business performance by providing operators the means to rapidly assess and respond to changing dynamics in their industry. This makes traceability essential to a global and competitive market (Gooch and Sterling 2013).

Therefore, implementing traceability can provide numerous benefits to companies that consistently apply these tools, both internally and within the context of an increasingly complex seafood value chain that requires effective information exchange among stakeholders (Blaha et al. 2023). Regulatory compliance has been a key driver of traceability systems. In previous decades, various food safety scandals prompted governments to act to restore consumer confidence in food supplies.

Furthermore, traceability tools have shown companies that they can significantly reduce costs by mitigating safety risks, sourcing necessary materials (e.g. feed) from reputable providers, and minimising stocks and wastes (Sterling et al. 2015).

Another outcome of traceability is to increase consumer confidence in the products they buy. Consumer perception associates traceability with provenance, food quality, and sustainability, which are important drivers across demographics (Sterling et al. 2015).

An effective traceability solution requires comprehensive and holistic standards around data collection, storing, and analysis, which allows companies to collaborate across the value chain (Sterling et al. 2015). Therefore, an integrated management system is crucial to effectively implement traceability framework. These require a systematic approach to data collection, storage, and analysis, and with the capacity to exchange this information with other business along the value chain. Studies demonstrate that an effective system can provide a competitive advantage that reinforces the smart use of traceability tools for increasing profitability of companies (Gooch and Sterling 2013).

This chapter evaluates the impact of traceability on improving aquaculture value chains performance. We discuss how traceability helps companies comply with legislative requirements, optimise energy, and reduce carbon emissions. Further, we provide information of case studies where traceability has shown to be an essential tool for aquaculture producers to achieve different objectives such as origin denomination, quality and sustainable production, improvement of internal processes and end-to-end traceability with blockchain technology.

8.2 Industry Adoption of Traceability Solutions

The proliferation of traceability solutions within an industry is driven by a combination of factors, including competitive differentiation, legislative requirements, consumer demands, and productivity improvements. In the context of the aquaculture industry, Blaha et al. (2023) classify traceability drivers into two main categories:

"Negative drivers" refer to actions that a firm feels compelled to take. These often involve compliance with traceability requirements mandated by legislation, adopted standards, or demands from buyers, customers, consumers, or the market. They also include efforts to mitigate risks, particularly those related to food safety and fraud, as well as initiatives aimed at short-term reductions in costs, resource use, and waste.

"Positive drivers" focus on leveraging traceability to create value. These include improving product quality, enhancing communication and information flow within the supply chain, and supporting brand trust, product differentiation, and storytelling. They also emphasise sustainability and ethical considerations that go beyond legal or contractual obligations.

8.2.1 Legislative Requirements

A key driver for the implementation of traceability systems by aquaculture stakeholders is the necessity to comply with regulations set by legislative bodies across international markets. Some of these regulations are influenced by the 1995 FAO *Code of Conduct*, which establishes international standards for responsible practices. This code aims to promote the sustainable and responsible use of fish resources, address international trade practices, and advocate for the implementation and monitoring of adequate fisheries management systems to ensure sustainability and traceability throughout the whole seafood value chain (FAO 1995).

Blaha et al. (2023) identify three main types of scenarios where legal requirements may apply: within organisations, between national organisations, and between different countries.

The European Union (EU) has established traceability requirements of the seafood products sold in member states. Objectives include enhancing sustainability and combating illegal, unreported, and unregulated (IUU) fishing through traceability systems. These systems enable the identification of key factors such as species, production methods (wild-caught or farmed), geographic origin, and, for some farmed species, the feed used in production (Control Regulation of the EU EC No 1224/2009, Common Market Organization Regulation EC No 1379/2013). The EU has also recently introduced Regulation EC No 2023/2842, effective from December 2023, which mandates improved monitoring of seafood using digital traceability along the supply chain. This regulation initially applies to fresh and frozen fishery and aquaculture products, both domestic and imported.

The United States has implemented the Seafood Import Monitoring Program (SIMP 2016) that requires traceability information for imported seafood, including aquaculture products, to detect and prevent the entry of illegal, unreported, and unregulated (IUU) fishing products into the U.S. market.

Traceability legal requirements for aquaculture products include the following items: documentation and record-keeping system, detailed records of the production process, farming practices, feed type and sources, welfare, and health management of the animals.

Traceability also has an essential role in managing aquaculture side streams by ensuring transparency, sustainability, and compliance throughout the production process. In the European Union, there are some pieces of legislation, such as the Water Framework Directive (2000/60/EC) and the Marine Strategy Framework Directive (2008/56/EC), that establish measures to protect water bodies from pollution and marine ecosystems, and this includes aquaculture installations that must monitor the release of nutrients, contaminants, and organic matter to the environment.

Moreover, EU traceability regulations for aquaculture emphasise detailed health monitoring of farmed animals. Operators must ensure compliance with health standards throughout the production process, maintaining transparency and accountability. Required traceability records include the following:

- Documentation for each batch of farmed animals to support health certifications
- Disease monitoring and surveillance records
- Detailed logs of treatments, such as antibiotics or vaccines, including dates, dosages, and reasons for use
- Records of mortality rates and causes of death
- Periodic health assessments conducted by qualified personnel to document the overall health status of the stock
- Environmental condition records, such as water quality parameters, temperature, and stocking density, which may affect animal health

These measures, outlined in Regulation EU 2016/429, aim to enhance the traceability, sustainability, and accountability of aquaculture production.

8.2.2 Consumers Requirements

As aquaculture becomes the primary source of seafood, meeting consumer's expectations will be crucial to addressing the growing demand for seafood. Studies indicate that consumers prioritise the freshness, price, taste, appearance, health and safety information, sustainability certification, country of origin, and nutritional information when making a purchase decision of seafood (Saidi et al. 2022; Kamiyama et al. 2024). While wild-caught fish has traditionally been preferred, recent studies indicate a growing perception of farmed fish as the more sustainable option. This shift is largely attributed to its potential for local production, which offers environmental benefits, supports local employment, promotes rural development, and provides a more affordable alternative compared to wild-caught fish (López-Mas et al. 2023).

Surfacing this information to consumers in a scalable manner demands an end-to-end traceability solution. In the EU market, the label of unprocessed aquaculture products typically shows the following information (Fig. 8.1): species commercial name, scientific name, production method (in the case of aquaculture products the word "farmed", should be indicated in the label), country where the animal has been farmed, batch number, expiry/best before date, and how the product should be prepared (EC1379/2013 and EC 1169/2011). However, if the farmed fish has been processed, such as in the case of canned salmon, some of this information is lost, since it is not mandatory to indicate the origin or whether it has been farmed or caught (example canned salmon).

Key factors that influence consumer purchase decisions are the location where the product was farmed and the sustainability standards achieved. Some origins are recognised by consumers as a quality mark, and traceability information are a key part of a premium branding strategy (Zander and Feucht 2018).

Besides origin, sustainable aquaculture certification is also an attribute with favourable consumer perceptions (van Osch et al. 2019), although studies indicate that it is less appreciated in Southern European countries than in Northern ones (Saidi et al. 2022). However, animal welfare and preservation efforts are becoming important to consumers across Western countries. This is particularly pronounced among younger consumers. The importance of sustainable certification schemes, such as those from the Aquaculture Stewardship Council (ASC), is growing in popularity among aquaculture companies globally. A recent study suggests that eco-certification of sustainable seafood production have a positive effect not only in fishing and aquaculture practices but also in consumer behaviour (Giacomarra et al. 2021). By setting rigorous standards for environmental and social performance, the ASC ensures that aquaculture practices minimise negative impacts on ecosystems and local communities ASC (Arton et al. 2020). This not only enhances the marketability of certified products but also drives industry-wide improvements by encouraging producers to adopt more sustainable practices.

8.2.3 Industry Requirements

The aquaculture industry faces growing demand for low-impact, energy-efficient seafood, with increasing attention to carbon footprint and animal welfare considerations. Transparency regarding feed ingredient sources and their environmental impact is becoming critical (Cooney et al. 2021).

Judging by Scope 1 and 2 emissions calculated using the Greenhouse Gas Protocol (GHG), aquaculture companies generally have a very low carbon footprint. However, the inclusion of Scope 3 drastically changes this perspective. For example, fish feed alone contributes up to 83% of the carbon footprint generated by

Fig. 8.1 Label for aquaculture products. The top figure presents an example of an unprocessed aquaculture fish while bottom is an example of a processed aquaculture product

Norwegian farmed Atlantic salmon (Skretting 2021), overshadowing improvements in Scope 1 and 2 emissions.

Historically, the largest aquaculture companies maintained a strictly operational focus, prioritising cost reduction, production planning, and delivery schedules. With farmed salmon already having a significantly lower carbon footprint than beef, lamb, and pork, the aquaculture industry has been slow to mandate environmental documentation and focused action from their suppliers.

Feed manufacturers—driven by customer demand for improved operational performance and cost efficiency—conduct extensive research to develop carbon-neutral feed. Their efforts focus on key elements such as ingredients, growth areas, and feed formulations. Recently, some aquaculture companies and feed manufacturers have expanded their focus to include environmental impacts. Combined with a focus on fish welfare, these companies are driving innovation on feed alternatives and operational changes that will improve fish welfare. Technology has become an essential tool in achieving these objectives.

Emerging evidence suggests that addressing feed composition, environmental footprint, and traceability not only reduces impacts but also increases consumer demand and margins. Retailers and processing companies serving the higher end consumer segments currently specify standards for production, documentation, or both. For instance, Helsingborg City Council pays premiums of up to 100% for seafood with documented lower environmental impacts. This shift aligns with the broader trend of responsible consumption, particularly among Gen Z, who is leading the movement toward sustainability-driven choices.

Unlocking and leveraging production data has become a cornerstone for driving operational improvements. Until recently, accessing and utilising production data were challenging. However, digitalisation initiatives have provided unique insights that enable significant advancements. Some companies now require data to accompany deliveries, creating opportunities for optimisation across the value chain.

For example, analysing why fish are downgraded from premium quality during slaughter offers valuable insights into improving production methods, genetics, and feed. Common reasons for downgrades include flesh colour variations, melanin spots, or visible external injuries. Addressing these issues helps ensure higher quality production and aligns with the industry's push toward sustainability and efficiency.

8.3 Traceability System

Analysing the elements and connections within an aquaculture value chain is crucial for implementing a functional traceability system. The geographic location where each process or company in the value chain operates plays a significant role as different regions may impose different administrative and legal requirements for trading and marketing products. The Global Dialogue on Seafood Traceability (GDST), a non-profit foundation, was established in 2017, addressing the complexity of global seafood trade. This trade involves multiple companies across different countries, each adhering to unique regulations. GDST recognises traceability as a critical tool to manage these challenges, emphasising the need for interoperability based on standardised data formats and harmonised communication protocols. Such approaches enable seamless data exchange and real-time collaboration across stakeholders in the seafood value chain.

According to Blaha et al. (2023), the aquaculture value chain generally comprises actors such as feed producers and distributors, hatcheries, farms, processing companies, distributors, and marketers (Table 8.1). Each of these stakeholders is responsible for providing information on Critical Tracking Events (CTEs), referred

Table 8.1 Examples of Critical Tracking Events (CTEs) and associated Key Data Elements (KDEs) required for traceability across key actors in the aquaculture value chain

Actor (value chain stage)	Critical Tracking Event (CTE)	Key Data Elements (KDEs)
Feed	Feed production	Feed producer details (name, unique operator identifier, location); production date; feed batch number; expiry date; unique seafood material identifier; ingredient list (species-specific); main additives used
	Feed commercialisation	Selling date; first buyer (name and details); linked batch information
Hatchery/nursery/seed growing	Broodstock reception	Broodstock identifier; reception date; species name; sanitary status
	Seed production and feeding	Hatchery identity (name and location); feed supplier details (name, brand, type, pellet size, feed batch number, expiry date, volume, additives used); veterinary drugs and chemicals used
	Harvest/seed transfer	Harvest date; harvest volume; batch number; seafood material identifier; transfer recipient; and transport details
Farms/growing areas	Stocking	Farm identity (name and location); pond/cage/tank/pen identifier; stocking date; stocking quantity; source batch number
	Feeding and treatments	Feed supplier details (name, brand, type, pellet size, feed batch number, expiry date, volume purchased); veterinary drugs and chemicals used
	Harvest	Harvest date; harvest volume/weight/quantity; batch number; seafood material identifier; destination details
Distribution and processing	Distribution event	Distributor name and details; movement date; transport vehicle identifier; destination
	Processing/transformation	Client details (distributor or processing plant); registration/approval number; processing plant identity; product origin; reception date; species name; net verified weight/volume

Adapted from Blaha et al. (2023) and GDST (2023)

to as Key Data Elements (KDEs), which are essential for traceability. Critical Tracking Events (CTEs) are specific points in the supply chain where data capture is required to maintain traceability and Key Data Elements (KDEs) are the data itself captured with specific format (GDST 2023). This ensures that data can be consistently captured, shared, and integrated across the supply chain.

8.3.1 Technical Challenges to Implementing Traceability Solutions

Achieving full traceability in aquaculture involves overcoming several technical challenges, including the following:

- *Data Availability*: Incomplete or inconsistent data capture along the value chain limits the ability to establish end-to-end traceability. Many aquaculture stakeholders lack robust systems for tracking and recording critical data.
- *Data Formats, Classification, and Integration*: Variability in data formats and classification standards across stakeholders poses significant integration challenges. Harmonising and consolidating data from diverse sources are essential for creating seamless traceability systems.
- *Contractual Legacy Restrictions on Data Exchange*: Existing contracts between stakeholders often restrict data sharing, impeding the transparency required for effective traceability. These restrictions can be a major barrier to implementing interoperable systems.
- *Legacy Systems with Limited Connectivity*: Many stakeholders operate legacy systems that are incompatible with modern, cloud-based data platforms. These systems often lack connectivity and fail to support real-time data exchange, hindering integration across the value chain.
- *Standards and Interoperability*: The absence of universally accepted standards for traceability complicates data exchange and collaboration. Initiatives like the Global Dialogue on Seafood Traceability (GDST) are working to address this by defining common standards for Key Data Elements (KDEs) and Critical Tracking Events (CTEs).
- *Data Exchange Mechanisms*: Establishing reliable and efficient mechanisms for data exchange is critical. Initiatives such as GDST and others aim to facilitate real-time, standardised data sharing to improve traceability. However, adoption of these mechanisms varies across regions and stakeholders, creating inconsistencies.

Addressing these technical challenges is essential for realising the full potential of traceability systems in aquaculture, ensuring compliance, sustainability, and transparency throughout the supply chain.

8.4 Case Studies

8.4.1 Case Study 1: Provenance and Quality Assurance Labelling

Galicia, in Northwestern Spain, is one of the most important mussel aquaculture regions in Europe, producing 250,000 tonnes of mussels annually. This makes Spain the third largest producer, after China and Chile, and the leading mussel producer in Europe. In fact, approximately half of all mussels produced in the EU are harvested in this Spanish region.

Mussels are cultivated in Galicia using a unique technique that started more than 70 years ago, using rafts or floating platforms. These rafts are made of eucalyptus wood framework with a maximum surface area of 550 m^2, supported by 4 or 6 metal floats and from which a maximum of 500 ropes of 12 m in length are hung (Fig. 8.2).

Cultivation is concentrated in five marine estuaries ("rías") where oceanographic conditions drive the upwelling of nutrients and promote the growth of primary producers (Figueiras et al. 2002). There are currently 3387 rafts and a fleet of 1061 boats dedicated to mussel farming in these estuaries. Furthermore, mussel farming in Galicia is a very important activity due to its contribution to ecosystem health, the quality and volume of the product, the economic and employment impacts, and its criticality to the sustainability of the region.

Fig. 8.2 Mussel raft used for aquaculture along the Galician coast

Mussel culture began in Galicia in rafts in 1946. After Spain joined the European Economic Community, foreign products entered traditional Galician mussel markets more easily. To differentiate their products, part of the Galician production sector initiated quality assurance strategies to highlight the premium nature of the produce. With this objective, at the end of the 1980s, a working group was set up within the production sector to seek differentiation through quality. This group was the seed of the current "Consello Regulador do Mexillón de Galicia" the entity that manages the PDO. Official recognition came in 1992 with the "Quality Galician Product" label (DOG 1992). Before this, the group lobbied for regulatory changes to include fishing, shellfish harvesting, and aquaculture under the Galician Quality label. With the approval of the European regulation on geographical indications (EC N° 2081/92), the Galician label was adapted to meet European standards. The Mexillón de Galicia PDO received provisional approval in 2000 and was finalised in 2007 after a lengthy process.

Attaining the quality label of Galician mussels marked a significant milestone in the history of mussel culture in Galicia. With the new PDO label, Mexillón de Galicia entered the Spanish and European gastronomy elite. Maintaining this position required continuous innovation with regard quality and sustainability of the Galician mussel cultures. This included adding value through new packaging, extended shelf life, and marketing campaigns underscoring the quality of the locally sourced produce. Achieving these objectives required enhanced traceability systems. Growing competition from mussels of other origins also necessitated measures to prevent the fraud of foreign mussels being marketed as Galician.

Quality foods with geographical indication labels (such as Mexillón de Galicia PDO) protect the genuine links between these foods and the natural and cultural resources in their areas. These labels help preserve the food heritage, contribute to food biodiversity, conserve world cultural heritage, promote sociocultural development and reduce poverty in rural production areas (Vandecandelaere et al. 2010). Geographical indications are intangible assets with the capacity to differentiate products, create added value, and extend their effects to areas related to the primary product for which the geographical indication is known.

Therefore, the label "Mexillón de Galicia" delimits a specific area of the territory where this protected product can be produced, processed, and transformed. This designation supports the establishment of companies and families within the value chain in the regions, ensuring that all the added value and the wealth generated remains in the territory.

The PDO Mexillon de Galicia pioneered the concept of indicator labelling for aquaculture products in Europe. The journey to achieve recognition and acceptance among most producers and industry players was fraught with challenges, notably the opposition from some canneries in Galicia. This industry often sourced raw materials from various global producing areas and insisted on the right to use geographical indications such as "mussels from Galician estuaries". This practice is very common, since consumers tend to prefer local, regional, or national products versus those produced in other remote areas (Carlucci et al. 2015). Therefore, over the years, the PDO Mexillon de Galicia has been strengthening

Table 8.2 Price comparison between quality assured PDO labelled Galician mussels and non-labelled produce highlighting the opportunity to obtain premium pricing with a premium label

	Price (€/kg)	
Size (number per can)	PDO canned mussels	Non-PDO canned mussels
Large (<8)	87.88 ($n = 8$)	79.75 ($n = 27$)
Medium (8–16)	43.70 ($n = 13$)	43.44 ($n = 54$)
Small (>16)	20.92 ($n = 29$)	17.97 ($n = 68$)

the traceability system by achieving complete document digitalisation, developing a pilot blockchain-based traceability solution to make available to its members, and intensifying and improving the analytical controls carried out on mussel batches throughout the entire value chain, among other actions, to monitor and control fraud in mussel labelling.

PDO labels are associated with increased market demand and a higher willingness to pay (WTP) among consumers. A recent meta-regression study estimated the relationship between WTP and product attributes, revealing that consumers from multiple countries are willing to pay a premium for farmed seafood with certain attributes such as environmental certification, processing standards, freshness, locally sourced, live sales, or domestic farming (Smetana et al. 2022).

Within the SEATRACES project, PDO Mexillon de Galicia conducted one key study focused on label authenticity and compared prices between canned PDO mussels and non-PDO mussels produced in Galicia. Results indicated that the average price for large-sized mussels was 10% higher, for small-sized mussels 16% higher, and nearly the same for medium-sized mussels, highlighting the economic benefits associated with the PDO label. Table 8.2 summarises findings from the study.

It is evident that the origin of a seafood product influences consumer's WTP and market demand. Consumers tend to place high importance on the origin of the products they purchase, often favouring those known for their proven high quality. This consumer preference directly impacts the marketing strategies for these products. Traceability systems are central to managing product labels, as they ensure the authenticity and reliability of the origin claims. Consequently, robust methods for label verification and provenance control are essential to maintain product authenticity and consumer trust.

8.4.2 Case Study 2: Pilot Action for Applying RFID Tags in Internal Traceability in Aquaculture

Oyster production has evolved greatly over the past 40 years in Southern Europe. Initially oyster cultivation involved capturing oyster spat from the natural environment and depositing it directly on estuary beds for growth and farming. This traditional method has developed into a more controlled process with oyster seed produced in specialised hatcheries. These oyster seeds are placed inside mesh bags and arranged on tables in intertidal zones of estuaries.

Throughout the oyster's growth cycle, which can range from 18 to 36 months depending on the farm's geographical location, the mesh bags are periodically spread out to accommodate the oysters' growth. This process ensures the oysters reach the commercial size of 100 g. Additionally, during the growth period, the oyster bags are regularly shaken and turned over to promote healthy and rapid growth. This method—known as the French method—has become a standard practice in modern oyster farming, reflecting the industry's shift toward more efficient and sustainable cultivation techniques (Fig. 8.3).

Simultaneously, the industry adopted another innovation, which involves growing oysters from a certain size (approximately 20–50 g) in hanging baskets. This process, called the Australian method (Fig. 8.3), has the advantage of avoiding manually turning and shaking the bags, as in the French method, but it involves a substantially higher investment.

Regardless of the method used to produce oysters, an essential aspect of their production chain is the sorting and calibrating process, which has become increasingly automated. At various stages of growth, oysters are sorted and calibrated. This involves removing the oysters from their bags or baskets, sorting them by size, and redistributing them. Oysters are calibrated to ensure uniform size within each batch. This helps to manage the growth process more effectively and ensures a consistent product for the market.

Depending on the sorting outcome, oysters are either directed for human consumption—typically involving transfer to purification and packaging centres if the cultivation zone is classified as B—or returned to the natural environment to continue growing.

In the final phase—return of the calibrated oyster to the oyster farm—internal traceability becomes particularly critical and demanding. This necessitates fast and efficient methods, such as those enabled by RFID technology.

EXPORSADO is an oyster farm consisting of three concessions in the Sado estuary in Setúbal, Portugal, encompassing a total farming area of about 60 ha. Currently, only 20 ha are being exploited and 2022 production volumes reached 190 tons of oysters. Given that most of the product is exported, primarily to France, a comprehensive traceability system is a crucial business differentiator.

RFID Tags in Seawater

In terms of management, an oyster farm can be likened to a warehouse. However, this warehouse possesses unique characteristics: the merchandise grows and dies at varying rates depending on natural conditions, batch quality, and handling practices. Additionally, growth and mortality also depend on the duration the oyster remains on the farm. There are fundamentally two types of warehouse management methods applicable to oyster farming:

Fig. 8.3 Systems of oyster culture: French method involving bags on tables (top) and the Australian method involving hanging baskets (bottom)

- Fixed allocation, where the space for a certain commodity, in this case a certain calibre and batch of oyster, is previously established and reserved.
- Chaotic, or random allocation, where any commodity is stored in whatever space is available.

These methods are adapted to accommodate the dynamic nature of oyster farming, ensuring efficient management of the growing and fluctuating stock. It is evident that the "chaotic allocation" allows a much greater efficiency in the use of the effective capacity of the warehouse than the "fixed allocation". However, when it comes time to pick oysters for the sorting machine, it is crucial to know the precise location of each sorting batch, including an estimate of the updated size and quantity of oysters in each bag. Without this information, the search process becomes impractical and highly inefficient.

Given the constant flow of oysters between the farm and the sorting machine, resulting in new batches being sorted during each sorting session, and the limited space available for a "fixed allocation" management approach, the oyster farmer often resorts to "chaotic allocation". The internal traceability of sorting batches is therefore critical for the day-to-day management of oyster farmers who practice "chaotic allocation" storage.

Several methods can be used for batch management and oyster redistribution during the sorting process. A common approach is applying coloured strips to oyster bags which meaning normally corresponds to homogeneous groups of sorting sizes. However, this method does not consider the age of the batch which hinders optimal management of oyster growth.

Internal digital traceability systems offer a more efficient approach to managing aquaculture operations. The state of the art in traceability involves the use of material requirements planning (MRP) systems supported by RFID technology. While RFID technology is extensively tested and widely used in the logistics of terrestrial industrial processes, its application in aquatic environments, particularly in aquaculture, is not as prevalent.

The pilot test conducted at EXPORSADO, as part of the European SEA-TRACES project, aimed to demonstrate the technical and financial feasibility of using RFID technology in a marine environment over an extended period—approximately 2 years (Fig. 8.4).

The pilot sought to investigate whether RFID reading could be reliably maintained over time considering the challenges inherent to the marine environment:

- Natural conditions: salt water, sun radiation, storms, temperature, algae signal obstruction
- Handling conditions: shocks, fatigue

Several models were successfully tested, and results indicate that while the tags demonstrated resistance to natural conditions, not all tags withstood the handling conditions. Further, results demonstrated variability in the performance of the readings with some returning spurious data.

Fig. 8.4 RFID tag and responder system tested on hanging bags at the Exporsado oyster farm

Selecting RFID Tags

The implementation of RFID technology in other oyster farms or similar marine environments should be evaluated on a cost-benefit basis. Specifically, the unit cost of each RFID tag tested in this pilot is approximately €2, while the cost of the reading transponder is around €300. The software and hardware required to manage the information generated by this technology can vary significantly, depending on the level of MRP integration desired.

To select the most suitable RFID tag model for the purpose to be used, the following features should be considered: frequency range, working environment, mounting surface, size, attachment method, read range, and custom printing and encoding.

Table 8.3 Results for the socio-economic impact of the implementation of RFID tags in marine culture (2022, Sea-Traces project, Interreg Atlantic Area, EU)

Dimension	Gain	Achievement	Socio-economic benefits
Survival indicator	1–5%	Competitiveness	Employment
Growth indicator	4–8%	Competitiveness	Employment
Mile indicator	11–13%	Competitiveness and Decarbonisation	Employment and Better Environment
Vacancy indicator	20–30%	Efficiency and Competitiveness	Employment
Engagement	10–20%	Awareness	Self-esteem
Carbon footprint	11–13%	Decarbonisation	Better environment

As highlighted in the socio-economic impact report of the European project SEA-TRACES, the use of RFID tags in oyster farming significantly enhances production efficiency. Table 8.3 summarises the key findings from the study. Results indicate that the growth rate increased by 4–8%, while the carbon footprint was reduced by 11–13%. This improvement results in increased competitiveness, enhanced efficiency, improved environmental performance, and the creation of more and better jobs.

Furthermore, RFID tags are integral to implementing blockchain technology in aquaculture activities, fostering greater transparency and strengthening commitment to consumers and the community at large. Results from the SEATRACES study indicate that customer engagement increased by between 10 and 20%.

8.4.3 Case Study 3: Blockchain-Backed Traceability Implementations

A blockchain is a distributed ledger technology that records transactions such as value, information, or digital events, which can be accessed and validated by participants without requiring a central authority. Blockchains are highly standardised, therefore facilitating interaction between many diverse actors. Any record added to the blockchain cannot be deleted which makes transactions immutable (Ganne 2018). Blockchains are increasingly used in aquaculture and fisheries to combat illegal, unregulated, and unreported (IUU) fishing, unsustainable fishing practices, and nefarious actors (Tolentino-Zondervan et al. 2023). The immutability, security, and distributed nature of blockchain make it an appealing solution for a transparent and trustworthy technology to manage distributed supply chains and enhance consumer trust and stakeholder engagement.

Several aquaculture farms and organisations have successfully piloted and implemented blockchain-based traceability systems, showcasing the potential of this technology in enhancing transparency and sustainability. For example, *Kvarøy Arctic*, a Norwegian salmon farm, partnered with IBM Food Trust to integrate blockchain into its supply chain. This system allows stakeholders and consumers to access detailed traceability information, including the salmon's origin, feed sources, and farm conditions, through QR codes on product packaging. Similarly, the *Sustainable Shrimp Partnership (SSP)* in Ecuador has adopted blockchain technology to ensure traceability and combat fraud in shrimp production. By integrating blockchain with a consumer-facing app, SSP provides end users with verified data on the shrimp's sustainability and quality. These initiatives demonstrate the practical application of blockchain in delivering transparent, verifiable supply chain information, meeting both regulatory requirements and growing consumer demands for ethical and sustainable seafood.

The implementation of blockchain-based traceability systems in fisheries and aquaculture involves a series of practical steps and considerations designed to enhance transparency, sustainability, and trust across the value chain. These systems leverage blockchain's decentralised, immutable ledger technology to create reliable records of the seafood journey, ensuring integrity and accountability.

A critical first step in the process is the design and configuration of the system. The choice of blockchain type—whether public, private, or hybrid—depends on the specific needs of stakeholders. Public blockchains offer high transparency but face scalability challenges, while private blockchains ensure faster transactions and greater privacy. Hybrid systems, which combine features of both, are often preferred for supply chains requiring selective data visibility. Standards and data consistency are also essential for effective implementation. Initiatives like the Global Dialogue on Seafood Traceability (GDST) have developed Key Data Elements (KDEs) and Critical Tracking Events (CTEs) to ensure consistent data capture and exchange across stakeholders. Integrating Internet of Things (IoT) devices, such as sensors and RFID tags, further enhances reliability by automating data collection and minimising human error (O'Donncha and Grant 2019).

However, any chain—whether physical or digital—is only as strong as its weakest link. When integrating with systems in the physical world, the application of blockchain's immutable, secure, and distributed properties is ultimately dependent on the quality and authenticity of external data it receives. This is because blockchains function as "digital walled gardens", unable to verify the veracity or quality of external data on their own. As a result, these off-chain components of blockchain-enhanced traceability systems, i.e., the original sources of external data, represent critical points of failure (Hassan et al. 2023; Lo et al. 2020). These components can undermine the system's sole purpose of providing definitive transparency and reliability for the products in question.

More specifically, a crucial but often weakest link in these systems is failing to establish a direct, secure, and unbreakable connection between the digital identity representing the products on-chain and the actual product being traced in the physical world. This is typically attempted by attaching a scannable or readable label/tag, commonly referred to as a "hardware oracle" in computer sciences, to the product (Al-Breiki et al. 2020). Common examples include QR codes, barcodes, and RFID tags, all of which are vulnerable to tampering and damage, lack versatility in how they can be applied to products at the farming stages, and are often cost-prohibitive for item-level traceability. Consequently, no additional traceability or transparency assurances can ultimately be provided when relying on such labels or tags.

For blockchain-enhanced traceability systems to reach their full potential, this crucial link must be significantly strengthened. The type of hardware oracle used must not only be securely bound to the product in a tamper-resistant manner; they must uphold the same security principles as those used in cryptographic hashing that enable blockchain's desirable properties in the first place.

A promising approach to strengthening this link, and establishing a definitive chain of custody, are developments of hardware oracles involving formation processes that are inherently unpredictable and result in a unique, secure fingerprint each time, otherwise known as physically unclonable functions (i.e. "PUFs"; Arppe and Sørensen 2017). For this to be scalable, the efforts and costs required to secure (i.e. "encrypt") every single oracle and its link to its product must be significantly less than the efforts and costs required to tamper with or break that connection. If such an oracle could be effectively integrated into aquaculture supply chains, it would serve as the closest physical counterpart to the same methods used in cryptographic hashing technology, thus enabling an end-to-end system that maintains authenticity and traceability of the product.

Stakeholder collaboration is another cornerstone of blockchain adoption. Successful implementation requires the participation of all actors in the value chain, from fishers and farmers to processors, distributors, and retailers. Early engagement and training ensure cooperation and understanding of the system. Providing tangible incentives can also drive adoption; for instance, projects like Fishcoin reward fishers with tokens as an incentive for providing data in the blockchain. These can be redeemed for benefits, such as phone credits.

Deployment often begins with pilot projects to test system functionality and identify areas for improvement. For example, the Norwegian salmon industry and the Sustainable Shrimp Partnership (SSP) have successfully piloted blockchain systems, demonstrating their value and feasibility. However, scalability remains a challenge, particularly for public blockchains that require significant computational resources. Hybrid solutions and second-layer technologies, such as Lightning Networks, can help address these scalability issues and reduce costs.

As an example, Narang et al. (2024) combined an IoT-based system together with a blockchain-based cryptographic algorithm to enhance the monitoring and management of mussel growth remotely. The system was equipped with location sensors and a load cell to sense the weight of mussel growth. The blockchain technology ensured the authenticity and integrity of data, facilitating optimisation in aquaculture practices and enabling data sharing in a multi-stakeholder environment.

Data security and authenticity are key benefits of blockchain technology. The system's cryptographic security ensures that records cannot be tampered with, fostering trust across the supply chain. Automated verification methods, such as IoT-enabled data collection and DNA testing, further enhance accuracy and reduce the risk of fraud. These technologies ensure that critical data—such as species authenticity and production conditions—are reliable and traceable.

Practical implementation also involves addressing regulatory and operational challenges. Different regions have varying traceability requirements, making cross-border implementation complex. Standardising data collection and exchange protocols is crucial to overcoming these challenges. A cost-benefit analysis ensures that the blockchain system delivers tangible value, particularly for small and medium enterprises with limited resources. Building trust among stakeholders is equally important, and this relies on clear communication of benefits and user-friendly interfaces.

Implementing blockchain-based traceability systems is a complex yet highly beneficial process. By addressing challenges such as scalability, regulatory compliance, and stakeholder trust, these systems can revolutionise the fisheries and aquaculture industries. The integration of blockchain with IoT and DeFi solutions further highlights its potential to enhance transparency, sustainability, and resilience across global seafood value chains.

8.5 Conclusions

Traceability systems are increasingly vital in addressing the challenges and opportunities within the aquaculture industry. As the demand for sustainable and high-quality seafood continues to grow, these systems provide a crucial mechanism for ensuring transparency, regulatory compliance, and consumer trust. The integration of advanced technologies such as blockchain has revolutionised traceability, enabling comprehensive data capture and real-time information exchange across the supply chain.

However, the implementation of traceability systems is not without challenges. Technical barriers, such as inconsistent data formats and legacy systems, as well as economic constraints, particularly for small and medium enterprises, limit widespread adoption. Furthermore, the variability in global regulations and consumer expectations necessitates a tailored approach to ensure inclusivity and efficiency.

Despite these obstacles, the potential benefits of traceability are undeniable. Beyond compliance, traceability systems offer opportunities for differentiation through sustainability certifications, enhanced product quality, and improved operational efficiency. They also empower consumers with the information needed to make ethical and informed purchasing decisions, fostering stronger connections between producers and end users.

Moving forward, key considerations include fostering collaboration among stakeholders, standardising data protocols, and reducing the cost of traceability technologies to enhance accessibility. By addressing these priorities, aquaculture can leverage traceability not only as a tool for compliance but also as a strategic advantage in creating a resilient, transparent, and sustainable global seafood supply chain. This transition will not only strengthen the industry's competitiveness but also contribute to broader goals of environmental stewardship and food security.

References

Al-Breiki H et al (2020) Trustworthy Blockchain Oracles: review, comparison, and open research challenges. IEEE Access 8:85675–85685. https://doi.org/10.1109/ACCESS.2020.2992698

Arppe R, Sørensen TJ (2017) Physical unclonable functions generated through chemical methods for anti-counterfeiting. Nat Rev Chem 1(4). https://doi.org/10.1038/s41570-017-0031

Arton A, Leiman A, Petrokofsky G, Toonen H, Longo CS (2020) What do we know about the impacts of the Marine Stewardship Council Seafood Ecolabelling Program? A systematic map. Environ Evid:1–20

Blaha F, Vincent A, Piedrahita Y (2023) Guidance document: advancing end-to-end traceability—critical tracking events and key data elements along capture fisheries and aquaculture value chains. FAO, Rome. https://doi.org/10.4060/cc5484en

Brenan T, Gjendemsjø AM, Silver C (2023) The next wave: alternative-seafood solutions. https://www.mckinsey.com/industries/agriculture/our-insights/the-next-wave-alternative-seafood-solutions

Carlucci D, Nocella G, De Devitiis B, Viscecchia R, Bimbo F, Nardone G (2015) Consumer purchasing behaviour towards fish and seafood products. Patterns and insights from a sample of international studies. Appetite 84:212–227. https://doi.org/10.1016/j.appet.2014.10.008

Cooney R, Wan AH, O'Donncha F, Clifford E (2021) Designing environmentally efficient aquafeeds through the use of multicriteria decision support tools. Curr Opin Environ Sci Health 23:100276

Council Regulation (EEC) No. 2081/92 of 14 July 1992 on the protection of geographical indications and designations of origin for agricultural products and foodstuffs. OJEU-L-1992-81207, DOCE No. 208, 24 July 1992

Dewali S, Sharma N, Melkani D, Arya M, Kathayat N, Panda AK, Bisht SS (2023) Aquaculture: contributions to global food security. In: Emerging solutions in sustainable food and nutrition security, pp 123–139

DOG (1992) Order of 26 March 1992 recognizing the designation of Galician quality product for Galician mussels and appointing its provisional governing council. DOG 69 of 9 April 1992. https://doi.org/10.1023/A:1021309222459

European Parliament and Council (2000) Directive 2000/60/EC of 23 October 2000 establishing a framework for Community action in the field of water policy. Off J Eur Commun L327:1–73

European Parliament and Council (2002) Regulation (EC) N° 178/2002 of 28 January 2002 Laying down the General Principles and Requirements of Food Law, Establishing the European Food Safety Authority and Laying down Procedures in Matters of Food Safety. Off J Eur Commun L31:1–24

European Parliament and Council (2016) Regulation (EU) 2016/429 of 9 March 2016 on transmissible animal diseases and amending and repealing certain acts in the area of animal health (Animal Health Law). Off J Eur Commun L84:1–208

European Parliament and Council (2023) Regulation (EU) 2023/2842 of 22 November 2023 amending Council Regulation (EC) No 1224/2009, and amending Council Regulations (EC) No 1967/2006 and (EC) No 1005/2008 and Regulations (EU) 2016/1139, (EU) 2017/2403 and (EU) 2019/473 of the European Parliament and of the Council as regards fisheries control. Off J Eur Commun L2842:1–105

European Parliament and Council Regulation (EC) (2009) No 1224/2009 of 20 November 2009 establishing a community control system for ensuring compliance with the rules of the common fisheries policy. Off J Eur Union L343:1–50

European Parliament and Council Regulation (EEC) (1992) No 2081/92 of 14 July 1992 on the protection of geographical indications and designations of origin for agricultural products and foodstuffs. Off J Eur Commun L208:1–8

European Parliament Regulation (EU) (2013) No 1379/2013 of the European Parliament and of the Council of 11 December 2013 on the common organisation of the markets in fishery and aquaculture products. Off J Eur Commun (L354):1–21

FAO (1995) Code of conduct for responsible fisheries. FAO, Rome, 41 p
FAO (2020) The state of world fisheries and aquaculture 2020: sustainability in action, Rome. https://doi.org/10.4060/ca9229en
FAO (2024) The state of world fisheries and aquaculture 2024—blue transformation in action, Rome. https://doi.org/10.4060/cd0683en
Figueiras FG, Labarta U, Reiriz MJ (2002) Coastal upwelling, primary production and mussel growth in the Rías Baixas of Galicia. Hydrobiologia 484(1–3):121–131
Ganne BE (2018) Can Blockchain revolutionize international trade? World Trade Organization report. https://www.wto.org/english/res_e/booksp_e/blockchainrev18_e.pdf
Giacomarra M, Crescimanno M, Vrontis D, Miret L, Galati A (2021) The ability of fish ecolabels to promote a change in the sustainability awareness. Mar Policy 123:104292
Global Dialogue on Seafood Traceability (GDST) (2023) https://thegdst.org/resources/standard/
Gooch M, Sterling B (2013) Traceability is free: competitive advantage of food traceability to value chain management, 23
Han K, Leem K, Choi YR, Chung K (2022) What drives a country's fish consumption? Market growth phase and the causal relations among fish consumption, production and income growth. Fish Res 254:106435
Hassan A et al (2023) From trust to truth: advancements in mitigating the Blockchain Oracle problem. J Netw Comput Appl 217:103672. https://doi.org/10.1016/j.jnca.2023.103672
https://www.regulations.gov/document/NOAA-NMFS-2015-0122-0111
Kamiyama R, Wakamatsu H, Seko T, Ishihara K (2024) Consumer preference for label presentations of freshness, taste, and serving suggestion on fresh fish packages of Japanese Flounder Paralichthys Olivaceus. Fish Sci 90(3):529–544
Lo SK et al (2020) Reliability analysis for blockchain oracles. Comput Electr Eng 83:106582. https://doi.org/10.1016/j.compeleceng.2020.106582
López-Mas L, Claret A, Arvisenet G, del Castillo RR, Kallas Z, Zuccaro M, Guerrero L (2023) European consumers' beliefs about the main pillars of the sustainability: a comparison between wild and farmed fish. Aquac Int 31(4):2035–2055
Narang A, Gupta R, Singh P (2024) Precision aquaculture framework for remote mussel growth monitoring with IoT, Blockchain, and Cloud Integration. In: 2024 IEEE international workshop on metrology for the sea; Learning to Measure Sea Health Parameters (MetroSea). IEEE. https://doi.org/10.1109/MetroSea.2024.1234567
O'Donncha F, Grant J (2019) Precision aquaculture. IEEE Internet of Things Mag 2(4):26–30
Olsen P, Borit M (2013) How to define traceability. Trends Food Sci Technol 29(2):142–150
Österblom H, Jouffray J-B, Folke C, Crona B, Troell M, Merrie A et al (2015) Transnational corporations as 'Keystone Actors' in marine ecosystems. PLoS One 10(5):e0127533. https://doi.org/10.1371/journal.pone.0127533
Saidi A, Sacchi G, Cavallo C, Cicia G, Di Monaco R, Puleo S, Del Giudice T (2022) Drivers of fish choice: an exploratory analysis in Mediterranean countries. Agric Food Econ. 10(1)
SIMP (2016) Seafood import monitoring program
Skretting (2021) Footprint report fish feed 2021. https://www.skretting.com/no/baerekraft/baerekraftsrapport/footprint-report-salmon-feed-2021/
Smetana K, Melstrom RT, Malone T (2022) A meta-regression analysis of consumer willingness to pay for aquaculture products. J Agric Appl Econ 54(3):480–495. https://doi.org/10.1017/aae.2022.28
Sperder WH, Stier RF (2010) Happy 50th birthday to HACCP: retrospective and prospective. Food Saf Mag:42–46
Sterling B, Gooch M, Dent B, Marenick N, Miller A, Sylvia G (2015) Assessing the value and role of seafood traceability from an entire value-chain perspective. Compr Rev Food Sci Food Saf 14(3):205–268
Tamm EE, Schiller L, Hanner RH (2016) Seafood traceability and consumer choice. Elsevier
Tolentino-Zondervan F, Thi P, Ngoc A, Luka J (2023) Use cases and future prospects of Blockchain applications in global fishery and aquaculture value chains. Aquaculture 565(February 2022):739158

van Osch S, Hynes S, Freeman S, O'Higgins T (2019) Estimating the public's preferences for sustainable aquaculture: a country comparison. Sustainability (Switzerland) 11(3):1–24. https://doi.org/10.3390/su11030569
Vandecandelaere E, Arfini F, Belletti G, Marescotti A (2010) Linking people, places and products. A guide for promoting quality linked to geographical origin and sustainable geographical indications, vol 153. FAO
Zander K, Feucht Y (2018) Consumers' willingness to pay for sustainable seafood made in Europe. J Int Food Agribus Mark 30(3):251–275

Chapter 9
Future of Intensive Pen Aquaculture

Heidi Moe Føre, Trine Thorvaldsen, Andreas Misund, and Hans V. Bjelland

Abstract Norwegian marine fish farming technology and innovations have had a great influence on technology development worldwide. While traditional net pens are the most common technology for fish farming in Norway, other production systems are emerging, making the industry more technologically diverse. This includes closed production units in sheltered areas, submersible farms, and large offshore structures. These innovations are motivated by the prospect of mitigating environmental concerns and the possibilities to install fish farms at sites further from the coast and in more exposed sea areas.

Fish production involves many different farming operations, including reception of juvenile fish, feeding, welfare inspections, parasite control, and cleaning to name a few. In future fish farming, many of these operations and sub-operations will benefit from automation and control through AI and digital twins. In addition, aquaculture workers face several hazards performing these operations and are at risk of occupational diseases and injuries. AI may provide safety relevant decision support systems that can improve safety and working conditions.

There is no doubt that automation and digitalization have the possibility to benefit future fish farming and solve some of the structural and technical challenges that have occurred during the latest years. These include challenges with structural integrity, interaction with vessels and other objects, station keeping, ballasting and stability, and water quality.

The future will probably bring more diverse pen farm designs, where the pens and their operation are tailored to different sites, species, and environments. This will include an increased use of automation, as many operations and daily rearing activities cannot be performed manually. There will be a need of new sensors, digital twins, AI, triggers, and robots to maintain both fish welfare and structural integrity of future fish farms.

H. M. Føre (✉) · T. Thorvaldsen · A. Misund · H. V. Bjelland
SINTEF Ocean, Trondheim, Norway
e-mail: Heidi.moe.fore@sintef.no

F. O'Donncha et al. (eds.), *Digital Twin for Marine Fish Farming*,
Aquatic Production Systems, https://doi.org/10.1007/978-3-032-10362-8_9

Keywords Marine fish farming technology · Innovations · Farming operations · Health and safety · Technological trends

9.1 Introduction

This chapter presents current trends and future directions for intensive pen aquaculture. The focus is mainly on industrialized sea-based aquaculture of fish (salmonids) in net pens. Intensive aquaculture requires farmers that actively stimulate the growth, condition, health, and welfare of the stock through inputs and operations such as high-quality feeds and feeding, parasite treatments, supply of oxygen, and light. Pen aquaculture means that the fish is kept in an enclosure at sea, typically in a net supported by a floating pen collar. Alternative approaches to intensive pen-based aquaculture, such as closed containers at sea or on land, where the fish farming volume is (semi-) closed from direct interaction with the sea, will also be briefly described in this chapter.

This chapter is written from a Norwegian perspective. Norway is the largest industrialized marine fin fish farming country. During the last decades, Norwegian fish farm innovations have had a great influence on technology development worldwide (Afewerki et al. 2022), supported by the rapid growth and development of the salmon (*Salmo salar*) farming industry. In addition to intensive pen aquaculture, fish is produced at large scale by other measures in seas, ponds, and lakes worldwide, which is not covered in this chapter.

In this chapter, we present recent innovations in Norwegian aquaculture and discuss possible future pen farm developments. We also give some examples of future trends in other areas of the world. In the beginning of this chapter, key drivers for technological developments in aquaculture are presented, followed by an overview of various production systems and innovations, ending with a section on future fish farming operations.

9.2 Farm Technology and Sites for Intensive Aquaculture

9.2.1 Different Production Systems for Fish Farming

While traditional net pens are the most common technology for fish farming in Norway, other production systems are emerging, making the industry more technologically diverse (Misund et al. 2024). Alternative farm technology has recently been suggested, and a few concepts have been implemented in prototype testing and commercial fish farming. In this chapter, the major fish farm design categories are defined as follows (Moe Føre et al. 2022):

Fig. 9.1 Conventional PE ring farm for aquaculture at sites. These may be applied in sites with significant wave height up to 6 m and currents up to 1.2 m/s (ill.: ScaleAQ)

- *PE rings*: Conventional surface floating flexible pens with a circular floater made of HDPE (high-density polyethylene) pipes and permeable nets (Fig. 9.1)
- *Rigid floaters*: Floating structures situated at the water surface holding a permeable net
- *Closed farms*: Closed or semi-closed sea-based fish farms that involve impermeable fish enclosures that separate the enclosed water volume from the external marine environment
- *Semi-submersible platforms (semi-subs)*: Rigid platforms with their buoyant elements below the water surface

Various types of fish farming structures have been designed and developed over the years. Currently, excluding homemade farms that are prevalent in certain regions, flexible circular pens (PE-rings) are a primary choice in the fish farming industry (Fig. 9.1, Halwart et al. 2007; Bjelland et al. 2024). These structures are made of polyethylene pipes that float on the water surface, with flexible nets that contain the fish (Chu et al. 2020). Additionally, floating steel collars with net pens are also common.

Industrialized marine fish farming has typically moved from utilizing sheltered locations to more exposed environments. This development has been followed by application of larger fish farming units designed to endure the increased environmental loads. Conventional fish farms are now capable of operating in even more exposed settings, with the industry providing PE rings certified for areas with a significant wave height of 6 m and currents reaching 1.2 m/s (Fig. 9.1). This represents a series of incremental innovations aimed at enhancing existing technologies, as opposed to the more radical innovations suggested through new fish farm designs (Bjelland et al. 2024).

9.2.1.1 Farming in Nets

Conventional fish farms consist of a net pen enclosing the fish, a pen collar keeping the fish pen afloat and providing a working platform, and a mooring system (Fig. 9.1). Pen collars are most commonly made of two parallel polyethylene pipes forming a horizontal circle on the surface (*PE-rings*); however, square-shaped steel pen collars also exist (*Rigid floaters*). The nets are made of Raschel knitted polymer fibres, with typical square mesh openings with sides of 15–30 mm (Moe et al. 2007). Weights are attached to the net to maintain its volume, for example, bottom rings containing steel wire or chain as shown in Fig. 9.1.

The main benefit of farming in nets is the natural exchange of water, which for most fish farms provide a suitable environment for the fish. However, at some farms, oxygen is added occasionally, and some sites may not be feasible for net pen aquaculture due to insufficient current, water quality, and high water temperatures. At the same time, the permeability of net pens may also be their main challenge, as they allow for free interaction with the sea and exchange of pathogens, larvae and parasites, algae, and dispersion of nutrients and organic matter (feed and faeces). This is a concern at sheltered and coastal locations, while such negative consequences of permeable nets may be less problematic at sites further away from the coastline. Innovative concepts for open ocean fish farming, like *semi-subs* (see Sect. 9.2.1), most often involve permeable nets (see Sect. 9.4.2). In addition to the benefits of natural exchange of water, nets will also involve smaller loads from water currents and waves acting on the structure, which is very beneficial in design to ensure structural integrity of the structure.

9.2.1.2 Closed Farming at Land and Sea

Land-based aquaculture involves rearing fish in tanks on land, which in Norway has been used by hatcheries for decades. In recent years, land-based aquaculture is also developed as an alternative production method for the whole lifespan of farmed marine fish (Fig. 9.2). In facilities that keep the fish on land until slaughter weight, common technologies are flow-through systems (FTS) or recirculating aquaculture systems (RAS) using sea water. Some companies use hybrid production technologies, combining elements from the two.

Production on land requires precise and continuous monitoring of environmental parameters, water quality, and welfare of the fish. In RAS, different types of water treatment remove particles, nitrogenous waste, and carbon dioxide while also adding oxygen. In FTS, the incoming water may be purified, and oxygen is added if necessary, but there is no reuse of water from the fish tanks. This makes the system less complex than RAS. However, FTS depends on high-quality water sources.

Closed farms in sea are designed to reduce sea lice infections, collect fish farming waste, and provide an optimal water environment for the fish (e.g. temperature and oxygen) and sheltered conditions for juveniles. While closed pens allow for increased control of the environment, maintaining water quality and fish welfare can

Fig. 9.2 Illustration of closed farm (semi-closed containment system) called Protectus with a steel collar and tarpaulin enclosure (FiiZK 2025)

be challenging, incurring high energy use and requiring backup systems. Although closed enclosures may be stronger than nets and enhance measures against fish escape, the increased environmental loads from water currents may cause structural challenges.

The technological principles for semi-closed production at sea are similar to those of flow-through systems (FTS) used in land-based facilities, but they are less complex (Aarhus et al. 2011). Water is often pumped from deeper waters (20–30 m) and fed to a coarse filter, while the wastewater is transported out of the unit for filtration and purification of sludge and feed residue. Control of the water quality (oxygen, carbon dioxide, ammonia, fish density, water exchange, nitrate, and nitrite) is essential to ensure the welfare of the fish (Rosten et al. 2011). Different closed farm designs and materials are used and suggested, and design concepts vary from flexible fabric enclosures (tarpaulins), via semi-flexible structures like fiberglass, to rigid structures of steel and concrete.

During fish production in closed facilities, there is potential to collect and utilize organic matter for other industries (e.g. biogas, fertilizers, and nutrients). Increased control of the environment can be positive for biosecurity, and contamination between facilities is reduced due to separated production units.

Challenges for land-based facilities often relate to fish welfare and water quality. For instance, formation of the poisonous gas hydrogen sulphide (H_2S) can cause acute fish mortality. Furthermore, energy consumption can become a limitation for both land-based facilities and closed farms at sea, as both are dependent on continuous water circulation systems. Closed pens respond differently to water currents and waves compared to traditional pens, resulting in increased mooring loads and risk of sloshing inside the closed pen (Kristiansen et al. 2018; Su et al. 2021).

9.2.2 Aquaculture Sites from Sheltered Areas to Open Ocean

In prominent salmon farming countries such as Norway, Chile, Scotland, and Canada, marine fish production occurs predominantly in sheltered coastal areas. However, there are not enough sheltered, highly suitable farming sites to meet demand from the industry, and aquaculture in such areas may conflict with competition from other interests and ocean uses, including local communities, wildlife, tourisms, fisheries, and shipping. Consequently, several nations are investigating the potential to establish fish farms farther from the coastline and in more exposed marine areas. In addition to increasing the potential of growth of existing industries, this may also bring new opportunities for local communities in rural coastal regions (Bjelland et al. 2024; Buck et al. 2024).

Both the terms offshore and open ocean aquaculture refer to aquaculture situated at a distance from the shore, further out at sea than conventional coastal sites (Buck et al. 2024; Holm et al. 2017). The term offshore is given different meaning and used in different ways across regions and sectors. Drumm defined offshore aquaculture in 2010 as follows: "In general offshore aquaculture may be defined as taking place in the open sea with significant exposure to wind and wave action, and where there is a requirement for equipment and servicing vessels to survive and operate in severe sea conditions from time to time. The issue of distance from the coast or from a safe harbour or shore base is often but not always a factor." Froehlich et al. (2017) wrote that there appears to be no consensus as to what measures commonly define an offshore site, and that this affects communication, regulatory processes, research and development, and understanding ecological implications of offshore farming.

Exposed aquaculture has been defined as the cultivation of aquatic animals or plants at exposed locations (Bjelland et al. 2025; Buck et al. 2024). Exposed locations are not fully sheltered or protected by their surroundings, meaning that fish farming structures will probably be exposed to increased environmental loads from high waves, strong currents, and strong winds during their lifetime. Aquaculture sites can be exposed in specific directions only, like in straits and fjords, whereas islands can offer protection in specific directions of otherwise exposed areas. It is common for an exposed site to be situated in open oceans and at a considerable distance from the shoreline (offshore). However, high exposure sites are also found close to shore in coastal areas facing open oceans which provide big waves and strong currents (Bjelland et al. 2025).

9.2.2.1 Level of Exposure

There is no established method to determine whether a site or farm is exposed or not or at what degree of exposure (Lojek et al. 2024; Heasman et al. 2024). The aquaculture industry, scientific community, and national authorities are seeking ways to describe and quantify the exposure of a given site for aquaculture (Bjelland et al. 2025).

The amount of energy inflicted on an aquaculture structure by the surroundings can be used as a measure of exposure. To quantify oceanic conditions and limit values for structural design and operation of aquaculture systems, six exposure indices (EI) were proposed in Lojek et al. (2024) that consider physical energy levels related to hydrodynamic forces at a site. This involves the seemingly impossible task of reflecting complex environmental conditions through a single number on a one-dimensional scale. This number with given thresholds could be useful both in regulation of the aquaculture industry and in planning of fish farming activities. The level of exposure could be associated with dedicated rules and regulations, certification and choice of equipment and vessels, and planning of various aquaculture operations (Bjelland et al. 2024).

In this section, one of the possible methods to quantify the exposure level of a specific site is applied on Norwegian aquaculture sites, and a corresponding categorization of exposure is proposed. Here, we suggest using structural design parameters as a basis for estimating exposure, i.e. critical design values for loads from waves and currents. This includes estimates of maximum expected wave heights, current velocities, and wind velocities for given periods. Typically, these extreme values are estimated for a return period of 10, 50 and/or 100 years, and combinations of these values are used in structural analysis, design, and dimensioning (Standards Norway 2021).

The environmental loads and exposure at a given site form a complex picture, varying both in time and space. Here, it is proposed that exposure level is measured by the estimated extreme horizontal water flow velocities at a given site and a water depth of 10 m (Table 9.1). This is equal to the "exposure velocity at reference depth" explained in detail in Lojek et al. (2024), applying 10 m instead of 5 m. The 10 m depth and the suggested exposure levels in Table 9.2 was chosen based on dialogue with Norwegian fish farmers, using some of their sites as examples. The fish farmer's practical assessment of exposure levels of their farms was used to set the limits in Table 9.2. This exposure velocity will be correlated with the environmental drag loads acting on a fish farm. It may also be considered as an estimate of the extreme velocities the fish may experience a few meters below the surface. Notice that the exposure velocity will act in single points in natural irregular waves; i.e. it will be local and last for a short time (Bjelland et al. 2024).

Table 9.1 Exposure velocity for various Norwegian example sites

Site characteristics	Sheltered	Coastal	Borderline	«Offshore»	Open ocean
Water current (m/s)	0.5	1	1	1.2	1
Significant wave height (m)	1	2.5	4	7	14
Peak period (s)	4	6	7	7	16
Exposure velocity (m/s):	0.6	1.4	1.8	2.6	3.0

Table 9.2 Suggested exposure levels (Bjelland et al. 2024)

Exposure level	Exposure velocity	Type of location
E0	<0.7 m/s	Sheltered
E1	0.7 – 1.5 m/s	Semi-sheltered
E2	1.5 – 2.5 m/s	Exposed
E3	2.5 – 3.5 m/s	Exposed
E4	>3.5 m/s	Extreme exposure

Table 9.1 shows typical parameters for salmon farms with different exposure levels. The exposure velocity can be calculated based on data found in standard site surveys, and values for 10- and 50-year return periods should be combined as explained earlier. Exposure velocity should be calculated for at least eight directions (north, north-east, east, etc.), and the highest resulting value is chosen as the exposure velocity.

The so called "borderline" location in Table 9.1 represents a common impression in Norwegian aquaculture industry of the practical limits for existing aquaculture technology and operations. Exposure may be even larger than at the exemplified Norwegian offshore and open ocean sites; e.g. un-sheltered near-shore locations may have exposure velocities well above 3.5 m/s. These sites are characterized by low water depths, which in combination with direct exposure to open ocean results in high waves and thus high energy sites.

Comparing the example sites, it is seen that the estimated exposure velocity increased from 0.6 m/s for the sheltered location, up to two to three for the more exposed sites (Table 9.2). Bjelland et al. (2024) used these values to establish five levels of exposure (E0–E4) as given in Table 9.2.

9.3 Drivers for Technological Developments in Aquaculture

During recent years, the most important drivers for technological developments have been the need to ensure sustainable production of farmed fish and utilize new areas for aquaculture. Sustainability includes many aspects such as environmental impact and economic and societal development. There is no doubt that the key motivation for the development of fish farming is a profitable production of fish for human consumption. Following growth in production numbers, environmental concerns such as the effect on wild fish stocks and interactions with local environment and ecology have become increasingly important. These concerns have curbed the growth of the aquaculture industry and led to an increased demand for technological solutions.

The past two decades have seen increasing concern of the impact of farmed escapees on wildlife. In salmon-producing countries, concerns include that salmon escapees could interfere with salmon breeding in rivers and that farmed fish may be sources of pathogens and parasites affecting wild salmon migrating between rivers and the sea. The parasitic salmon lice (*Lepeophtheirus salmonis*) is a major welfare issue of both farmed and wild fish. In farmed fish, welfare concerns are often related to the stress associated with salmon lice removal, while migrating wild salmon smolt may suffer from extensive lice infections. Moving fish farms further from the coast and reducing fish farming waste in near-shore areas are possible sustainability measures that are of great current interest internationally.

In Norway, authorities have used the licensing system to encourage technology development that may reduce environmental risks (Osmundsen et al. 2022). Exposed and open ocean sites may require new technological solutions. In Norway, technology development for open ocean sites was boosted by special purpose development licenses (Moe Føre et al. 2022). While farm concepts for open ocean aquaculture exist, many are not yet placed in open ocean areas as the process of establishing a regulatory regime is still ongoing.

Comparing aquaculture regions worldwide, it is found that challenges affecting establishment and growth of fish farming may differ. For instance, while sea lice and escapes are a major concern in Norway, other parts of the world with no local wild stocks of salmon are not as concerned with these issues. However, several challenges driving innovation in aquaculture are similar between regions, such as the need to reduce environmental impacts, the need for area economic development, and acceptance for fish farming from the local communities, which are prominent in Canada, the USA, Japan, the UK, and New Zealand. Still, regulations and drivers vary between countries and even within countries. For instance, Canada has seen a stronger pressure for land-based solutions in the west than at the east coast and introduction of closed fish farms.

Of international concern is also negative interaction with wildlife, benthic environment, invasive species, and fish welfare. Societal impact and acceptance and the use of ocean areas are primary concerns in many areas of the world, in addition to micro and macro plastic waste. Areas identified for aquaculture facilities are often also of interest for other industries, or the local communities can be opposed to the establishment of fish farms. In some parts of the world, organized multi-use of sea areas are requested and even a requirement for establishments.

This section will elaborate on key drivers for technology development, with focus on Norwegian salmon farming.

9.3.1 Environmental Concerns and the Need for New Areas

In Norwegian aquaculture, challenges with salmon lice and escape of fish have led to a stagnation in production, making the regulations related to these environmental concerns key drivers for technological innovation. Additionally, the need for new areas has become a significant driver of development in recent years, also serving as a potential measure to reduce sea lice infections. These three drivers are expected to be prominent also in the years to come.

Salmon lice is a parasite that feed of the mucus, skin, and blood of salmon. They multiply at a high rate and may become a welfare issue for the host. Their larvae spread with the water currents and find numerous hosts at fish farms.

In Norway, there are regulations related to lice management that aim to reduce both the negative effects on wild salmon stocks and ensure an ethical and efficient production of farmed salmon. Farmers must ensure that the number of lice is below set limits. Lice numbers at each fish farm are closely monitored and reported, and new technologies have provided the farmers with the opportunity to count the lice automatically without handling the fish (for more discussion on fish welfare monitoring see Chap. 1).

In the current regulatory regime, sea lice are identified as a key indicator for sustainability. The number of lice in 13 different production areas determines the allowed production growth in each area through the so-called traffic light system (TLS) introduced in 2017 (Sønvisen and Vik 2021). This involves production in net pens at sea, while closed pens and land-based aquaculture are exempted from these limitations. Each production area is given the colour green, yellow, or red. In green areas, increased production is allowed, in yellow no production growth is allowed, while in red areas salmon production must be reduced.

Several different methods are used to prevent sea lice infections or de-lice the farmed fish. Since lice are most abundant high up in the water column, a common preventive measure is to apply so-called lice skirts, consisting of tarpaulins that cover the upper metres (typically 5–10 m) of the net pen wall. There has been a downward trend in using chemotherapeutants and an increase in mechanical de-lousing treatments (e.g. brushing and flushing) and thermal treatments. Fish are pumped into a vessel for parasite removal, which can result in challenging fish welfare and increased mortality. Mechanical and thermal treatments involve rougher handling of the fish, often causing ulcers and skin damage. The prevalence of these methods is mostly due to the lice becoming resistant to chemotherapeutants, and controversies surrounding the release of chemotherapeutants into the marine environment. However, mechanical and thermal treatments have also received much negative attention, and gentle parasite control is now a major driver for technical innovations in Norway. Some farms also use cleaner fish, i.e. fish species that feed on sea lice. It is however debated whether this is an effective measure, and there are also concerns with the welfare of cleaner fish.

Escape of fish is another key driver for technology development in the Norwegian context and has received high attention from both regulators and media over the years. Preventing fish escape is required by law, and in 2003, a technological standard for fish farms (NS9415) was introduced and enforced by law through the NYTEK-regulations (Føre and Thorvaldsen 2021). Causal analysis has shown that most escapes are directly caused by technological factors such as holes in the nets, but underlying causes may also relate to human and organisational factors.

In recent years, the need for *new areas* for aquaculture has boosted technological developments aimed at designing different systems for specific locations. One major motivation for fish farming at more exposed and offshore locations is to exploit the natural abundant supply of clean water. This water exchange helps to reduce risks of disease outbreaks, salmon lice infections, and organic matter accumulation beneath the farm. Aquaculture operations at sites exposed to large ocean waves, strong currents, and winds is therefore becoming increasingly relevant both in Norway and internationally. This includes locations close to open oceans, offshore locations, and some near-shore locations. More extensive use of such high energy areas opens for new opportunities and added benefits for farming of finfish, shellfish (i.e., mussels, oysters), seaweed, and other aquatic organisms. Societies worldwide seek to expand aquaculture operations to exposed locations as this both facilitates growth in existing industries and generates new opportunities for communities located in rough coastal environments (Bjelland et al. 2024).

While exposed locations offer beneficial conditions for fish growth, they also introduce challenges that need to be addressed through new solutions within farming operations, farming technology, and governmental regulation. The move to more exposed sites has led to the establishment of new standards and requirements for safety measures, farm design, logistics, and operations and even the capabilities of the cultured species. The narrow operational windows at exposed sites means that the enforcement and accommodation of these new demands will require increased use of remote control and automation. Moreover, moving operations into new areas also entails increased potential for conflicts with local wildlife (e.g. large mammals, wild fishes) but also through the potential of facilitating establishment of invasive species. In addition, aquaculture will need to find its place among other industries including oil and gas, maritime traffic, fisheries, energy production, and tourism.

Increased utilization of sheltered areas may be possible but may require a closed fish farm preventing negative interactions between fish farming and the surroundings. This includes fish farming waste and pollution and spread of diseases and parasites but also suboptimal water quality for the farmed fish including risk of low oxygen levels and high temperatures. On-shore locations are also considered as an opportunity in fish farming.

9.3.2 Special Purpose Licenses

In Norway, fish farming is governed by licenses that are both highly valuable and in great demand. Ordinary licenses have requirements set by the preliminary fish farming act of 1973 (Hersoug et al. 2019). The first regulations focused on economic profitability, increased employment in rural areas, and limiting environmental impacts. In the last decade, several special purpose licenses have been designed to achieve new technologies, knowledge, and innovations (Hersoug 2022).

Green licenses, implemented in 2013, had stricter requirements regarding salmon lice and escapes. The objective of the "development licenses", introduced in 2015, was to accelerate the development of innovative fish farming technologies that could address environmental issues associated with fish farming, especially in terms of sea lice, preventing escapes, and facilitating access to areas for production. These licenses encouraged applicants to propose significant innovations and original ideas rather than further development of current solutions (Bjelland et al. 2024).

9.4 Future Fish Farms

9.4.1 Designing for Exposed and Offshore Sites

Designing fish farms for use at exposed locations introduces both challenges and opportunities in structural design. For instance, new open ocean sites may be more suitable for larger production units and new design concepts that might otherwise not fit with conventional sites. This type of systems includes solutions ranging from those that mostly represent an upscaling of the structures used today, through new structures inspired by systems used in the offshore industry, to mobile systems that are more similar to vessels (Bjelland et al. 2016; Moe Føre et al. 2022). Moving operations to more exposed sites will also require revising the design of other fish farming components and technology such as barges for feeding systems, storage, and personnel facilities.

Utilizing new and more exposed areas will also require the systems to be designed to cope with larger environmental loads from the prevailing wind, current, and waves at such sites. Fish escape is a major concern of the industry, and the combination of larger environmental loads and new farm designs may increase the risk of incidents that may lead to escape events. Although a knowledge base exists in established industries such as shipping and oil and gas, aquaculture sites may require new and more advanced analytical methods to cope with the more complex hydrodynamics at fish farms. Furthermore, the unique structures found in fish farms may require a refinement of existing numerical models for analysing marine structures. A concrete example of this is that most net pens consist of highly flexible netting structures and ropes and are permeable to the environment allowing flux of water from the outside into the fish enclosure.

Large and exposed facilities also require more precise descriptions of wave and current loadings and the structural response to environmental loads. Moreover, it is important to ensure that fish farms are safe and reliable from a structural perspective while ensuring economic viability. However, these are not the only design criteria to consider for fish farming structures, as the conditions on site will

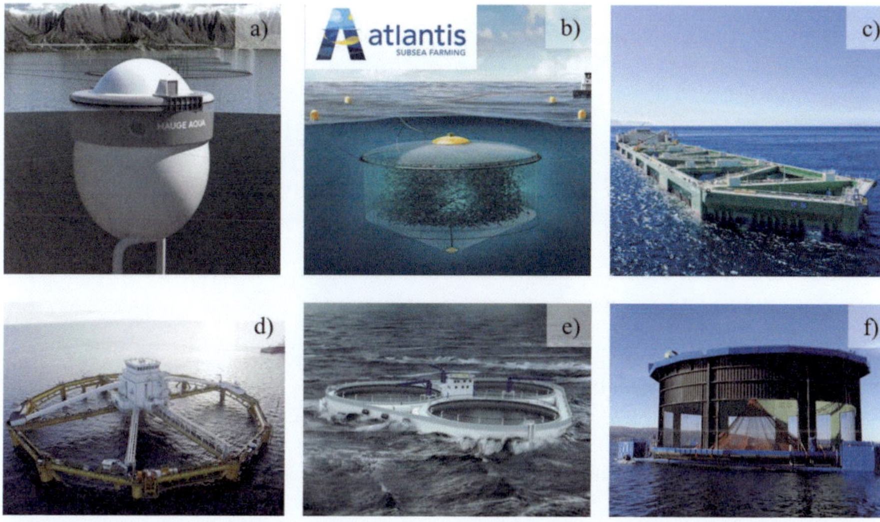

Fig. 9.3 Different farm concepts that have been awarded development licenses (Moe Føre et al. 2022). (**a**) Closed pen called "The egg" (ill.: Hauge Aqua), (**b**) submersible PE-rings called Atlantis (ill.: Atlantis Subsea Farming), (**c**) semi-subs called Havfarm (photo: Nordlaks/Deadline Media), (**d**) semi-subs called Ocean Farm (photo: Ocean Farming), (**e**) rigid floater called Øymerd (ill.: Astafjord Ocean Salmon), (**f**) partly closed farm called Aquatraz by Seafarming systems (photo: Steinar Johansen/MNH)

also pose a challenge for the personnel that are needed to perform farm operation. The design must thus be such that the farming crew are able to carry out essential tasks safely, efficiently, and cost-effectively. In addition, the structures need to provide a good habitat for the fish in the sense that they can be fed regularly, are kept at adequate stocking densities, and are reared under beneficial environmental conditions (Bjelland et al. 2024).

The environmental conditions will also challenge the sites' accessibility for humans during periods of time. This can impact possibilities for daily inspections and repairs and day-to-day management of the farm, which is inconvenient when aspiring to maintain continuous and efficient production. Moreover, as they are located further from shore, these locations may experience an increased risk of collisions with other vessels, wildlife (large mammals in particular), and even icing and drifting ice.

One of the main development tracks toward exposed farming has been the incremental innovation of PE rings which has led to PE-based systems designed also for exposed sites (Fig. 9.1 and Tables 9.1, 9.2). However, potentially radical innovations that may open completely new possibilities for exposed fish farming have emerged in parallel with this development. Units designed to withstand waves of 10–15 m significant wave height are currently in development (Fig. 9.3). In addition, there have been steps in the direction of new fish farm structures based on principles from the offshore industry, including the Havfarm and Ocean Farming

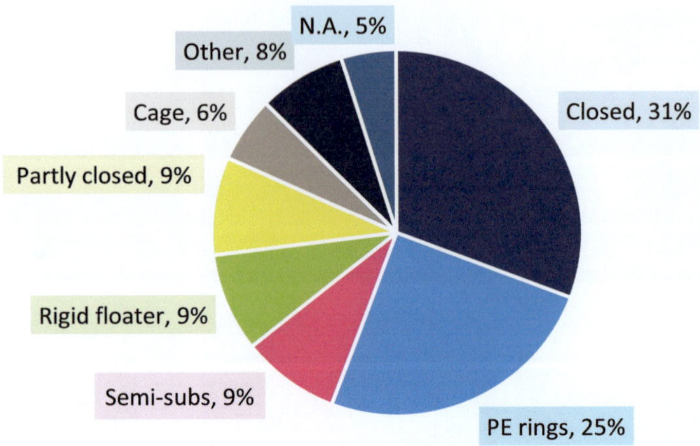

Fig. 9.4 Farm concepts found in applications for development licenses

1 (Fig. 9.4). While these systems are designed specifically for holding fish, they are clearly inspired by ocean going vessels and semi-submersible platforms commonly used in the oil and gas industry.

Open ocean aquaculture, which is a version of exposed farming where operations are moved even further from shore, has also received attention of late, leading to the suggestion of several new offshore fish farming concepts. While some of these have already been built or are close to realization, yet others are still only on a concept stage. One of the obstacles in this area is that the legislation needed to establish an offshore farming industry has yet to be determined, but considering the worldwide interest in new sustainable aquaculture concepts for industrial scale farming at exposed locations, this industry is likely to grow in the coming years.

The Norwegian development license arrangement was a government-supported initiative that initiated and supported the development of several new concepts for fish production including those aimed at open ocean fish farming. Nine of these that were granted licenses (per March 14, 2023, Moe Føre et al. 2022) were particularly relevant concepts featuring four semi-submersible platforms with buoyant elements below the water surface and rigid platforms above the waterline, four concepts based on rigid floaters placed at the water surface and one based on PE rings (see Sect. 9.2.1). There also exist other concepts internationally that fall within the same categories, including a semi-submersible rigid system called Shenlan (China), a submersible REFA Tension leg Cage (Italy), and the SeaStation concept (USA) to mention a few.

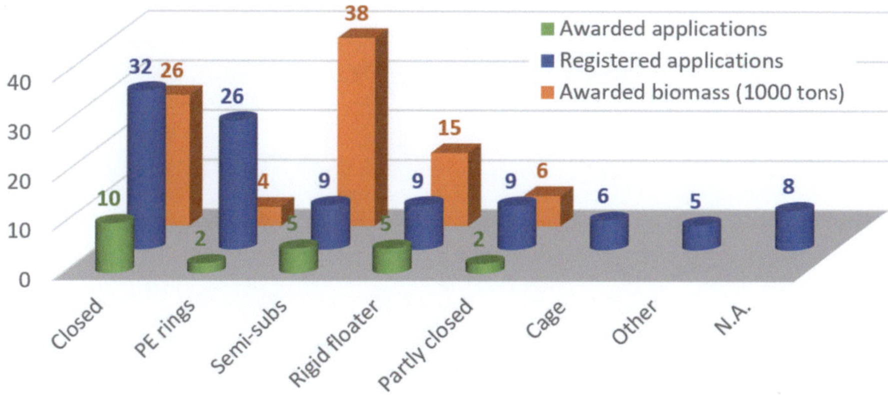

Fig. 9.5 Number of registered (blue) and awarded (green) fish farm concepts and awarded biomass (orange) associated with development licenses

9.4.2 New Fish Farm Concepts Supported by Development Licenses in Norway

This chapter gives an overview of technological innovations suggested through the Norwegian *development licenses*, with focus on the innovations that were rewarded licenses (see Sect. 9.3.2 and Moe Føre et al. 2022). The Norwegian Directorate of Fisheries (NDF) received 104 applications requesting in total 884 licences to farm 780 tonnes of fish each. By March 2023, 24 applications had been awarded, 79 had been rejected, and 1 application had not received a final decision by the NDF. Since Moe Føre et al. (2022) published the status per September 2021, several complaints have been processed resulting in one additional rewarded application, and while one application will be given a new assessment, several others have received a final rejection. In this section, we present an updated summary of the findings in Moe Føre et al. (2022). By March 14, 2023, the awarded applications corresponded to 118 production licenses with a total maximum allowed biomass of 89,309 tonnes.

Most applications claimed to involve an innovative fish farm design (Fig. 9.3). Most of the various designs fell within the already mentioned categories *PE rings*, *rigid floater*, *closed pens*, and *semi-subs* (see Sect. 9.2.1). In addition, a few *partly closed* concepts were also suggested, involving partly closed enclosures or combined use of nets and closed pens, while the *pen* category included permeable spherical and cylindrical net volumes without a pen collar.[1] Results presented in Figs. 9.4 and 9.5 show that the largest categories were closed farms (31% of the applications) and PE rings (25%). And while ten of the closed farm applications were awarded, only two concepts involving conventional PE rings were awarded. The latter included the submersible Atlantis in Fig. 9.3b, and the iFarm sensor

[1] The terms cage collar, platform, and floater are often interchangeable.

Fig. 9.6 Service vessel AQS Ravn at a PE ring with a snorkel pen. (Photo: AQS)

system which involves development of a sensor unit for monitoring of individual fish in a submerged volume using machine vision (so-called snorkel pen with a submerged net roof, Fig. 9.6). Four of the awarded concepts included submergence; i.e. the fish will be permanently or periodically submerged below the water surface to avoid parasite infections and large waves (Moe Føre et al. 2022).

Although 42% of the awards involved closed farms, they were in total given only 30% of the allocated biomass. Many of these concepts had relatively small units, which according to the authorities did not require a high number of licenses to test and validate their technology. Two of the concepts were particularly successful in the allocation process; five of nine applications of both semi-submersible platforms and rigid floaters were given licenses. In general, the semi-subs were large farms for exposed, open ocean conditions representing 21% of the rewarded applications and 43% of the total allocated biomass. These structures are inspired by oil and gas industry, with the permeable nets as a special feature.

The estimated investment costs of the awarded concepts varied from 80 million NOK for the submerged PE-ring *Atlantis* to 2044 million NOK for the closed facility *Eidsfjord Giant*. Of the 24 awarded projects, 13 had an estimated cost above 400 million NOK.

Size and Shape of Innovative Fish Farms

The awarded concepts included 12 circular floater concepts, 6 polygon-shaped floaters (1 of them square), 4 oblong floaters, and 2 triangular floaters with rounded corners (Fig. 9.3). Both small and large units are seen as a part of future aquaculture.

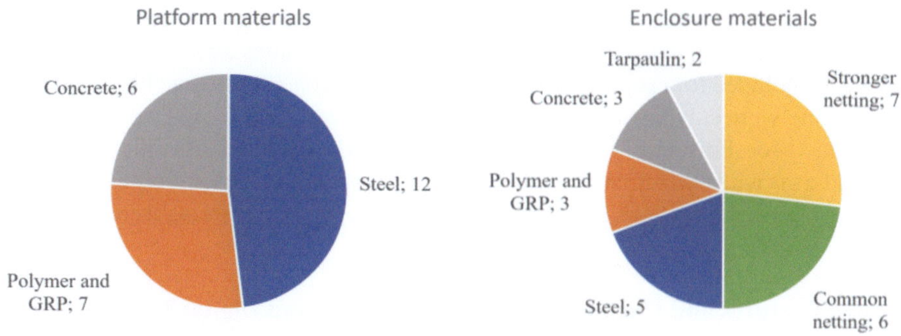

Fig. 9.7 Applied materials in floaters and platforms (left) and shape of floaters and platforms (right) associated with the 24 awarded innovative concepts

Floater size has been defined as the largest horizontal main dimension, like the diameter of a circular floater or the length of a production ship (Moe Føre et al. 2022). Of the awarded innovations, 4 had small floaters with a size less than 35 m, 7 had a size in the range of 35–60 m corresponding to a typical Norwegian PE ring, while 13 of the floaters were larger. Eight concepts involved fish encloser with a volume above 60,000 m^3, i.e. a larger volume than typical Norwegian net pen.

Materials in Platforms and Fish Enclosures

An overview of the applied construction materials in the awarded concepts is shown in Fig. 9.7. Half of the innovations involved the use of steel in platform (pen collar) structures, while the other half used concrete or polymers, and in one case a combination of the two. These three materials and tarpaulins (also polymers) were found in fish enclosures of closed farms, while netting materials (i.e. nets) were equally common. Seven concepts planned to apply a stronger netting material than the conventional knitted nylon netting. Thirteen concepts involved double enclosures to prevent escape of fish, and 2 involved multiple material use. The latter include the awarded concepts iFarm, combining common and stronger netting in a double net, and the partly closed "production tank" by Hydro Salmon Company that combine a steel lice skirt with stronger netting below (Moe Føre et al. 2022).

Technological Measures for Enhanced Sustainability

All awarded concepts had measures against sea lice. Eighteen involved shielding the fish from sea lice, including closed farms and lice skirts of various materials and shapes and four involved submergence. Further, four applications claimed that the use of open ocean areas will reduce the infection pressure. Most awarded concepts (22) involved measures to prevent escape, with stronger and double enclosures as the most common measure (21). Six concepts also involved safer operations. Ten of the concepts involved collection of waste, including all the six concepts designed for sheltered sites. Nine concepts were designed for coastal areas (significant wave height of 2–4 m), while nine concepts were designed for exposed and open ocean sites with significant wave heights above 4 m. The nine awarded open-ocean concepts included four semi-sub platforms, five rigid floaters, and one PE ring (three of the nine were submersible), while in sheltered areas, only closed concepts were awarded (Moe Føre et al. 2022). Concepts were designed for specific locations: closed farms with collection of waste for sheltered areas, designs promoting reduced environmental interactions in coastal areas, while open ocean farms were designed to withstand large waves.

Design Features and Challenges of Relevance to AI and Digital Twin
Innovative fish farm designs and open ocean sites have several features and challenges of relevance to AI and digital twins. Only 1 among the 24 awarded concepts have AI and digital twins as their main innovative feature (iFarm). Eight designs, especially submersible structures and large enclosures, are planning to implement underwater feeding, either as an alternative to or alongside the traditional method of surface feeding distribution. One of the semi-sub designs (Havfarm, Fig. 9.3) is considering the use of dynamic positioning for station keeping, and another (Smart Fish Farm) are planning for continuous surveillance of the fish and automated feeding, and cleaning and inspections of the net.

There is no doubt that automation and digitalization have the possibility to benefit future fish farming and solve some of the challenges that have occurred during the latest years. These include challenges with net integrity, interaction with vessels (boarding of personnel and various operations, see Sect. 9.5.1), orientation of farms with single-point moorings, ballasting and stability, water quality and last, but not least, challenging working conditions.

9.5 Future Fish Farming Operations at Sea

9.5.1 Fish Farming Operations in Pen Aquaculture

Net pen aquaculture involves several primary operations, some of which occur daily, while others may take place less frequent than once a year. Here we define fish farming operations as operations directly involving farming of fish, disregarding technical operations as installation of equipment, inspections, and repairs.

In the future, artificial intelligence and digital twins may play important roles during fish farming operations, and some examples and suggestions are given below.

Fig. 9.8 Handling of net. (Photo: AQS)

9.5.1.1 Reception of Juvenile Fish

Juvenile fish are commonly transported to the fish pen by a well boat. At conventional pens, the well boat will be fastened by ropes to the fish farm, most often directly to the pen collar that will receive the fish (Figs. 9.8 and 9.9). The fish is unloaded into the pen through flexible pipes going from the vessel to the inside of the pen. Then follows the so-called on-growing period until the fish has reached the requested size (approx. 14–22 months for salmon) or other circumstances require this phase to end (e.g. diseases or harmful algae blooms).

Well boat operations may be challenging, and even more at future exposed and open ocean sites. Mooring of well boats in high energy environments (large waves, strong winds, and currents) will require control of displacements of the various farm components, in particular net deformations, and loads in pen collars and moorings. This is a perfect task for a digital twin. In high-energy environments, the well boat may not be fastened to the fish farm. Dynamic positioning systems with AI may contribute to increased safety and facilitate operations in weather conditions that limit the use of existing solutions. Launching of fish in challenging weather conditions may require new solutions for compensating relative movements between vessel and fish farm.

9.5.1.2 Daily Operations

During on-growing, daily chores for personnel involve feeding of fish, health and welfare inspections, and removal of weak fish and mortalities. Live fish are captured

Fig. 9.9 Crowding and loading of fish into well boat. (Photo: SINTEF Ocean)

using landing nets for inspections. Mortalities will sink to the bottom of the net pen, and it is common to have a cone-shaped bottom that naturally collects them in the bottom centre. Mortalities are then pumped up to a boat through a hose that may or may not be connected to a stationary collector. Alternatively, a landing net is kept in this area and raised to the surface to remove mortalities. Every 14th day, the average number of sea lice per fish must be found, either through manual inspections or specialized automatic counting equipment (e.g. utilizing machine vision).

Feeding fish is a top priority. In conventional farms, feed is stored at the feed barge and transported to the pens through individual feed pipes. At the pen, the feed is normally distributed over the surface through a rotating dispenser. In recent years, methods for under water feeding have also been developed.

At high-energy sites, many of these operations may need to be automated as secure manual operations may be impossible. Already, equipment is developed to use images and AI to count lice and inspect fish. The requirements for daily removal of mortalities have been mentioned as a key problem with the current technology and operational procedures. AI and digital twins will also be important tools in optimizing feeding, ensuring that the fish are fed when hungry at the appropriate place in the pen volume, and at the same time avoiding spilling of feed through pellet detection. Important logistics, including deliverance of feed, personnel and other necessities, and the corresponding fleet management, will be challenged by unpredictable and harsh weather conditions.

9.5.1.3 Biofouling Mitigation, Parasite Control, and Moving of Fish

In periods, biofouling control will also require frequent cleaning operations (Bloecher and Floerl 2021). Nets without a functioning anti-fouling paint will be colonized by various organisms, including algae, hydroids, and mussels. The biofouling organisms will occlude the net openings, reduce the water flow through the net, and increase the hydrodynamic loads on the pens and moorings, which result in deformations and thereby reduced net pen volume. This may negatively affect the water quality inside the pen and increase the risk of structural damages to the farm. To mitigate this, cleaning operations are often performed with a regularity up to once every week in late summer with high water temperatures and long hours of sunlight. Typically, specialized cleaning equipment with rotating discs with nozzles discharging pressurized water is applied.

In farming of Atlantic salmon, the fish is routinely vaccinated against several diseases (Adams 2019), but should the fish stock be infected with diseases or parasites, they may be treated using medicated feed (antibiotics are seldomly used). At many fish farms, the fish will occasionally need to be treated for parasite infections in specialized treatment barges (see Sect. 9.3.1) or well boats. These treatments involve several sub-operations and vessels. To prepare for de-lousing, the volume of the fish enclosure is typically reduced to crowd the fish. For a conventional net pen facility, this involves lifting of weights attached to the net pen, followed by lifting of the net pen (Fig. 9.8) itself and then reducing the net volume using a buoy rope to crowd the fish (Fig. 9.9). Cranes, winches, and other mechanical lifting equipment are frequently used in all phases of these operations, which are known to involve a relatively high risk of damages to the net enclosure and consequently escape of fish (Føre and Thorvaldsen 2021).

Delousing often involves moving fish from one pen to another, i.e. infected fish are pumped into the treatment vessel and then into a different pen. The same type of operation is used if fish need to be moved, for instance, during sorting of fish. Finally, the fish is pumped into a well boat and delivered at a land-based facility for slaughter (Fig. 9.9).

In future fish farming, many of these operations and sub-operations will benefit from automation and control through monitoring of important parameters such as biofouling, net pen integrity and position, loads in critical structural components, water quality, and fish welfare.

9.5.2 Occupational Health and Safety

Globally, aquaculture workers face several hazards and are at risk of occupational diseases and injuries. Still, occupational health and safety have not been an important driver for technological innovation in the industry.

For fish farm and vessel employees, the most reported types of accidents are fall and crush/impact. Fatalities during the last 10 years have mainly happened in

Fig. 9.10 Causes for work-related sick leave (self-reported) given in an HSE-Survey among fish farm employees (Thorvaldsen et al. 2020)

relation to lifting and maintenance operations (Holen and Holmen 2025; Holen et al. 2018a, b). Working at sea involves impact from wind, waves, and current that makes it challenging to perform manual work tasks and operations.

Alongside acute injuries, employees may be exposed to ergonomic risk factors including lifting with a twisted or bent upper body, repetitive and monotonous tasks, and heavy lifting that can lead to musculoskeletal pain. Employee surveys have revealed that pain in the neck, shoulder, arm, back, and hand and wrist are prevalent work-related health complaints (Fig. 9.10). Strain injuries are the leading reason for work-related worries and sick leave in the surveys (Thorvaldsen et al. 2020, 2025).

Different risk dimensions described in the Norwegian context of salmon farming include risks to personnel, fish welfare and health, material assets, environment, and food safety (Yang et al. 2019).

The organisation of work or organisational factors is crucial for safe operations. Examples of such factors include sufficient staffing, adequate training, proper planning, work hours, and sufficient time for maintenance work. Conflicting objectives related to protection versus production may be challenging, as workers at fish farms and vessels must handle a range of risks in their work (Føre and Thorvaldsen 2021; Thorvaldsen et al. 2025).

Moving forward, more must be done to protect personnel, by implementing safety in design, aiming to control hazards by removing or substituting them. Furthermore, reduction of straining tasks through a higher degree of automatization can be beneficial for workers' health. Training of personnel in safe environments using digital twins and AI can also be beneficial. Similarly, using these new tools can provide important decision support for planning and conducting operations.

9.5.2.1 Operational Limits and Decision Support

Weather and environmental conditions are important for safe operations, especially at exposed and offshore sites. Strong currents affect equipment under water, and waves impact crane-based operations involving personnel. Bad weather also affects operations involving different types of vessels. The weather forecast and knowledge about the conditions at a given site are therefore very important in planning of operations.

During operations, the weather is assessed continuously based on visual observations from the employees. The fish is also under observation to ensure fish welfare and health. Decisions to call off an operation due to conditions of concern are based on experience, visual observations, and assessments by the people running the operations. This is supported by the following quote from Thorvaldsen et al. (2015):

> If strong winds, excessively strong currents or high waves occur during operations, the work should be called off. This is, first and foremost, to ensure the safety of employees, but also to minimise the risk of damaging fish and the risk of escape. Previous experience remains crucial in decision-making processes. (Thorvaldsen et al. 2015)

Since personnel safety is affected by the sea state, objective measurements of waves and wind and monitoring of farm movements may be helpful during operations to avoid uncomfortable and unsafe working conditions.

Specific operational limits are used in other industries and may be a possible strategy to ensure safe operations in the aquaculture industry. The use of AI may provide safety relevant decision support related to operational limits if relevant data is provided. Looking into the potential use of machine learning techniques for decision support in high-risk aquaculture operations, one study suggests that criteria for operational limits may relate to weather and sea conditions (wind, wave height), type of operation, visibility, daylights, vessel size and age, age of fish farm, as well as organizational and human factors (Yang et al. 2019).

9.5.2.2 Emergency Preparedness

Establishing a robust emergency preparedness plan is important for safe operations in all types of locations and should be based on thorough risk assessments involving operational personnel (Holmen et al. 2018).

Regulations state that emergency preparedness plans should cover scenarios related to personnel, fish, food safety, environment, and structures (Thorvaldsen et al. 2023). This includes hazardous events such as serious personal injury, mass mortality of fish due to algae, jelly fish or other causes, escape of fish, pollution and breakdown of facilities, fleets, boats, or equipment.

Offshore fish farming involves increased distance from shore and increased number of fish in each farm, probably more automatization, and increasingly demanding logistics. New risk scenarios relate to more extreme weather conditions as well as helicopter transport accidents, sabotage, and loss of power. Cybersecurity will become crucial for ensuring safe remote operations and automatization.

9.5.3 Fish Farming Operations at Exposed and Offshore Sites

Although extreme exposure levels may be high at a given site, as discussed in Sect. 9.2.2.1, most days will bring lower environmental loads which will represent operational conditions at which routinely fish farming operations must be performed. Thus, foreseeing the duration and frequency of various exposure levels will be important knowledge to plan for operations ranging from daily maintenance and rearing to larger operations involving the use of multiple vessels. While waves may be less critical for large well boats, waves are highly critical for crane and lifting operations on smaller service vessels. Operations should be planned to take place in calmer periods, and the possibility for suitable weather windows at frequent enough intervals is often of higher interest than extreme conditions (Bjelland et al. 2024).

In conclusion, estimating the extreme exposure level is mainly of interest in structural design, while the probability and duration of different exposure conditions are of great interest during the operational phase. In the future, it is envisioned that the exposure velocity (Sect. 9.2.2.1) will be given in the site survey report. Specific values will be given for at least eight directions (north, northeast, east and so forth) based on combinations of different wave and current design value return periods as stated in NS9415.

Rose plots may be applied to illustrate the distribution and direction of the exposure velocity (Fig. 9.11). The rose plots utilize a polar coordinate system indicating frequency and direction of exposure velocities, with different colour bands for given intervals of velocities. The highest triangle represents the most common direction of the velocity (Bjelland et al. 2024). In addition, weather windows, i.e. time intervals with environmental conditions below a specified limit, are of interest. These require statistical estimates, which can be based on the rose plot data, but will also involve other sources such as weather forecasts.

As an example of an exposure velocity rose plot, Bjelland et al. (2024) applied 18 months of data from measurements of current and waves at the location called *Salatskjæra* (Kristiansen et al. 2017), which is found just within the Norwegian sea boundary, defined by the uttermost points of land, west of the city Trondheim (Fig. 9.11). The maximum measured current was found to be 0.59 m/s toward east (104 degrees), and a maximum significant wave height was found to be 2.03 m from north-east (84 degrees). The rose plots give the directional distribution of exposure velocity in time. For example, 23% of the time, the main environmental loads came from a north-eastern direction, while the maximum exposure velocity for the 18-month period was found to be 1.4 m/s coming from east. However, 95% of the time the exposure velocity was below 0.7 m/s and 70% of the time, and the exposure velocity was below 0.4 m/s.

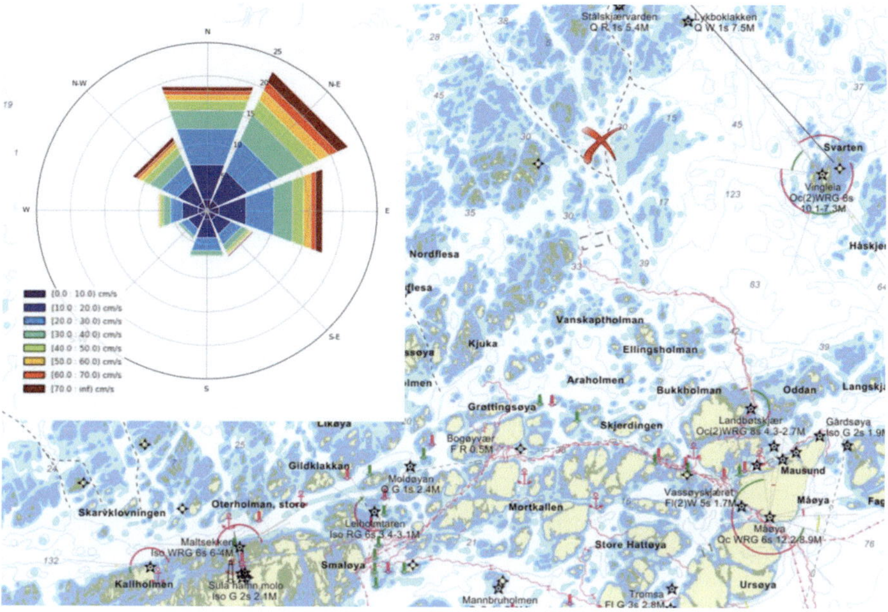

Fig. 9.11 Exposure velocity rose plot based on wave and current measurements at a given location on the outer coast of Norway (Bjelland et al. 2024)

In the example in Fig. 9.11, we use the velocity in the crest; however, the wave through velocity should also be considered where its direction is similar to the water currents. Another aspect that will be of importance with regard to experienced exposure is from how many directions the farm may experience high waves and environmental loads. As stated by one fish farmer during a discussion on exposure: “The most demanding sites today have long fetch length over a large angle above 180 degrees”.

9.6 Overall Trends and Future Outlook

While traditional net pens are the most common technology for pen fish farming, other production systems are emerging, making the industry more technologically diverse. Alternative farm designs are introduced, and a few concepts have been implemented in prototype testing and commercial fish farming. This includes both modified and scaled-up conventional farms, new big and strong farms for exposed locations, and closed farms for sheltered locations. Submerged pens are currently introduced at locations with high parasite pressure and large waves.

Although there have been governmental incentives in Norway to support development of new farm designs, which have been embraced by the aquaculture industry, most farms are still similar and of the conventional type. There are also challenges with new types of farms that must be addressed, both relating to economic feasibility, technical robustness, and pen environment. Thus, this development may take some time.

However, everything points to the fact that the future will bring more diverse pen farm designs, where the pens and their operation are tailored to different sites, species, and environments. This will lead to an increased use of automation, as many operations including feeding, net cleaning, maintenance and repairs, and daily rearing activities cannot be performed manually. There will be a need of new sensors, digital twins, AI, triggers, and robots to maintain both fish welfare and structural integrity of future fish farms.

References

Aarhus I, Høy E, Fredheim A, Winther U (2011) Kartlegging av ulike teknologiske løsninger for å møte de miljømessige utfordringene i havbruksnæringen

Adams A (2019) Progress, challenges and opportunities in fish vaccine development. Fish Shellf Immunol 90:210–214. https://doi.org/10.1016/j.fsi.2019.04.066

Afewerki S, Asche F, Misund B, Thorvaldsen T, Tveteras R (2022) Innovation in the Norwegian aquaculture industry. Rev Aquac. https://doi.org/10.1111/raq.12755erences

Bjelland HV, Føre M, Lader P, Kristiansen D, Holmen IM, Fredheim A, Grøtli EI, Fathi DE, Oppedal F, Utne IB, Schjølberg I (2016) Exposed aquaculture in Norway: Technologies for robust operation in rough conditions. In: Proceedings of MTS/IEEE OCEANS 15, 19–22 October 2015. Washington, USA

Bjelland HV, Størkersen KV, Føre HM, Thorvaldsen T, Folkedal O, Holmen I, Lona E, Grøtli EI, Selvik Ø, Fagertun JT, Endresen PC, Ohrem S, Slette HT, Gutsch M, Mathisen BM, Hvas M, Klebert P (2024) Safe, robust, and productive exposed aquaculture operations – challenges and solutions. SINTEF Ocean AS, report no. 2023:00755 A. https://hdl.handle.net/11250/3148412

Bjelland HV, Folkedal O, Føre HM, Grøtli EI, Holmen IM, Lona E, Slette HT, Størkersen KV, Thorvaldsen T (2025) Exposed aquaculture operations: strategies for safety and fish welfare. Rev Aquac 17:e12964. https://doi.org/10.1111/raq.12964

Bloecher N, Floerl O (2021) Towards cost-effective biofouling management in salmon aquaculture: a strategic outlook. Rev Aquac 13(2):783–795. https://doi.org/10.1111/RAQ.12498

Buck BH, Bjelland HV, Bockus A, Chambers M, Costa-Pierce BA, Dewhurst T, Ferreira JG, Føre HM, Fredriksson DW, Goseberg N, Holmyard J, Isbert W, Krause G, Markus T, Papandroulakis N, Sclodnick T, Silkes B, Strand Å, Troell M, Wieczorek D, van den Burg SWK, Heasman KG (2024) Resolving the term “offshore aquaculture” by decoupling “exposed” and “distance from the coast”. Front Aquac 3:1428056. https://doi.org/10.3389/faquc.2024.1428056

Chu YI, Wang CM, Park JC, Lader PF (2020) Review of cage and containment tank designs for offshore fish farming. Aquaculture 519

Drumm A (2010) Evaluation of the promotion of offshore aquaculture through a technology platform (OATP). Marine Institute, Rinville, p 46

Føre HM, Thorvaldsen T (2021) Causal analysis of escape of Atlantic salmon and rainbow trout from Norwegian fish farms during 2010–2018. Aquaculture 532:736002. https://doi.org/10.1016/j.aquaculture.2020.736002

Froehlich HE, Smith A, Gentry RR, Halpern BS (2017) Offshore aquaculture: I know it when I see it. Front Mar Sci 4:154. https://doi.org/10.3389/fmars.2017.00154

Halwart M, Soto D, Arthur JR (eds) (2007) Cage aquaculture: regional reviews and global overview, vol 498. FAO

Heasman KG, Sclodnick T, Goseberg N, Scott N, Chambers M, Dewhurst T, Rickerich S, Føre HM, Buck BH (2024) Utilisation of the site assessment energy indices for aquaculture in exposed waters: biology, technology, operations and maintenance. Front Aquac 3. https://doi.org/10.3389/faquc.2024.1427168

Hersoug B (2022) "One country, ten systems" – the use of different licensing systems in Norwegian aquaculture. Mar Policy 137:104902. ISSN 0308-597X. https://doi.org/10.1016/j.marpol.2021.104902

Hersoug B, Mikkelsen E, Karlsen KM (2019) "Great expectations" – allocating licenses with special requirements in Norwegian salmon farming. Mar Policy 100:152–162. https://doi.org/10.1016/j.marpol.2018.11.019

Holen S, Holmen IM (2025) Occupational injuries and fatilities in Norwegian Fish Farming. J Agromed 30(2):319–329

Holen S, Utne IB, Holmen IM, Aasjord H (2018a) Occupational safety in aquaculture – Part 1: Injuries in Norway. Mar Policy 96:184–192

Holen S, Utne IB, Holmen IM, Aasjord H (2018b) Occupational safety in aquaculture – Part 2: Fatalities in Norway 1982–2015. Mar Policy 96:193–199

Holm P, Buck BH, Langan R (2017) Introduction: new approaches to sustainable offshore food production and the development of offshore platforms. In: Buck BH, Langan R (eds) Aquaculture perspective of multi-use sites in the open ocean: the untapped potential for marine resources in the Anthropocene. ISBN(1): 978-3-319-51157-3 – ISBN(2) 978-3-319-51159-7 (eBook). 1–20

Holmen et al (2018) Risk assessments in the Norwegian aquaculture industry: status and improved practice. Aquac Eng 83:65–75

Kristiansen D, Aksnes V, Su B, Lader P, Bjelland HV (2017) Environmental description in the design of fish farms at exposed locations. In: ASME 2017 36th international conference on Ocean, offshore and Arctic Engineering. American Society of Mechanical Engineers Digital Collection. OMAE2017-61531, Trondheim, Norway

Kristiansen D, Lader P, Endresen PC, Aksnes V (2018) Numerical and experimental study on the seakeeping behavior of floating closed rigid fish cages. In: Presented at the ASME 2018 37th international conference on ocean, offshore and Arctic engineering. American Society of Mechanical Engineers Digital Collection. https://doi.org/10.1115/OMAE2018-77254

Lojek O, Goseberg N, Føre HM, Dewhurst T, Bölker T, Heasman KG, Buck BH, Fredriksson DW, Rickerich S (2024) Hydrodynamic exposure – on the quest to deriving quantitative metrics for mariculture sites. Front Aquac 3. https://doi.org/10.3389/faquc.2024.1388280

Misund A, Thorvaldsen T, Strand AV, Oftebro TL, Dahle SW (2024) Opportunities and challenges in new production systems for salmon farming in Norway – industry perspective. Mar Policy. https://doi.org/10.1016/j.marpol.2024.106394

Moe Føre H, Thorvaldsen T, Osmundsen TC, Asche F, Tveterås R, Fagertun JT, Bjelland HV (2022) Technological innovations promoting sustainable salmon (Salmo salar) aquaculture in Norway. Aquac Rep 24:101115. https://doi.org/10.1016/j.aqrep.2022.101115

Moe H, Olsen A, Hopperstad OS, Jensen Ø, Fredheim A (2007) Tensile properties for netting materials used in aquaculture net cages. Aquac Eng 37:252–265. https://doi.org/10.1016/j.aquaeng.2007.08.001

Osmundsen TC, Olsen MS, Gauteplass A, Asche F (2022) Aquaculture policy: designing licenses for environmental regulation. Mar Policy 138:104978. https://doi.org/10.1016/j.marpol.2022.104978

Rosten TW, Ulgenes Y, Henriksen K, Terjesen BF, Biering E, Winther U (2011) Oppdrett av laks og ørret i lukkede anlegg-forprosjekt. SINTEF Trondheim

Sønvisen SA, Vik C (2021) Shaping aquaculture management – an interest Tug O' War. Sustainability 13:8853. https://doi.org/10.3390/su13168853

Standards Norway (2021) NS 9415:2021, Floating aquaculture farms – site survey, design, execution and use. Standard Online. https://online.standard.no/en/ns-9415-2021. Accessed 4.4.25

Su B, Tsarau A, Endresen PC, Kristiansen D, Lader PF (2021) Numerical study of closed rigid fish cages in waves and comparison with experimental data. Ocean Eng. 233:109210. https://doi.org/10.1016/j.oceaneng.2021.109210

Thorvaldsen T, Holmen IM, Moe HK (2015) The escape of fish from Norwegian fish farms. Causes, risks and the influence of organisational aspects. Mar Policy 55:33–378. https://doi.org/10.1016/j.marpol.2015.01.008

Thorvaldsen T, Kongsvik T, Holmen IM, Størkersen K, Salomonsen C, Sandsund M, Bjelland HV (2020) Occupational health, safety and work environments in Norwegian fish farming – employee perspective. Aquaculture 524

Thorvaldsen T, Salomonsen C, Ranum SA, Trædal T, Misund A, Holmen IM (2023) Prepared for the worst? Emergency preparedness in Norwegian fish farming – status and further improvements. Aquaculture. https://doi.org/10.1016/j.aquaculture.2023.739921

Thorvaldsen T, Kongsvik T, Holmen IM, Størkersen K, Holen S (2025) Revisiting occupational health and safety in Norwegian fish farming – results from an employee survey. Aquaculture 603

Yang X, Utne IB, Holmen IM (2019) Methodology for hazard identification in aquaculture operations (MHIAO). Saf Sci 121:430–450

Conclusion

Martin Føre • Jon Grant • Fearghal O'Donncha

Conclusion: From Precision Aquaculture to Ocean Intelligence—Realising the Potential of Digital Twins

Aquaculture stands at a pivotal juncture, uniquely positioned to address the global demand for nutritious food yet constrained by the very complexities of the marine environments it inhabits. As this book has demonstrated, the sector's growth is hampered by deep-rooted challenges in observability, controllability, and operational efficiency. The confluence of climate change, resource scarcity, and evolving consumer expectations no longer constitutes a distant risk but an immediate operational reality. This is not merely a time for refinement; it is a moment that demands structural transformation in how aquaculture interfaces with technology and the environment. The digital twin, as explored throughout these chapters, serves as a foundational architecture for that transformation. It is a necessary evolution from reactive farming to predictive, intelligent, and resilient food production.

Synthesising the Digital Twin Framework

Our journey began *On the Farm* (Part I), confronting the challenges of collecting reliable, high-resolution data on fish health and welfare in the dynamic ocean environment. We saw how emerging robotics and autonomous systems are extending our senses, generating the empirical data streams required to construct and validate a digital twin. We then shifted to *the Cloud* (Part II), exploring how this raw data is forged into actionable insight. Through advanced simulation, hybrid analysis, and

F. O'Donncha et al. (eds.), *Digital Twin for Marine Fish Farming*,
Aquatic Production Systems, https://doi.org/10.1007/978-3-032-10362-8

machine learning, we can now model the intricate interplay between biology and environment, enabling us to forecast operational dynamics and elucidate complex system interactions in ways that were previously out of reach.

This fusion of on-farm sensing and cloud-based analytics culminates in the vision of *Precision Fish Farming* (Part III). The digital twin is not merely a data dashboard; it is an integrated ecosystem for decision-making. It is the framework that allows us to manage systemic risks, from the immediate threat of an algal bloom to the long-term, existential challenge of climate change. It provides the verifiable data needed to build transparent supply chains that meet modern consumer demands for provenance and sustainability. And fundamentally, it gives us the tools to reimagine the future of aquaculture itself.

Aquaculture as a Testbed for a Digital Ocean

The challenges that make aquaculture a difficult domain for data collection also make it the ideal catalyst for innovation. The sector's unique concentration of economic, biological, and logistical assets within a constrained marine space makes it a powerful beachhead for developing ocean intelligence. The technologies honed in the crucible of fish farming—from autonomous underwater vehicles to predictive ecological models—have applications that extend well beyond aquaculture, informing broader marine domains. By solving the problems of aquaculture, we are simultaneously building the technological stack for broader ocean observation, conservation, and resource management. In this sense, a digital twin for a fish farm is a microcosm of a digital twin for the ocean itself.

The Future of Work and the Human-in-the-Loop

The path to a digitised and automated future is not about replacing human expertise but augmenting it. Digital twins empower remote operations, automate repetitive tasks, and provide decision support that can mitigate a growing labour gap and enhance worker safety. However, this transition demands a parallel evolution in our workforce. The farmer of the future must evolve into a data interpreter, a systems manager, and an overseer of autonomous fleets. Cultivating these new skillsets, establishing institutional governance frameworks that address accountability, validation, and ethical use of machine-generated insights, and fostering trust through human-in-the-loop oversight are paramount. The success of these technologies is reliant on the people who use them.

Looking Ahead: Emerging AI Paradigms

While much of this book has focused on the current digital technologies driving transformation in aquaculture, it is worth noting the recent advances in generative AI and agent-based systems. These approaches are beginning to reshape decision-making and automation in other industries through enhanced reasoning, planning, and human-machine interaction. Although still nascent in aquaculture, such tools may in time augment digital twin frameworks—particularly in supporting adaptive control, constructing synthetic scenarios to support rare-event modelling, forecasting, and risk mitigation, or enabling more autonomous farm operations. Their successful adoption will depend on targeted research that adapts these general-purpose technologies to the specific constraints and data realities of aquaculture.

A Call to Action: Charting the Course Forward

This book has provided a map of the technological landscape and a strategic vision for the future of aquaculture. A map, however, is only useful when it guides action. Realising the full potential of the digital twin requires a concerted and collaborative effort.

- *For researchers and technologists*: The challenge is to continue pushing the boundaries of sensor technology, communication, and predictive modelling to overcome the ocean's opacity.
- *For industry leaders and investors*: The opportunity is to champion this transition, fostering a culture of data-driven decision-making and investing in the infrastructure and skills required to make precision aquaculture a reality.
- *For policy-makers and regulators*: The task is to create supportive frameworks that encourage innovation while ensuring data ownership, interoperability, and the responsible stewardship of our marine resources.

If implemented with foresight and collaboration, digital twins will not only elevate aquaculture to new heights of productivity and sustainability but will also position the sector as a leader in the digital transformation of our oceans. The journey is ambitious, but the destination—a resilient, transparent, and intelligent food system in harmony with the marine world—is within our reach.

Looking Ahead: Recovery AI Paradigms

MIX
Papier aus verantwortungsvollen Quellen
Paper from responsible sources
FSC® C105338

If you have any concerns about our products,
you can contact us on
ProductSafety@springernature.com

In case Publisher is established outside the EU,
the EU authorized representative is:
Springer Nature Customer Service Center GmbH
Europaplatz 3, 69115 Heidelberg, Germany

Printed by Libri Plureos GmbH
in Hamburg, Germany